ROMAN DECORATIVE STONE COLLECTIONS

in the Kelsey Museum of Archaeology

Kelsey Museum Studies

Roman Decorative Stone Collections in the Kelsey Museum of Archaeology, J. Clayton Fant, Leah E. Long, and Lynley J. McAlpine

Latin Inscriptions in the Kelsey Museum: The Dennison and De Criscio Collections, Steven L. Tuck

Indian Block-Printed Cotton Fragments in the Kelsey Museum, Ruth Barnes

Roman Brick Stamps in the Kelsey Museum, John P. Bodel

Pottery from Karanis: Excavations of the University of Michigan, Barbara Johnson

Karanis Excavations of the University of Michigan in Egypt, 1928–1935: Topography and Architecture, Elinor M. Husselman

Latin Inscriptions in the Kelsey Museum: The Dennison Collection, Edited by Martha Welborn Baldwin and Mario Torelli

Terracotta Lamps from Karanis, Egypt: Excavations of the University of Michigan, Louise A. Shier

The Gospel of John in Fayumic Coptic (*P. Mich. Inv. 3521*), Edited by Elinor M. Husselman

Funerary Stelae from Kom Abou Billou, Finley A. Hooper

ROMAN DECORATIVE STONE COLLECTIONS

in the Kelsey Museum of Archaeology

Kelsey Museum Studies
Volume 10

J. Clayton Fant,
Leah E. Long, and Lynley J. McAlpine

The Kelsey Museum of Archaeology

University of Michigan Press
Ann Arbor, Michigan

Published in the United States of America by the
University of Michigan Press
Manufactured in the United States of America
Printed on acid-free paper

First published August 2024

A CIP catalog record for this book is available from the British Library.

Library of Congress Control Number: 2024014077

ISBN 978-0-472-13195-2 (hardcover : alk. paper)
ISBN 978-0-472-12675-0 (e-book)

Contents

Digital materials related to this title can be found on the Fulcrum platform via the following citable URL: https://doi.org/10.3998/mpub.11534228

Tables

Figures

Acknowledgments

In a project that began in 2011, we have inevitably piled up considerable debts. First, Elaine Gazda, curator of Hellenistic and Roman Collections at the Kelsey Museum of Archaeology, sponsored and guided us from the beginning. The Kelsey staff made our work possible and provided constant enlightenment when we got bogged down in accessions issues, so we gratefully acknowledge Sebastián E. Encina, collections manager, and Michelle Fontenot, registrar. Lorene Sterner, graphic artist, provided invaluable support not only on the creation and organization of the illustrations, but also helped us navigate photography and shepherded the catalogue entries. Sharon Herbert, Christopher Ratté, Terry Wilfong, and Nicola Terrenato supported the project throughout their tenures as directors of the Kelsey Museum. In general, the museum as an institution devoted considerable resources in hours and talent throughout. Jamie Newhall, senior media producer at the University of Akron, executed the photographs of the artifacts during a long campaign in academic year 2017–18. Leslie Schramer, editor, coped with a large manuscript with three different authorial imprints, and Ellen Bauerle at the University of Michigan Press oversaw the production.

We have had significant financial support. Two Faculty Research Grants by Virginia Commonwealth University in Qatar were awarded, in 2014 and 2015, for the research, travel, and scientific analysis of the collection. The laboratory analysis was conducted by Dott. Fabrizio Antonelli in the Laboratorio di Analisi Materiali dei Antichi of the Istituto Universitario di Architettura di Venezia. The Loeb Classical Library Foundation provided a generous grant in 2016 to defray research and production expenses and professional photography; we thank them not only for that but also for their unbureaucratic and flexible collaboration.

The authors are listed in alphabetical order because this was truly a collaborative effort. Over four years we inventoried the 704 pieces as we found time together. Over the next three years, we undertook the writing, starting with an outline drawn up in common. The chief writing assignments emerged organically: Fant with the catalogue since he had the broadest knowledge of marbles and quarry sources, Long with the archaeological side because she had the depth of field experience from work in Turkey, and McAlpine with the archival sleuthing, for which it quickly became clear that she has a special aptitude. Fant led drafting for several grant applications, including the generous Loeb Classical Library Foundation grant of 2016 which supports the overall publication and the professional photography, which he also oversaw. Long coordinated tasks to ensure the project's completion, also drafted the Loeb grant, secured several grants to fund the laboratory analysis which enriches our understanding of the white marble samples in the collections as reflected in chapter 3. McAlpine was the steward of the database.

Many scholars helped us with specific advice, guidance, and criticism; our appeals for help with identifications and often with bibliography were always met with generosity. Dott. Federico Guidobaldi and Antonio Dell'Acqua gave the archaeological chapter close attention. We drew upon the expertise of Drs. Simon Barker, Matthias Bruno, Patrizio Pensabene, Simona Perna, Anna Gutiérrez Garcia-Moreno, and especially Lorenzo Lazzarini (Laboratory for Analysis of Ancient Materials at Iuav University of Venice) and James A. Harrell (Department of Environmental Science, University of Toledo, Ohio), who kindly not only answered emailed questions but made several visits to the collections in person. We also wish to thank the anonymous readers for the University of Michigan Press.

J. Clayton Fant, Leah E. Long, and Lynley J. McAlpine
August 2023

Abbreviations

BCE	before the common era
C.	Celsius
c.	century
ca.	*circa*, approximately
cat.	catalogue
CE	common era
cf.	*confer*, compare
ch(s).	chapter(s)
cm	centimeter(s)
cont.	continued
cu.	cubic
D.	diameter
et al.	*et alii*, and others
e.g.	*exempli gratia*, for example
esp.	especially
etc.	*et cetera*, and so forth
fig(s).	figure(s)
forth.	forthcoming
g	gram(s)
i.e.	*id est*, that is
kg	kilogram(s)
KM	Kelsey Museum
km	kilometer(s)
L.	length
l.	*lapis*
m	meter(s)
m.	*marmor*
max.	maximum
MGS	maximum grain size
mm	millimeter(s)
n(n).	note(s)
no(s).	number(s)
n.p.	no publisher
p(p).	page(s)
pl(s).	plate(s)
q.v.	*quod vide*, which see
s.v.	*sub verbo*, under the word
trans.	translation
vol(s).	volume(s)
W.	width
CIL	*Corpus Inscriptionum Latinarum* (1863–)
MUSNAF	Museo di Storia Naturale dell'Accademia dei Fisiocritici, www.museofisiocritici.it/collezione-marmi.asp (the very helpful search functions are available only in the Italian version)
OUMNH	The Corsi collection at the Oxford Museum of Natural History, www.oum.ox.ac.uk/corsi

Chapter I

THE COLLECTIONS AND THE COLLECTORS

Lynley J. McAlpine, Leah E. Long, and J. Clayton Fant

A Panorama of the Roman Stone Trade

Both the wide variety of lithotypes and the multiplication of examples of some dozen are accurate reflections of the nature of the ancient Roman trade in decorative stones. In the Greek and Hellenistic worlds, most building materials were quarried locally; most projects made a pragmatic compromise between proximity (less expensive transport) and appearance. Exceptions, like Kleisthenes' rebuilding of the temple of Apollo at Delphi in imported Parian marble (Herodotus, *Histories* 5.62) attracted wide attention, as intended.

The growth of fine stone imports tracked the expansion of Roman hegemony in the eastern Mediterranean beginning in the second century BCE.[1] Successful commanders rewarded their tutelary gods and advertised themselves with commemorative temples in Rome built *e manubiis*, from the spoils of their campaigns. Employing Greek architects and importing Aegean marbles (no Italian marble resources were available then) simply followed the same logic.[2] In the first century BCE, imports of colored marbles became more frequent, still linked more often than not to imperial expansion and dynastic politics. Traditional (white) marbles were named in Greek for their sources, for example, Pentelic and Parian. This practice was continued in Latin nomenclature but took on a more imperial connotation as the range of place names spread.[3] "Phrygian" marble alludes to the Roman alliance with the kingdom of Pergamon and then its reconstitution as the Roman province of Asia. "Numidian" points to the Romans' Numidian allies against Hannibal and, ultimately, to the rump Numidian state folded into Roma as Africa Nova in 46 BCE. *Marmor Luculleum* was unique in trumpeting an explicit connection to a Roman general, Lucullus, whose successes in Asia Minor were notable but later eclipsed by his successor, Pompey.[4] The defeat of Cleopatra in 31 BCE by the future Augustus and creation of the special imperial province of Aegyptus opened the floodgates to novel as well as traditional Egyptian stones bearing evocative and exotic names from the Nile and Eastern Desert. Simultaneously, the vast resources of the Apuan Alps from Carrara south to Pietrasanta became accessible, and the opening up of the Luna quarries allowed Augustus to boast of transforming Rome into a city of marble (Suetonius, *Life of Augustus* 28.3). Subsequent exportation westward carried the new imperial styles to southern France, Spain, and even North Africa.[5]

As Roman imports grew during the long peace, quest for novelty and assiduous imitation of imperial metropolitan trends remained constants. Indeed, they fueled each other, keeping pressure on imperial builders to stay ahead. Thus we see constant prospection in the Eastern Desert of Egypt throughout the first century and into the early third century CE, when Aurelius Heraclidas brags about his discovery of new quarries in the vicinity of the Aswan granite quarries of Egypt's First Cataract (*novae lapidiciae adinventae,*

1. A fuller overview is provided in Pensabene 2002. See Russell 2013 for a novel survey and analysis of the full scope of the Roman stone trade.
2. Gros 1973 and 1976.
3. Lazzarini 2007 provides extensive discussions of nomenclature for all the marble varieties that fall inside the boundaries of the modern Greek state.
4. Fant 1988 and 1999.
5. Pensabene 2013, 426–28.

CIL III 75). Even in Italy itself, new lithotypes from smaller quarries continued to be sought out.[6]

Collectors of Roman fine decorative stones, mostly based in Rome in the nineteenth century, were only emulating ancient practice in seeking to include the rarest lithotypes as well as to fully represent the subvarieties of the most widely used polychrome stones.

A Brief History of the Collections

Since the Renaissance, collectors have assembled samples of exotic marbles, granites, and alabasters to represent the varieties of decorative stone in order to emphasize the rarity and beauty of the earth's natural resources. In Italy, this aesthetic exercise was paralleled by the production of tables with inset samples, such as Florentine *pietre dure* (fig. 1.1). Typically, the samples were cut into plaques of uniform size and polished to bring out their color and veining. Most European collections of decorative stone have found homes in institutions and universities, and a number of them have recently been published and lavishly illustrated in catalogues.[7] While the samples in European collections came from ancient monuments in the Middle East, Egypt, Greece, and Rome, in many cases their provenances are now lost and their original architectural characteristics have been obliterated by later reworking.

In the United States, it was not until the mid-nineteenth century that the study of the material culture of ancient civilizations was established as a field. As museums opened their collections to the public and universities expanded their Greek and Latin departments to include classical archaeologists, Francis W. Kelsey, professor of Latin at the University of Michigan, undertook major educational initiatives. He spearheaded excavations in Karanis (Egypt), Carthage (Tunisia), and Pisidian Antioch (Turkey), and founded a museum that now bears his name. The Kelsey Museum of Archaeology today houses more than 100,000 ancient and medieval objects from Egyptian, Middle Eastern, Greek, Roman, Coptic, and Islamic cultures. Between Kelsey's arrival at the University of Michigan in 1889 and his death in 1927, he amassed large collections of many types of artifacts to broaden the teaching of ancient studies at the university. Among these was a collection of stone samples that date to the Roman period. Professor Kelsey not only collected many of the samples himself but also energetically solicited donations from other leading scholars and wealthy Detroit industrialists. The resulting collection, numbering some 700 samples, is probably the largest collection

Figure 1.1. Semiprecious stone-cut table from the Opificio delle pietre dure, Florence, Italy.

6. Bruno and Lazzarini 1999, Bruno 2002c.
7. Mielsch 1985; Gnoli 1988; Borghini 1989; Dolci and Nista 1992; Price 2007.

outside Europe and is one of the richest and most varied anywhere.

In contrast to most European collections, the architectural fragments of the Kelsey Museum were collected as archaeological objects, often from well-known ancient buildings. Kelsey solicited his colleagues working abroad to ship them to the University of Michigan where they could be used for the purposes of teaching and research. He began to collect seriously in 1893 on visits to Tunisia, Sicily, Naples, and Rome, routinely buying antiquities of all materials and states of preservation. It was in 1900–1901, during his leave for an appointment as professor at the American School of Classical Studies in Rome, that he set his mind to purchasing decorative stone in bulk. Like the industrialists of his age, Kelsey amassed wealth, but his was in archaeological riches. He seems to have bought marbles at every opportunity, and at what must have been a dizzying pace, from foremen of archaeological excavations, dealers, quarrymen, and even a local child at the train station in La Malka, Tunisia. Given Kelsey's exhaustive pursuits in championing the study of antiquity, our work, which should be read in tandem with other studies, represents a small and opaque window on his vision of collecting.[8]

Kelsey did not compile the present collection alone. Thirteen other individuals gathered, donated, or bequeathed marbles to the museum. Major donors include Esther Boise Van Deman, a pioneering female archaeologist and founding scholar of Roman construction, and Canon Giuseppe De Criscio, a parish priest from Pozzuoli, Italy, who had amassed a large collection of antiquities and was persuaded to sell them to the museum. J. D. Candler, founder of a roofing company in Detroit responsible for many major projects in the city, donated "curios" from his travels to the Mediterranean. Each collector was motivated by a different set of circumstances. Francis Kelsey's collection showcases the breadth and quantity of artifacts from the ancient world, Esther Van Deman's for the full range of materials used in Roman architecture, De Criscio was motivated by his passion for local history, and J. D. Candler by his interest in collecting Grand Tour souvenirs. The contributions that this fascinating array of characters made to the Kelsey collection is our subject here.

Biographies of the Collectors

Francis Willey Kelsey

The Kelsey Museum of Archaeology is named for Francis Willey Kelsey, professor of Latin at the University of Michigan from 1889 to 1927 (fig. 1.2). Kelsey held a number of prominent appointments throughout his career, such as president of both the American Philological Association and the Archaeological Institute of America, and was influential in the fields of philology and archaeology, as well as in American college education in general.[9] He started to collect antiquities early in his academic career, and these eventually made their way into the holdings of the archaeological museum named after him. He began in Carthage in 1893, obtaining more than one thousand artifacts of various types from dealers and archaeologists in Tunisia that year. In Timgad, Algeria, also in 1893, Kelsey picked up at least thirteen marble fragments, probably surface finds as they are still encrusted with dirt and are difficult to identify (cat. no. 701). These North African artifacts were the first to be accessioned into the museum's collection when it was established in 1928.

From the early autumn of 1900 to summer of 1901, Kelsey took a leave from the University of Michigan to take an appointment as professor at the American School of Classical Studies in Rome and spent much of his time abroad touring Italy, Greece, and North Africa.[10] He resided for most of that period in Rome, and his personal diaries indicate that he circulated with the leading archaeologists of the day. Rodolfo

8. Important for our research were Kelsey's diaries, now housed in the Bentley Historical Library at the University of Michigan. John Pedley's 2012 biography, *The Life and Work of Francis Willey Kelsey*, focuses on Kelsey as a seminal figure in the history of classical archaeology in the United States, and helped us enormously. So too did the accounts of the other donors documented in Talalay and Root 2015.
9. A detailed account of Kelsey's life and contributions can be found in Pedley 2012.
10. See Pedley 2012, 64–77.

Figure 1.2. Francis W. Kelsey and his wife, Isabelle, in Palermo, Sicily, ca. 1925. Francis Willey Kelsey papers, Bentley Historical Library, University of Michigan, no. HS5259.

Lanciani, known for his work on the topography of ancient Rome, and August Mau, scholar of Pompeian culture and art, accompanied Kelsey on trips to Ostia and Pompeii, respectively. Giacomo Boni, famed for his excavations in the Roman Forum, provided consultation on the nomenclature of ancient marbles Kelsey acquired in Italy and Greece.

Kelsey not only studied the archaeology of the ancient Mediterranean during his 1900–1901 stay, he also set out in earnest once again to collect artifacts of many kinds, amassing a comprehensive teaching collection that includes the architectural marbles catalogued here. His methods for collecting were varied and opportunistic. In some instances, he seems to have gathered surface artifacts himself, while at other times he sought out antiquities dealers. On occasion he also approached—or was approached by—local people who had an ancient object or two to offer him.

Soon after arriving in Italy for his appointment, in September 1900, Kelsey made an excursion to the ancient site of Volterra and the nearby quarries at Carrara on the northwest coast of Italy. At Volterra he acquired a fragment of white marble purchased from a workman on site (cat. no. 303), and at Carrara a piece of the white marble typical of that quarry area (cat. no. 559).

According to Kelsey's notes, many of the pieces of colorful stone he collected were obtained at the Baths of Diocletian in Rome (see fig. 2.2, map of Rome). Kelsey's stay in Rome coincided with the major construction project of the Piazza dell'Esedra (now the Piazza della Repubblica) at the end of the Via Nazionale, near the Baths of Diocletian. This project uncovered the remains of a Flavian monument,[11] many fragments of which are now divided between the Museo Nazionale Romano and the Kelsey Museum. Pieces of paving and veneer must also have been unearthed by workers on that site—indeed, workmen there reportedly stole artifacts and sold them.[12] The Baths of Diocletian were themselves adorned with lavish decorative stone of many types, so perhaps fragments now at the Kelsey Museum once covered its walls and floors.[13]

11. Perhaps the *Templum Gentis Flaviae*; Gazda and Haeckl 1996, 8.
12. Gazda and Haeckl 1996, 8.
13. On the stone used to decorate the Baths of Diocletian, see Paribeni 1920.

Kelsey's notes, as transcribed into the museum's accession books after his death, provide a "Locality" for each object, but it is not always clear if this refers to the site at which Kelsey himself obtained the artifact, or to the site where it was originally found by him or the person from whom he acquired it. The process of accessioning the items is described thus in the first accession book, at the bottom of page 80:

> The catalogue of marbles was made up from the notes made during the trip to Rome 1900 '01. Unless some definite place and date was given for the stone no entry is made. The original material is [in] an envelope marked "marbles" in the Collections file. In some cases the stone cutter had numbered the stones also and cut off the original nos. This will explain many discrepancies.
>
> In 1930 the material contained in several lots of loose sheets and cards was copied here to save the information. Had it not been for his long illness in 1902 Professor Kelsey would have brought all this together himself. The various titles seem to have been added by various stone experts as the writing varies. The one used first here was the one which stood 1st on the slip.

If Kelsey's original notes were retained after they were copied, their current location is unknown.

In the case of the Baths of Diocletian, the objects may have been found on site at the Baths or merely purchased there. Around ninety stone artifacts were bought from one F. Romano (also spelled Romanò), who is described as a "marmorista [marble worker] in new bldg. on part of site of Baths of Diocletian," and as a "dealer whose shop was in area of old Baths of Diocletian" (accession book, notes at KM 2090 and 2028).[14] According to the accession books, fifteen of these come from the Baths themselves, ten from Ostia, three from the Roman Forum, and one from Centocelle, ca. 5 km east of the city center (see maps, figs. 2.1 and 2.2). The rest are listed only as from "Rome" or even "Italy" or have no provenance at all.

While Kelsey purchased artifacts from Romano on several separate occasions in 1901 (February; March 2, 9, 10, 13, and August 10), he had also acquired objects from the Baths of Diocletian in October 1900. The 1900 group, which includes a fragment of a sarcophagus as well as a piece of *sperone* or *lapis Gabinus* (cat. no. 595) from the "bottom of trench for foundations of new building" and a fragment of Pentelic marble from a "bit of concrete" (cat. no. 499), were apparently not purchases. It seems likely, therefore, that Kelsey found some artifacts himself at the Baths of Diocletian in October 1900, before or after conducting business with Romano at his nearby shop.

Most of Romano's marble fragments were cut and polished (Kelsey's notes specify that one fragment, cat. no. 272, was "cut with copper wheel by Romano who polished it").[15] Marble fragments Kelsey collected on his trip to Greece in spring 1901 (described below) were also polished once they had arrived back in Rome with Kelsey.

Widespread construction in Rome at the turn of the century furnished Kelsey with a number of opportunities for collecting newly unearthed artifacts. Several marble samples collected in February 1901 (cat. nos. 554 and 562) are recorded as having originated from the "tunnel under Quirinal," which according to his diary he visited "with Prof. Wright."[16] This tunnel must have been the Traforo Umberto I, which connects Via Nazionale and Via Tritone. Construction of the tunnel began in June 1900, just in time for Kelsey's arrival in September. A humbler construction project in front of a palazzo on the Quirinal Hill yielded a smaller trove of architectural fragments for Kelsey in early 1901. On January 4, he discovered a fragment of a *cipollino* paving or veneer in the "street in front of 38 Via Firenze" (cat.

14. Ninety-two pieces are securely attributed to Romano. Four others are likely, based on date and place of collection, but not explicitly stated to have been purchased from Romano. See Index D, Collectors, on page 195.

15. A coating of wax polish is found on many of Kelsey's pieces, mostly accessioned the year after his tenure at the American School of Classical Studies in Rome, and on Van Deman's pieces, accessioned in 1938, along with more Kelsey pieces found after his death in 1927. In 2017, Kelsey Museum conservator Caroline Roberts examined several fragments from the collections of Kelsey, Van Deman, and Candler under a longwave UV lamp. She said, "As I expected, the coated surfaces fluoresce much like beeswax; if you look at the edges of the applied coating you'll see that it glows a yellow-white color similar to that of the beeswax in the sample jar. The coating is also soluble in mineral spirits, which is typical of beeswax. I think it's safe to say that this is what was used on these objects. The wax was likely applied after the stone was sampled or broken, since the edges of the coating drip onto break edges." The wax must once have intensified the colors and patterns in the stone, but today it has darkened and has the opposite effect. Application of wax to at least one surface seems to have been standard practice at the Kelsey Museum during the early twentieth century.

16. This is presumably Professor J. H. Wright of Harvard University.

no. 188).[17] On January 21 and 23, several additional items were unearthed from that location when "an excavation for laying electric light brought [them] to light."[18] According to Kelsey's diary, these items were obtained while he and his family were in quarantine for scarlet fever elsewhere in the city. The quarantine lasted until January 25.[19]

On February 13, Kelsey wrote in his diary that he "went with Professor Wright and his son and Orlandi to identify marbles in S. Maria Maggiore and S. John Lateran."[20] No marble fragments were obtained that day, it seems, but Orlandi's identifications of stone varieties accompany many entries, alongside Boni's, in the Kelsey Museum accession books. This man may have been Aristide Orlandi, perhaps a collector of marbles and/or an artist who built inlaid furniture similar to the table in figure 1.1.[21]

On March 13, 1901, Kelsey purchased at least eleven marble fragments from an "old man" in Rome.[22] At an unknown date he purchased twenty-one fragments from "Vecchio."[23] Most of Vecchio's fragments have no provenance. Three are said to have come from Ostia (cat. nos. 251, 319, 694), another from nearby Tor Paterno (cat. no. 101), and one from Rome (cat. no. 356) (see fig. 2.1b).

Other specimens gathered in Italy may have been acquired during trips sponsored by the American School in 1901 (see fig. 2.1b, map of Italy). Kelsey and his students visited Ostia with Lanciani on February 2, where he acquired two fragments not catalogued here (KM 2286 and 2300); several other marble fragments are noted as having come from "Tor Paterno, Castel Fusano," located a few kilometers from Ostia, though the date of these acquisitions is not recorded in Kelsey's notes or diaries.[24] On February 15, Kelsey acquired fragments from "place marked Rudei [Ruderi] on staff map 145-III (Cassoli) S.E. of Roviano & E. of Anticoli Corrado near Mola Nuova" (cat. nos. 219, 452, 600). On March 1–5, the group visited Palestrina, guided by Thomas Ashby.[25] Some fragments came from "Hadrian's Villa and adjoining sites, Palestrina" (cat. nos. 80, 107, 195, 198, 327, 399, 457, 584). Other fragments were found at Astura, the site of Roman imperial villas on the coast of Latium (cat. nos. 61, 176, 357, 509, 540, 585). Kelsey also acquired a fragment of limestone (cat. no. 598) from a quarry near the town of Subiaco well outside of Rome. Kelsey's notes usually do not tell us whether he purchased these objects or if they were surface finds that he collected himself—with the exception of a lozenge-shaped fragment of *palombino opus sectile* purchased from a farmer near Arricia on February 18, 1901 (cat. no. 582).

In mid-March 1901, Kelsey left Rome to lead students and colleagues on an arduous tour of Greece, during which he made a point of stopping at various ancient quarry sites and obtaining marble samples (see fig. 2.1c).[26] Some of these quarries were still in use and he was able to consult managers and workers. For example, Kelsey's diary for April 17 notes that he "visited new quarries of Pentelic marble as guest of Mr. Arthur Hill, President of 'Marmor Limited,'"[27] which ran a quarrying operation on Mt. Pentelikon about 20 km northeast of Athens. Kelsey recorded in his notes the specific location within the quarry from which many samples were taken, much like a modern geologist does when taking samples for scientific

17. The palazzo at 38 Via Firenze, located on the site of an ancient temple, now houses the Evangelical Methodist Church of Italy, and has since the early twentieth century. It was also a common residence for American students and scholars in Rome during the first few decades of the last century; see, e.g., Spencer 1938. Though Kelsey, an active member of the Presbyterian church (Pedley 2012, 114–15), did not reside there, in his diary he mentioned visiting missions and attending a Methodist church service on February 17.

18. January 21: cat. nos. 121, 122, 225, 451, 561; January 23: cat. nos. 123, 124, 125, 126, 510. The source of the quote is the accession book at KM 2256.

19. Spencer (1938, 271) claims that this is palazzo was built "on the site of a late church, which had in turn been built over an ancient temple." Spencer was a member of Methodist Missionary Society in 1910.

20. See n. 16 above.

21. A similar table from the late nineteenth century in a private collection is signed by an Aristide Orlandi (Grieco 2008, fig. 471). In addition, a fragment of "lumachella cenere" was purchased from an "Orlandi Aristide" from the Via Sistina in Rome (Sackler Museum, object no. 1977.216.88).

22. Only seven of these were brought back to Michigan and accessioned: cat. nos. 13, 21, 89, 163, 422, 517, 518. See accession book 1, insert at p. 81.

23. While it is possible that these two figures are one and the same, the purchases from "old man" are dated to March 13, 1901, while those from Vecchio are not dated.

24. One fragment from Tor Paterno, cat. no. 101, was "purchased of Vecchio."

25. Pedley 2012, 66.

26. For a detailed account of this trip, see Pedley 2012, 67–73.

27. On Arthur Hill, also see chapter 2, n. 9.

analysis. For example: "new quarry, north side" (cat. no. 501), "new quarry, above monastery" (cat. no. 502), and "old quarry, w. of entrance to cave" (cat. no. 503). One fragment (cat. no. 500) was a "piece from block cut for monument of Von Moltke, Berlin." On April 28, Kelsey visited Larissa, where he wrote in his diary, "afternoon went out to quarries of Verde Antico with the manager, who had come in for the day, and to whom I had a letter of introduction from Mr. Hill, manager of the marble company." Elsewhere, pieces of *verde antico* were taken from sites only recorded as "New Quarry" (cat. nos. 685, 686, 687) and "Quarry whence columns for Westminster Cathedral are being taken" (cat. no. 685). Other fragments were purchased or picked up at archaeological sites all over Greece, such as the Acropolis, Areopagus, and Olympeium in Athens, the theater in Sparta, the odeum in Patras, and the temple of Athena Alea in Tegea (see Index C, Places Where Acquired, p. 193).

Immediately upon the group's return to Italy in May, Kelsey arranged for August Mau to lecture for them on site in Pompeii, though Kelsey appears not to have obtained any marble fragments from that site.[28] He then spent the summer touring students in Italy, at sites such as Veii and Astura, and getting ready to return to Michigan. Presumably with the help of his Italian colleagues, Kelsey was given permission to ship his archaeological materials to the United States. A diary entry from Tuesday, August 27, 1901, records a "shipment of 17 cases of brickstamps, ex-votos, and architectural specimens [sent] to the university."[29] All in all, Kelsey shipped 437 building stones, colored marbles, and architectural decorations back home.[30]

Kelsey continued to collect artifacts throughout his life, though the bulk of his architectural and decorative marble fragments were amassed during that early year in Rome. Occasional objects that must have seemed particularly interesting to Kelsey were acquired later. He collected a piece of *pavonazzetto* during his visit to Herod's Palace at Jericho in 1920 (cat. no. 177), and six surface finds of decorative marble were brought back from the excavations at Pisidian Antioch in 1924 (cat. nos. 50, 179, 258, 487, 488, 489). The main task of building a complete study collection of ancient stone varieties, however, was taken up by Kelsey's student Esther Boise Van Deman, whose fascination with architectural materials surpassed even Kelsey's.

Esther Boise Van Deman

Esther Boise Van Deman (fig. 1.3) was Kelsey's student at the University of Michigan, where she received her BA in 1891. She went on to the University of Chicago, becoming the first woman there to earn a doctorate in Latin.[31] Both Van Deman and Kelsey appear to have aimed at creating comprehensive teaching and research collections, whether for use at the University of Michigan or at the American Academy in Rome. Van Deman focused especially on obtaining building materials, among them examples of the many varieties of stone used in Roman architecture.

Van Deman was a student at the American School of Classical Studies in Rome in 1901, the same year Kelsey was in residence there, and she remained until 1903.[32] It may have been then, as Kelsey sought out samples of decorative marble and other architectural fragments, that she began her own collection of building materials. It was certainly at this time that Van Deman shifted her research interests from Latin to archaeology, embarking in her pioneering fashion on the road to becoming one of the leading experts in Roman construction.[33] Perhaps she was present on occasion when Kelsey consulted Italian experts to identify stone types; certainly she observed the excavations being carried out in the Roman Forum by Giacomo Boni (whom she described in a letter as "patronizing"

28. Kelsey did pick up a specimen of tufa (KM 2445, not catalogued here), and a few other miscellaneous artifacts from Pompeii in 1901.
29. See also Pedley 2012, 75.
30. Gazda and Haeckl 1996, 59.
31. Talalay and Root 2015, 77.
32. Welch 2004, 71.
33. Talalay and Root 2015, 77–78. Van Deman's role as an influential archaeologist and uncompromising scholar in a time a when few women achieved her prominence has led to a number of biographical profiles in addition to Talalay and Root's. Among them are profiles by M.E. Blake in the *Dictionary of American Biography* (1958), and by L. Shoe Meritt in *Notable American Women 1607–1950: A Biographical Dictionary* (1971). Other sources of biographical information are Bull-Simonsen Einaudi 1983; Einaudi 1988; and, most thorough, Welch 2004.

Figure 1.3. Esther Boise Van Deman. Self-portrait taken on the deck of the steamer *Frankfurt*, 1901. Courtesy of the American Academy in Rome, image: VanDeman.2275.

to women),[34] and in the Atrium Vestae, on which she published her first book.[35]

Van Deman had a keen interest in the nomenclature of archaeological stones (as the letter to Orma Butler quoted below demonstrates). Most of her samples now in the Kelsey Museum's collection are labeled in her distinctive handwriting with the name of the stone type and her initials (fig. 1.4). The names she provided are precise and often idiosyncratic. For example, she distinguishes among a number of varieties of alabaster, such as, "alabastro palombara attosoni," "alabastro di Civitavecchia bruciato," "alabastro a tartaruga giallo e rosso," and "alabastro amythisto dendritico."

Figure 1.4. Stone sample (cat. no. 286) labeled by Esther Boise Van Deman.

During her visit to Rome at the turn of the century, Van Deman began her life's project studying Roman architecture and construction methods. She was especially interested in their application for dating Roman buildings and their development. In addition, Van Deman began to make a list of "the more important decorative materials, namely, marbles, granites, basalts, and porphyries, with a brief description of each," for her Roman construction volume, presumably collecting samples when possible for study.[36] Unfortunately, she did not live to complete that project, but her colleague Marion Blake used Van Deman's extensive research materials to compose the volume.[37]

Van Deman held various appointments at American institutions during her career, including Bryn Mawr, Wellesley, Mount Holyoke, and the Women's College of Baltimore.[38] She evidently preferred to spend as much time as possible in Rome, however; in 1906, she won a Carnegie fellowship for the School of Classical Studies in Rome from 1906 to 1909, and after this period spent most of the rest of her life in Rome, including her longest phases of residence from 1922 until her death in 1937, with brief visits to the U.S. to lecture.[39] During her time abroad, in addition to becoming a

34. Welch 2004, 74.
35. Van Deman 1909.
36. Van Deman 1925.
37. Blake 1947. For a thorough critique of Van Deman's research, see Welch 2004.
38. Welch 2004, 70–71.
39. Welch 2004, 71–72; Van Deman 1925.

leading expert and teacher of archaeology, she taught herself photography, documenting ruins from all over the ancient world.[40] While in Rome she amassed a large collection of samples of archaeological materials, focusing particularly on construction materials. Some of these she sent to the Kelsey Museum in Ann Arbor for storage, and others to Johns Hopkins University, where she came to feel they were not properly appreciated or cared for. She also collected objects specifically for the Kelsey's collection at the request of Orma Butler, the curator of the Kelsey Museum from 1928 to 1938. Below is her response to a request from Butler to send some specimens of tufa for an exhibition.

> August 1st, 1930
> My dear Dr. Butler,
>
> It's too bad about the specimens. Sanders handed me the list and I should have gotten them ready and put them in his study at once for I was sick much of these last months and could not ever go to the Forum for them—where I could have picked tiny pieces. Now I am at the foot of Monte Bianco and can not see Rome until the end of September, Alas! In fact, I'm bothered by your names, for they are the doubtful choice of Tenney Frank and are largely useless. "Gray pozzolana" can be gotten in a hundred places but must be treated as a general term. "Grotta Oscura" tufa is senseless, for this is only one quarry, which he liked, out of many, in other sections—it even occurs in the city. "Fidenae tufa" means only a theory of his own & cannot be made a generic term—at least until he's proved his theory. "Monte Verde" may pass, but it's doubtful. Better use Lanciani's divisions or those of a Geologist. "Anio tufa" is blind, for his "Grotta Oscura" tufa is found in abundance in the same sections. "Silex" is a real name. When I'm back in Roma, I can ask Stevens what he thinks about the names—I think I shall stick to colors in my work, with maybe Monte Verde, since it's still in use. I am sorry Frank made our job harder by doing unscientific work in that book. It has not helped his reputation, in several ways. I can send you the stones, of course, which I suppose he means, when I'm back in Rome.
>
> My work goes slowly. The "Aqueducts" apart, except that the very last part, with Mr. Sanders. I am here to get well enough to get the last thirty or forty pages finished in order to begin my "Roman Construction."
>
> Regretting my seeming neglect—for I was too ill to lift the stones in their quarries—and trusting you are resting now.
>
> Very cordially yours,
> Esther B. Van Deman[41]

A conflict with Tenney Frank of Johns Hopkins may have led her to leave her collection to the University of Michigan when she died in 1937.[42] An inventory of the 227 objects then stored at Johns Hopkins was made, including pottery, lamps, bronze jewelry, lead, weapons, bone and ivory, stucco, tile and brick, and "miscellaneous." The largest category, however, was marble, consisting of seventy-one objects.[43] Her will stipulated that, "All my books and antiquities now in the museum of Johns Hopkins University which I have not disposed of in my lifetime I give to the University of Michigan. My books and antiquities that are in Europe at the time of my death I give and bequeath to the American Academy in Rome, provided that, if at any time the Classical School should be separated from the American Academy, then such books and antiquities shall pass to the Classical School." She seems not to have revealed her wishes to the faculty at Johns Hopkins, as archaeologist David M. Robinson wrote in a letter of December 21, 1937, to John G. Winter at Michigan that he was

> surprised that Miss Van Deman left to the University of Michigan the few antiquities which she had put long years ago in our museum at Johns Hopkins. She had always said she would leave them to us (and some to me personally), and we have taken care of them and given them a good place. They are not valuable, worth $300 or $400, and I am publishing with Miss Van Deman's consent some of the minor vases in my next fascicule, now in print. But of course I am very glad to follow the wishes of my good friend, Miss Van Deman, whose grave in the English cemetery in Rome I recently visited and decorated. I know she was enraged with Tenney Frank and for this reason she probably changed her intentions....[44]

Van Deman's objects enrich the variety and completeness of the collection of architectural marbles in the Kelsey Museum, though unfortunately we have

40. Welch 2004, 73.
41. Van Deman Collection, Kelsey Museum of Archaeology Archives.
42. Talalay and Root 2015, 79.
43. Will of Esther Boise Van Deman, October 18, 1936. Van Deman Collection, Kelsey Museum of Archaeology Archives.
44. Van Deman Collection, Kelsey Museum of Archaeology Archives.

no records of where individual pieces of her collection were gathered.[45] Some objects, such as a large *opus sectile* rondel (cat. no. 420; see fig. 2.9d), composed of seven fragments glued back together, seem likely to have been found in situ, or at least near their original location of use, but the Kelsey Museum's accession books tell us only that Van Deman's collection presumably came from Rome.

Canon Giuseppe De Criscio

Figure 1.5. Canon Giuseppe De Criscio of Pozzuoli, Italy. Kelsey Museum Archives.

While F. W. Kelsey was back in Michigan between his trips abroad in 1893 and 1900, he continued to acquire antiquities through gifts and purchases. One of the most important such purchases was the collection accumulated by Giuseppe De Criscio (fig. 1.5), a parish priest, scholar, and collector of local antiquities in Pozzuoli (ancient Puteoli), near Naples, Italy. Walter Dennison, a student of Kelsey and later professor of Latin, met with De Criscio in 1897 in order to study some of his inscriptions. De Criscio informed him that he was trying to find a suitable home for his collection, and so Dennison wrote to Kelsey about the matter. Kelsey was able to obtain funding as a gift from Henry P. Glover of Ypsilanti to purchase a portion of the collection, namely, 276 inscriptions.[46] More objects were received by the University of Michigan in 1905 and 1909, and after De Criscio's death in 1911, his sister contacted the university about buying the remainder of the collection. Those last objects, among which were more fragments of decorative stone, came to Michigan in 1922. A note in the accession book (at KM 2691) on this final batch reads, "Here begins the second purchase from the Collection of Canon DeCriscio, purchased by the University in February, 1922. [...] Purchase arranged by J. G. Winter." As for their provenance: "Puteoli or vicinity so far as known. No definite data."

In all, the Kelsey Museum has 1,060 artifacts once owned by Giuseppe De Criscio. The best-known component of De Criscio's collection are the many inscriptions from Pozzuoli and neighboring areas, but he also collected objects of other types, including architectural components. Some are things that he or his parishioners[47] perhaps gathered from surface debris in the local region, such as the bags full of small, dusty chunks of decorative stone, whose condition makes them difficult to identify (cat. nos. 702, 703, 704).

James Deforest (J. D.) Candler

An explanatory note in the Kelsey accession books at the beginning of the Candler entries (at KM 29575) reads as follows:

45. A note in the accession book at KM 6602 states in part, "By the will of Esther Boise Van Deman her collection of objects, presumably made in Rome, was left to the Museum of Classical Archaeology." We have listed (Italy) as the provenance for all of Van Deman's materials, with the understanding that many could be from Rome and its vicinity.
46. Pedley 2012, 52–53, 87; Talalay and Root 2015, 52.
47. Talalay and Root 2015, 52.

The Candler Collection
This collection consists mostly of "curios" garnered from unlikely places (stone from top of pyramid of Cheops, Gibraltar, etc.). Those pieces seeming to have a definite place in the collections of the Kelsey Museum have been individually accessioned, and any accompanying information recorded. Mr. Candler was evidently a prosperous Detroit businessman. Most dated items show the dates 1860–1870, but one or two are marked 1925. How these items came to be given to the Kelsey Museum is not known, or when.
J.A. Brown – Jan. 17, 1964

Little is known about how the small but remarkable collection of marble fragments and other artifacts donated to the Kelsey Museum by James D. Candler was amassed.[48] Candler (fig. 1.6) was the owner of J.D. Candler Roofing Company of Detroit and his collection may have passed to the Kelsey Museum upon his death in 1925 at age 69, though, as is evident from the note above, it was not accessioned until 1964, when it was discovered in storage. Most of the "accompanying information," with the exception of a few handwritten labels still affixed to objects, has been lost in the intervening years since John G. Winter accessioned the collection.

Figure 1.6. J. D. Candler, ca. 1925. Courtesy of J. D. Candler Roofing, Livonia, Michigan.

It seems possible that the collection was passed down to J.D. Candler by his father, William Candler, who had an interest in antiquities and particularly in collecting mineral specimens. William's collection of mineral specimens was given to the University of Michigan in 1922 by his daughter Miriam. James's brother Henry wrote in his biography of this father that William was an avid traveler—though he makes no mention of travels to the Mediterranean, where these objects were collected.[49] J.D. Candler's son Claude made a monetary gift to the American School of Classical Studies in Rome in 1900, demonstrating a particular interest in classical antiquity and probably participation in American School programs in Rome. Various other objects, archaeological and geological, were donated to the university in decades following by the Candlers, though there is no record of a specific gift to the Kelsey Museum early in the twentieth century. These tantalizing fragments of information suggest that several members of the Candler family in the late nineteenth and early twentieth centuries were interested in both the ancient world generally and in collecting stone specimens in particular, though precisely who gathered the objects now at the Kelsey and who donated them is not known.

Among the Candler objects now in the Kelsey Museum are many stone fragments (and other artifacts) with handwritten labels affixed that indicate where and when they were collected. These objects seem to be souvenirs pocketed during a Grand Tour that took the collector to Italy, Greece, Turkey, and Egypt (fig. 2.1a–c). In Italy, a piece of the floor of St. Peter's Basilica was acquired (cat. no. 181), as well as bricks from the Roman Forum, Colosseum, and Hadrian's Villa at Tivoli. At Olympia, Greece, the same traveler collected a piece of *greco scritto* (cat. no. 514). Several stone fragments (cat. nos. 182, 263, 291, 567) were collected at Ephesus from the Church of St. John, the Odeion, and the Great Theater, the latter dated

48. We are grateful to Daniel P. Diffendale for allowing us to consult his detailed research on J.D. Candler.
49. Candler 1933.

September 6, 1867. Moving on to Cairo, on October 4, 1867, a piece of gray sandstone was picked up from "the Great Pyramid 'Cheops'," (cat. no. 603), and a fragment of granite from "the Tombs near the great Sphinx" (cat. no. 24). Other pieces of marble, numbered between 11 and 96 on the surface of the stone in pencil, are not labeled with their origins, though some perhaps had been when Winter discovered them in the Kelsey Museum, as is suggested by notes of uncertain origin in the accession book. Several of Candler's pieces are small, delicately carved fragments of *opus sectile* in various colors and shapes (cat. nos. 142, 143, 144, 145, 146, 147, 148, 149, 150, 180; see ch. 2). When and where these were collected is not known.

Since J. D. Candler would have been only eleven or twelve years old in 1867, it seems most likely that these objects were collected by his father, William, or another older family member. James, however, traveled abroad himself and obtained antiquities en route, including a piece of mummy cloth, now in the Kelsey Museum's collection, which was removed from a mummy that he and his companions had paid to be unwrapped for them 1890.[50] Other artifacts, including a murex shell and a small marble head of Pan from a sarcophagus, both from Carthage, were evidently collected in 1925, the year of James's death. Kelsey made an expedition to Carthage in spring of 1925, but J. D. Candler had been ill since 1923, so it is unlikely that he was involved in that project, unless to fund it. How these latest dated pieces came to be among the Candlers' collection remains a mystery.[51]

Other Collectors

Small numbers of marble fragments in the Kelsey's collection were donated by a handful of other collectors: Ms. Pauline Emerson Burton, Elisha Jones, J. G. Winter, Cyrus L. W. Eidlitz, Arthur Hill (the president of Marmor Ltd., whose quarries in Greece Kelsey visited in 1901), Arthur E. R. Boak, Eugene S. McCartney of the University of Michigan, C. L. Meader, Miss Wanere, and C. P. Sellards (who gave to F. W. Kelsey a modern plaque of pink marble that his father had received from J. P. Morgan). As the last five people only provided an object or two to the marble collection, we focus here on the first four.

Ms. Pauline Olson Emerson Burton, an alumna of the University of Michigan (BA 1916, AM 1938) was the chairman of the foreign language department and a Latin teacher at Libbey High School in Toledo, Ohio.[52] According to the 1960 Libbey High yearbook, the school's Latin students had a tradition of visiting the University of Michigan to enhance their studies. It seems very likely that they would have toured the Kelsey Museum during these excursions. Burton was a highly acclaimed teacher and active in the field of classical studies. Though she bequeathed her collections to the Kelsey Museum in her will, as well as establishing an endowment for the University of Michigan in her name, almost $120,000 of her estate remained mistakenly untouched until it was discovered in 2013, almost forty-seven years after her death in 1965.[53] She had intended most of this money to go to the University of Michigan, in particular for student scholarships in Latin and for the Kelsey Museum, which has now created the Pauline Olson Emerson Burton Acquisition Fund. While she seems to have had an interest in decorative stone, not all of her stone fragments are actually ancient—several samples come from North American quarries.

Elisha Jones was associate professor of classical studies, Greek, and Latin at the University of Michigan during the 1870s and 1880s.[54] A note in the Kelsey accession book at KM 1975 states that his artifacts were collected in "1885–86(?)" and received by the Kelsey in "1912(?)." "Given to H. R. Cross by Miss Caroline Pattengill after the death of Mrs. Jones, then turned over to Latin Department 11/29/1912." Most of these objects seem to have been collected by Jones during trips to Italy. The provenances for many are recorded in the accession book, such as "Tomb of Lucilla, Via Salaria," "Palatine," "Roman Forum," "Tablinum of

50. Talalay and Root 2015, 120–21.
51. What is more, a fragment of wall painting from the F. W. Kelsey collection joins with a fragment of wall painting said to have been part of the Candler collection. Clearly, the origins of the objects in the Candler collection are muddled.
52. Alumni Association of the University of Michigan 1944, 222.
53. "Libbey Latin Teacher Donates 120K to UM 47 Years after Death," *The Blade* (Toledo, OH), January 28, 2013. (http://www.toledoblade.com/Education/2013/01/28/Libbey-Latin-teacher-donates-120K-to-UM-47-years-after-death.html)
54. Hinsdale and Demmon 1906, 262–63.

House of Vestals," and "St. John Lateran" (cat. nos. 49, 152, 326, 331, 378).

John G. Winter was the director of the Kelsey Museum from the time it opened in 1928 until his retirement in 1951. It was he who coordinated the acquisition of the last portion of De Criscio's collection after the priest died in 1911, before the museum was established. The origin of most of his decorative marble fragments is unknown, with the exception of a fragment labeled, "Porphyry near Arch of Titus, 1922" (cat. no. 51). Objects from his personal collection of antiquities were given to the Kelsey Museum by his widow in November 1969.

Cyrus L.W. Eidlitz was an architect in New York who through his profession was imbued with an appreciation of ancient building materials. He took pride in presenting, presumably to F.W. Kelsey himself, three "specimens of Numidian marble" in June 1898 that were "used by him in the decoration of the entrance hallway of the Washington Life Insurance Company building, 145 Broadway."[55] Eidlitz was a prominent architect who is best known for designing One Times Square, formerly the home of the New York Times. Perhaps Kelsey met him on one of his frequent visits to New York, or when Eidlitz designed the Michigan Central Railroad Stations in Detroit and Kalamazoo, Michigan, built in 1884 and 1887, respectively.

55. Kelsey Museum accession book at KM 1089–1091.

Chapter 2

ARCHAEOLOGICAL ASPECTS OF THE COLLECTIONS

Materials, Provenance, and Decoration

Leah E. Long

In keeping with the nature of the Kelsey Museum of Archaeology's collections, the thrust of this study is archaeological. In the following pages, each constituent class of material is presented as a single entity with attention paid to the objects' history of use and reuse. In-depth examination provides the opportunity to highlight unique pieces in the collection, add comparanda to existing studies, and place the Kelsey marbles in the wider context of the Roman world. The chapter is not intended to provide an interpretation of marble use in the Roman world, but rather to examine decorative surface ornamentation and understand how these objects were used in their original contexts. As one of the first studies published in English on Roman decorative stone, this chapter is also intended as a guide for identification and analysis.

Project Methods, Goals, and Problems

The project to study the Kelsey's collection of decorative marbles began during the 2011 academic year. Over the course of four years, Clayton Fant, Lynley McAlpine, and Leah Long investigated and catalogued over seven hundred samples. The most immediate goal was to identify the lithotype and original function of each of the samples. For cataloguing, we adopted the schema used in the Grassi marble collection in the Museo Nazionale Romano.[1] Each piece was carefully examined and many of its characteristics were entered into a FileMaker database according to a number of archaeological classification fields. Those entries included object identification and history of use (original object, ancient reuse, modern re/working), provenance (by site and country), measurements (thickness, maximum and minimum dimensions as represented by preserved height and width), material/variety (Italian name, ancient name, geological description), and treatment of the flat surfaces (principal faces, number of edges of the original object preserved, shape, polishing). Crucial to the project were the Kelsey Museum accession books. Information about the marbles recorded in the accession books by the museum curators was copied directly and entered into the database, and much of that information is cited here.

When we first convened, we realized that the task of deciding what to include in the collection would be a difficult one. Other special studies made on the Kelsey's holdings focus on relatively well-defined collections, such as Roman glass, brick stamps, and inscriptions.[2] Whereas European collectors specialized in assembling collections of the spectrum of marbles used by the Romans, there is no sign that this was Kelsey's goal, and we have been unable to find an explanation that defines his aims. Kelsey's diaries refer to the stones variously as "architectural specimens" or "building materials." The category "marble samples" was also in the minds of the curators who arranged the collections into their current form. An "explanatory note" bound into the first accession book at page 80 explains that in 1924 the museum curators accessioned the "building materials," relying on "a bundle of slips prepared by Professor Kelsey in 1900–'01 in Rome, and some sheets of hasty notes, poorly written

1. Dolci and Nista 1992.
2. Root 1982; Bodel 1983; Tuck 2005.

and badly spelled made by various Italian stone cutters were turned over to help in the work of entering the specimens." Curatorial decisions in assigning materials to cabinets in the museum gave us a reasonably well-framed corpus of objects of colorful marble or kindred stone, mostly small and mostly lacking in clear identity. Included among the cabinets were lime deposits from aqueducts, fieldstones, and an assortment of volcanic stones. With ambiguities in "building materials," "architectural specimens," and "marble samples," the term "decorative stone" was chosen so as to define the collection as artifacts intended for decorative use and display.[3] Within these parameters, utilitarian, structural, architectural, and unworked geological materials were omitted. Objects in the collection of decorative stone consist of fragments of floor paving, wall veneer, figural and geometric *opus sectile*, mosaic tesserae, reworked architecture, decorative water basins and furniture, reused sarcophagi and sculpture, unidentifiable objects of exotic stone classified as "other," debris from ancient quarries, and commercial plaques from modern quarries.

The long post-antique history of ancient marble has resulted in a confusing nomenclature.[4] The ancient definition of marble included alabasters, breccias, granites, porphyries, diorites, basalts, and finer-quality limestones. Pliny's *Natural History*, Strabo's *Geography*, and Diocletian's Edict on Maximum Prices are the main sources for the names of the most popular marbles. Latin terms usually referred to the geographic origin of the stone (e.g., *marmor Numidicum*). Less commonly, they were bestowed to honor a person (e.g., *marmor Luculleum*, *marmor Claudianum*), or describe the stone's appearance (such as the snakeskin-like *lapis ophytes*). Italian craftsmen working in the Renaissance and Baroque periods coined their own sets of terms, adding complexity to an already confusing vocabulary. Italian names for stone tend to describe its physical characteristics or the place it was quarried. No example is more emblematic than *pavonazzetto*, a term that comes from *paonazzo*, an Italian word meaning "violet." The Romans, who spent enormous resources to obtain this purple-and-white marble from deep within Turkey, had various designations for it. Strabo called it *marmor Dokimeion* for its source next to the ancient city of the same name. Claudian called it *marmor Synnadicum* after the city of Synnada, which neighbored Dokimeion. Both cities were in the province of Phrygia, and Juvenal referred to the stone as *marmor Phrygium*. To add to the confusion, the term *pavonazzetto* can also be applied to similar stones from entirely different sources. Italians stonecutters used *pavonazzetto* to refer to another purple-and-white marble from the Apuan Alps of Italy. The Kelsey Museum curators clearly fretted over identification, as the explanatory note dealing with accession indicates (see p. 9). Esther Van Deman's personal correspondences show she was perplexed, if not downright annoyed, by scholars using "unscientific" names (see Van Deman's letter to Dr. Butler, quoted on p. 9).

Studies on ancient marble have been on the rise in recent decades, making contributions on topics as varied as the discovery of new quarry sources, scientific analyses, workshop production, and the marble trade, and these investigations have informed our interpretation of the materials here. Some studies illustrate the diversity and range of luxurious marbles available to the Roman elite.[5] Others focus on the interpretation of specific objects, such as Roman *labra* or *opus sectile*.[6] Also invaluable are the archaeological studies of marble revetment discovered in situ, which have revealed much about the decorative programs as well as technology for installation.[7] While existing scholarship forms the basis for understanding the materials, the Kelsey collection is unique and idiosyncratic in that it does not represent every object made from colorful and exotic stones, but rather is a reflection on each collector's own interests, ambitions, and practices. In the following sections, aspects of production, distribution, original use, display, materiality, and technology will be addressed.

---

3. As in the title of Price 2007, *The Sourcebook of Decorative Stone.*
4. For clarification on terminology, Appendix A, Gazetteer of Quarries, provides both the Italian and Latin names for decorative marbles and other types of stones represented in the Kelsey collection.
5. De Nuccio and Ungaro 2002.
6. Ambrogi 2005; Guidobaldi and Angelelli 2002; see the chapter on *opus sectile* in Dunbabin 1999, 254–68.
7. Barker and Fant 2019a; İlhan 2014; Barker et al. 2014; Ball 2002.

Provenance and Materials

About half the decorative marbles in the Kelsey Museum collection have some sort of attested provenance. While the accession books often record the country, city, or building from where a collector acquired an object, an exact excavation findspot is never provided. The means by which Kelsey and his associates gathered the decorative stones now in the museum reflect attitudes toward collecting at the turn of the twentieth century. Their methods of purchasing materials from dealers, workmen, and even the archaeologists in charge of excavations are certainly questionable by today's standards of ethics and scientific rigor. Some pieces were taken from well-known buildings, such as Olympia's stadium in Greece, the summer palace of Herod at Jericho, and the Baths of Diocletian in Rome, but our ability to add new information about architectural decoration and design of these structures is hampered by the absence of precise archaeological context. Nevertheless, broader trends in art history, museum studies, and more recently in classical archaeology have embraced publications on the history of museum collections and object biographies, even in cases where the artifacts have dubious origins.[8] The Kelsey Museum's decorative stones have for the most part been held in storage cabinets since they were accessioned; we believe they deserve to be brought to light.

The geographical scope of the sources of the Kelsey marbles is vast and reflects the travels of the collectors as they visited archaeological sites and quarries across the Mediterranean region (figs. 2.1 and A.1). For example, on his extensive travels in Italy and Greece during his year teaching in Rome (1900–1901), Kelsey toured many archaeological sites and active quarries, and often obtained marble samples from the dig directors and quarry owners with whom he met. Almost half of the decorative stones in the Kelsey collection (over 300) were collected in Italy, and nearly half of those in Rome (fig. 2.2). The Baths of Diocletian were a favorite collection point for Kelsey; the sixty-seven pieces of marble from the Baths represent the largest collection from any single building in the collection, although it is uncertain if the objects were a part of the Baths themselves or merely purchased there. Figure A.1 shows the geographic locations of the quarries represented in the collection. Included on the map are only the ancient quarries where our attestations are secure and the locations known. The gazetteer of quarries (appendix A) provides a comprehensive list, with the majority being ancient Roman quarries. Others that were not opened

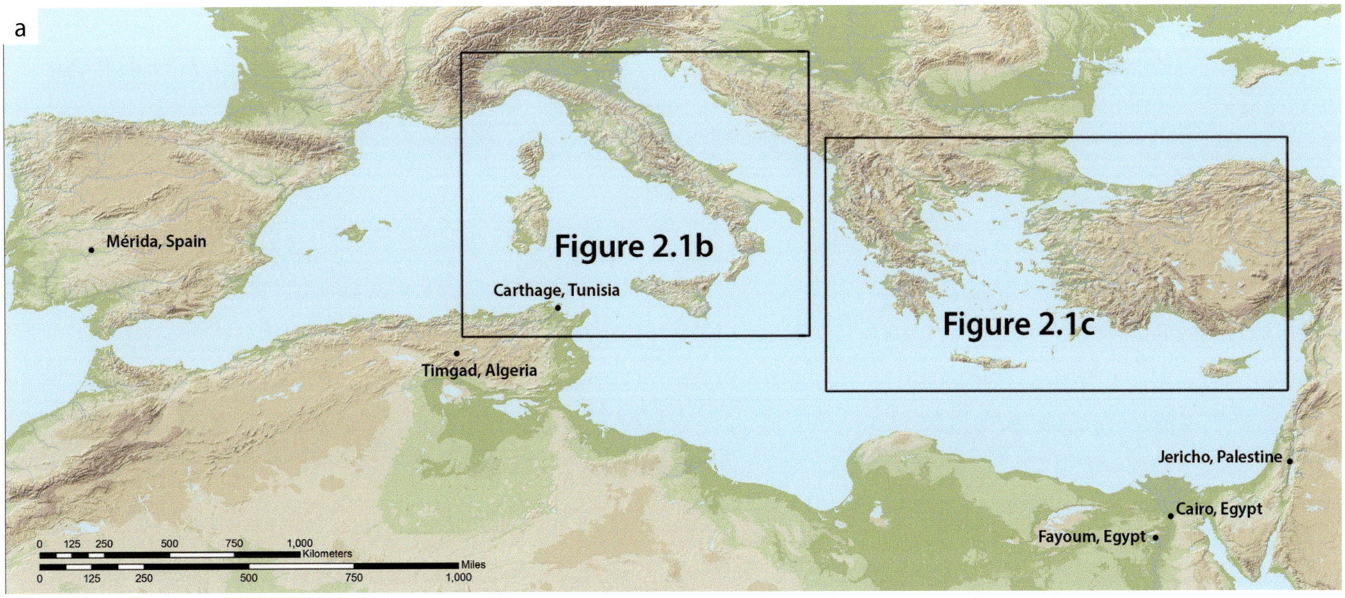

Figure 2.1a–c. Maps showing archaeological sites across the Mediterranean from where decorative stones were taken. Maps by Lorene Sterner.

8. Osborne 2015.

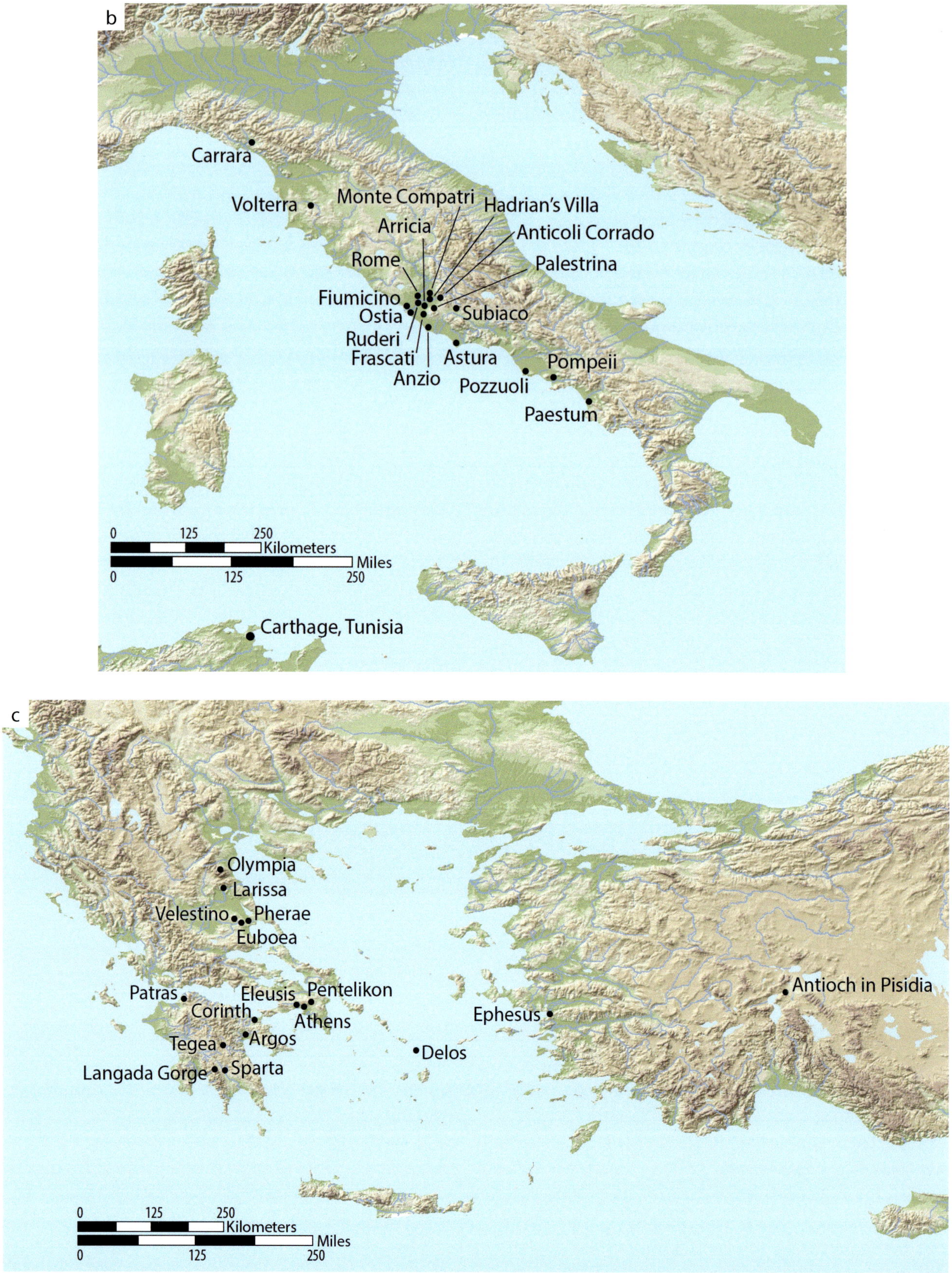

Figure 2.1a–c. Maps showing archaeological sites across the Mediterranean from where decorative stones were taken. Maps by Lorene Sterner.

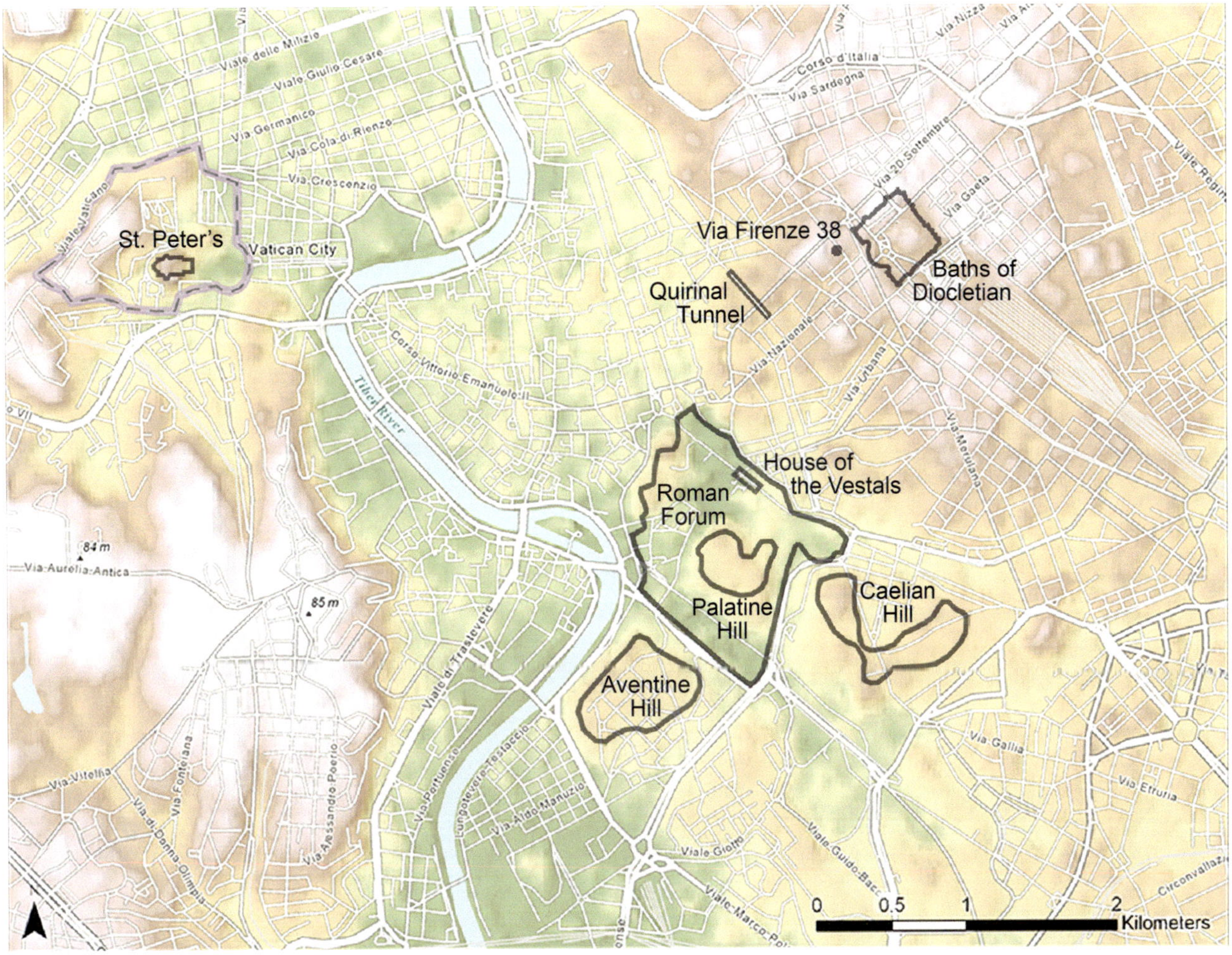

Figure 2.2. Map showing archaeological sites in Rome from where decorative stones were taken. Map by Lorene Sterner.

until the medieval or Renaissance period, such as the *verde di Prato* quarry near Florence, are listed alongside the pieces determined to be modern.

There is an extraordinary range of lithotypes among the ancient pieces, with over sixty-four varieties and subvarieties from well-known ancient quarries. Adding in the fragments designated as modern, the number climbs to over eighty separate lithotypes from the Mediterranean. While a good teaching collection should be representative and diachronic, those in the Kelsey show a marked bias for the most prized Roman imperial marbles. A breakdown of the pieces by lithotype reveals that the collectors were drawn to the exotic and colorful marbles used in the imperial art and architecture of Rome, with the most numerous being *portasanta*, *giallo antico*, *porfido rosso*, and *pavonazzetto* (fig. 2.3). For example, the sixty-four pieces of *portasanta* Kelsey collected, which make up 9 percent of the entire collection, were not necessary to represent its half dozen subvarieties. That the rarest and most valuable stones from Egypt were collected as single specimens underscores an impetus toward collecting for representation, yet on the other hand, common Carraran white and *bardiglio* were purchased in bulk. Although Rome was a focal point, the materials with which Republican Rome was built and decorated are a curious omission. Kelsey almost completely ignored Republican-era lithotypes, buying only two pieces of the local tufas like *sperone* and *peperino*, used for four centuries into the time of Cicero; a few examples of travertine from Tivoli; basalt for paving quarried south of Rome; and a fine, buff-colored limestone called *palombino*. The building stones of Pompeii are neglected entirely. The presence of white marbles is also significant at around fifty pieces. Similarly, having a complete record of the major white

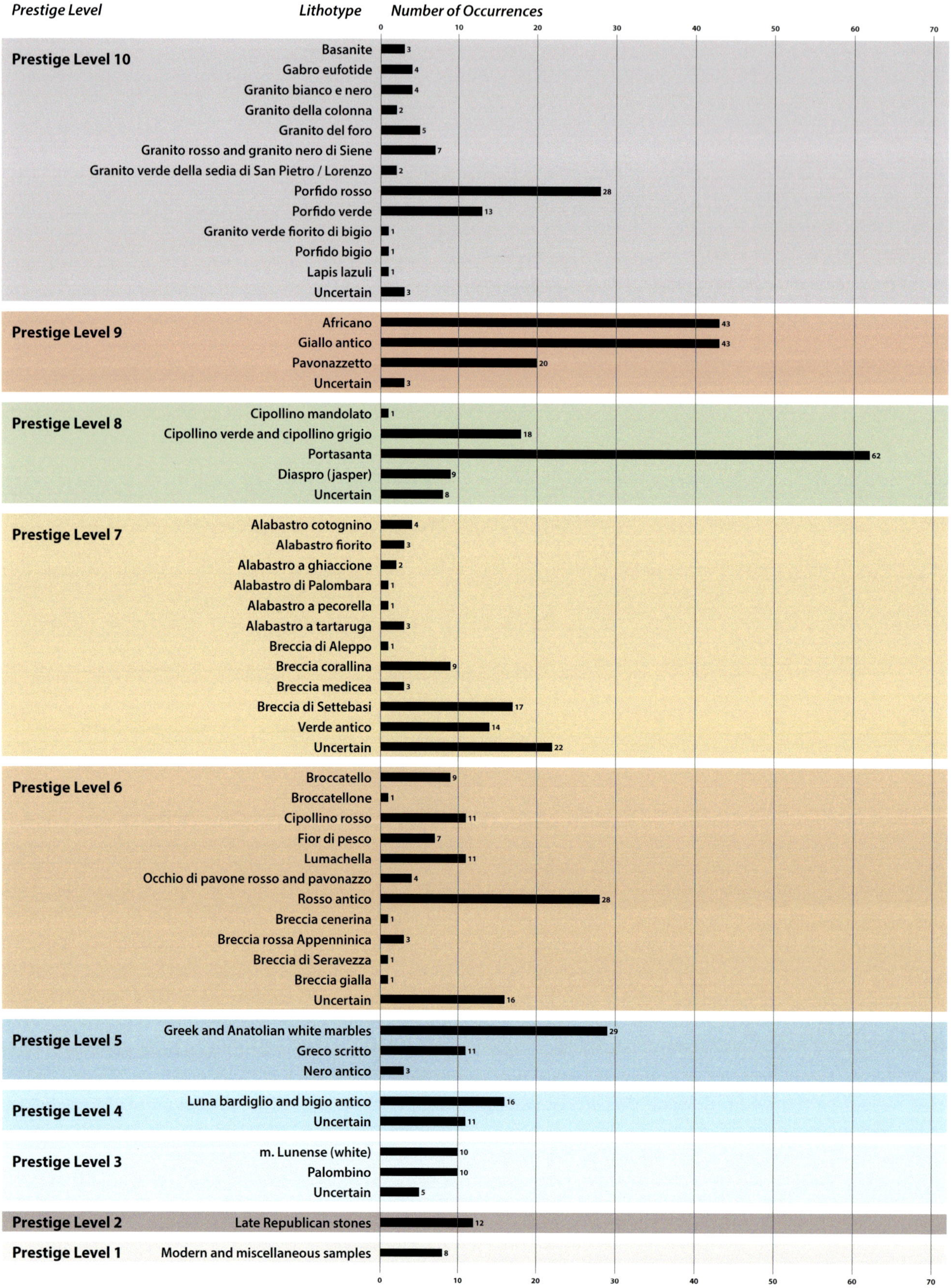

Figure 2.3. Distribution of lithotypes in the Kelsey decorative stone collection by Prestige Level.

marbles quarried in antiquity was clearly not part of the procurement strategy—if there was one—of any of the collectors. While Carraran and "Island" white marbles are present, it is unclear if some of the most important white marbles, such as Paros and Proconnesian, were included. For this reason, isotopic and petrographic analysis was desirable and was performed on select artifacts with known provenance. The results are reported in chapter 3.

Materials from Quarries

Quarries were of great interest to Kelsey, and upon his arrival in Italy he wasted no time in visiting them. The accession books record that a trip was made to Carrara in September of 1900, where he acquired at least three pieces (cat. nos. 557, 558, 559). While in Greece, in the spring of 1901, Kelsey made excursions to Mt. Pentelikon, 20 km northeast of Athens. There he met and was accompanied by Arthur Hill,[9] the president of Athens-based Marmor Limited, which quarried the pure white marble of Pentelikon. Kelsey collected debris from the ancient quarries "w. of entrance to cave" (cat. no. 503), the "new quarry, north side" (cat. no. 501), a "new quarry, above monastery" (cat. no. 502), as well as a "piece from block cut for monument of Von Moltke, Berlin, North quarry, Pentelicus" (cat. no. 500). Presumably this was when the German sculptor Joseph Uphues was beginning work on his statue of Helmuth von Moltke, who served a long tenure as chief of staff of the Prussian Army. The sculpture, entitled *Der große Schweiger* ("the Great Silent One"), was dedicated in 1905 in the Square of the Republic in Berlin, some five years after Kelsey's visit to Mt. Pentelikon. As Kelsey and his group of students traveled into northern and central Greece, Hill helped to arrange a stop at the *verde antico* quarry in Larissa, Thessaly, by sending a letter of introduction to the quarry manager. There they collected debris from the new quarry "whence columns for Westminster Cathedral are being taken" (cat. nos. 685, 686, 687).

Original Function of Kelsey Marbles

Paving or Veneer

The 454 pieces of floor pavement and wall veneer comprise well over half (64%) of the Kelsey's collection of decorative marble (fig. 2.4). A preliminary definition of paving and veneer and how it is distinct from *opus sectile* and architectural revetment may be helpful here. Paving and veneer have been worked in such a way as to have two or more flat and parallel faces. Usually the exposed face was smoothed, while the posterior was left at an intermediate stage of finish. In their original form, most pieces were rectangular or square, while more complicated shapes such as triangles, circles, and lozenges should be classified as *opus sectile*. Architectural revetment imitates three-dimensional architecture in two-dimensional form. Technically, all surface adornment is architectural, as it is part of architectural decoration, but here architectural revetment is distinguished from plain as having a profile or relief inspired by classical architecture.

Criteria particularly helpful in labeling the marbles as paving or veneer are measurements of thickness, height, and width; lithotype; and treatment of the flat surfaces (principal faces, number of edges of the original object preserved, shape). Ancient workings, such as tool marks or polishing, were also documented. Around thirty pieces of paving and veneer preserve technical aspects, confirming their use in buildings from antiquity. In particular, traces of mortar encrusted on a fragment reveal that it was set in a mortar bedding as part

9. Hill was an active member of the American School in Rome and British School in Athens. Shortly after his dealings with Kelsey, both institutions report his death in 1902 (Baker-Penoyre 1902–03, 396; West and Norton 1903, 31). This is not the same Arthur Hill who was a University of Michigan Regent and the benefactor of Hill Auditorium on the University of Michigan central campus in Ann Arbor.

of the preparatory foundation. Likewise, the presence of metal dowels or empty holes for the insertion of metal clamps is evidence of the mounting system used for attachment to a wall.

A number of factors make it impossible to distinguish the fragments as either floor paving or wall veneer. Many pieces lack the normal features associated with their original surface and have been reworked either in antiquity or in modern times. Although over 310 pieces of paving or veneer have some sort of provenance, very few were found and removed in situ, and most have an indeterminable origin. To avoid erroneous classification, the label "paving or veneer" was chosen. Additionally, few publications deal with marble coverings, partially because it is unusual to discover them in situ, but also because they are often treated as inferior archaeological materials. Studies on wall veneer and floor paving have attempted classification where context is known, but given that the Kelsey artifacts are largely decontextualized, incontrovertible proof for attribution as either paving or veneer is not possible. The strongest indications of use as wall veneer are cat. nos. 478 and 479, which have holes for metal pins that would have secured them to a vertical surface (fig. 2.5).

Roman writers used a host of terms to refer to the thin slabs of colorful marble that decorated the floors and walls of public and private buildings, and engaged in vigorous debates (and often scathing condemnation) over the Roman elite's proclivity to spend lavishly on decorative stone.[10] The artifacts in the Kelsey Museum's collection underscore the extent to which Romans loved decorating with exotic and beautiful stones. Difficulties abound when using the written sources to understand the material evidence, since they can be vague, varied, and imprecise. Cicero and Suetonius refer to *pavimenta* for marble floor paving, while Vitruvius mentions floors paved with slabs of stone cut in geometric shapes as *sectilia pavimenta*.[11] This distinction between floor and wall decoration is evident elsewhere in the sources. Seneca the Elder describes the variegated stone cut to cover walls (*parietes*) with a thin façade (*varius ille secatur*

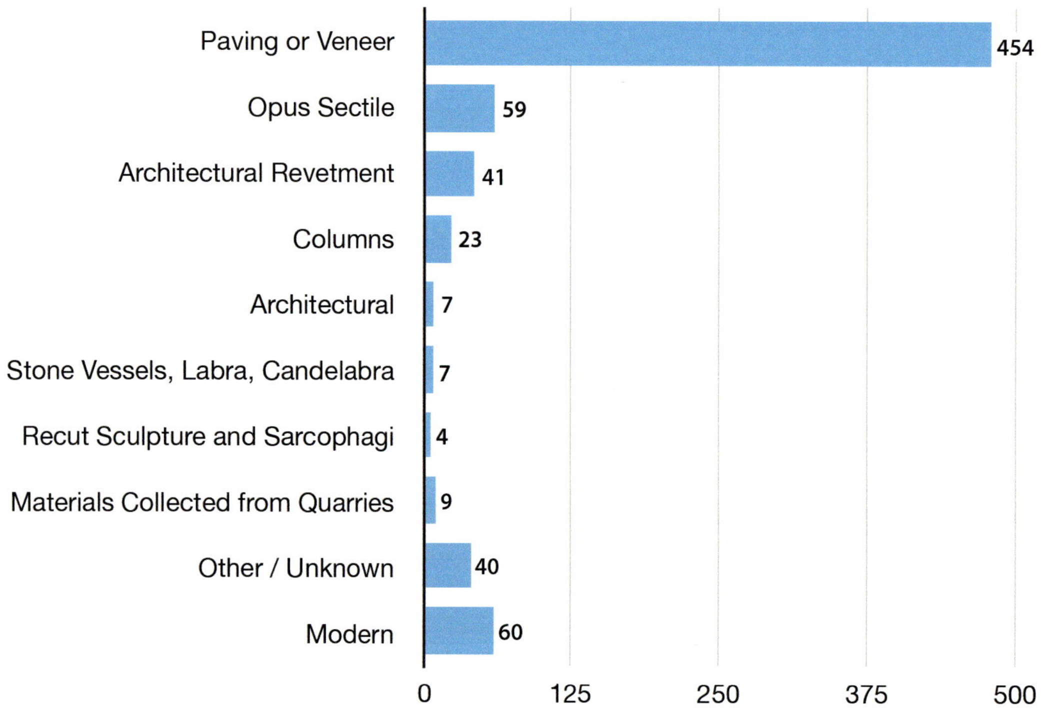

Figure 2.4. Types of archaeological materials in the Kelsey decorative stone collection.

10. Dubois-Pélerin 2007; Barker et al. 2014, n. 14.

11. Cicero, *De Legibus* 2.2.5; Suetonius, *Augustus* 72; Vitruvius, *On Architecture* 7.1.3–4.

Figure 2.5. Holes for metal pins used for wall attachment. Cat. nos. (a) 479 and (b) 478.

lapis, ut tenui fronte parietem tegat).[12] The Late Antique lexicographer Nonius Marcellus uses the term *pavimenta et parietes incrustatos*, referring either to stone paving, veneer, or mosaics. The most common Latin term for veneer was *crustae*.[13] According to Pliny, the earliest attestation of the custom of lining walls with slabs of marble was documented in Rome in the first century BCE,[14] when Mamurra, Caesar's chief engineer, became the first person to cover the walls in his house with *crusta marmoris*. Vitruvius tells us that the earliest type of wall painting was meant to imitate marble veneer (*crustae marmoreae*),[15] showing that the trend for embellishing interiors with colored marble had become so widespread that it branched out to other media. For the most part, modern scholars try to follow the ancient vocabulary, or use invented terms like *opus sectile* to describe more elaborately cut veneer or paving.[16]

The Kelsey pieces of paving and veneer are notable for their substantial range of thicknesses, a metric that has proven to be an interesting criterion for investigation. One would assume that the thinnest pieces in the collection would not have been placed on floors, being unable to bear much weight, yet studies show this was not necessarily the case. In proposing a classification method, Guidobaldi and Angelelli noted that variability in thickness may not be relevant for distinguishing between paving and veneer, but rather between private and public settings, since studies have shown that slabs in residential settings tend to be thinner than those from monumental buildings.[17] In the Late Antique villa of Cazzanello, near Rome, Angelelli recorded floor pavings over 3 cm thick, while pilasters, cornices, moldings, and vegetal and figural *opus sectile* used as wall decoration ranged in thickness from 1.5 to 5.5 cm.[18] Fant, Russell, and Barker's study on the marble

12. Seneca the Elder, *Controversiae* 2.1.12.
13. *CIL* VI 10237; Vitruvius, *On Architecture* 7.5.1; Pliny, *Natural History* 35.2, 35.3, 36.48, 36.49; Seneca, *Epistulae* 86.6; Lucan 10.111–117; Sidonius Apollinaris, *Carmina* 22.146.
14. Pliny, *Natural History* 36.48.
15. Vitruvius, *On Architecture* 7.5.1.
16. On this subject, see Becatti 1969; Asimakopoulou-Atzaka 1980; Guidobaldi 1990; Guidobaldi 2000a; Guidobaldi 2000b.
17. Guidobaldi and Angelelli 2002.
18. Angelelli 2000, 90, 96.

bars at Pompeii and Herculaneum showed that *opus sectile* flooring was typically 0.5 to 2 cm thick, while wall veneer ranged from 1.5 to 6 cm in thickness.[19] The intact decoration from Villa A ("of Poppaea") at Oplontis displays how wall veneer and floor paving could be variously very thin or very thick, and that assignment to either category on the basis of thickness alone, especially with small pieces, is difficult.[20] In comparison to previous research, the Kelsey artifacts show an even larger range of measurements. Fragments of *opus sectile* range from 0.38 to 8.9 cm thick, with an average of 1.91 cm and a median of 0.97 cm. This relative thinness underscores the delicate craftsmanship typical of *opus sectile*. Overall, the 481 fragments categorized as paving or veneer varied from 0.4 to 13.1 cm, with an average of 2.82 cm and a median of 2.12 cm.

Some fragments defy classification, falling within one or more categories. Cat. nos. 47, 201, 291, 513, 518, and 589 have indications of a geometric or curvilinear shape normally associated with *opus sectile*, but because they bear only hints of their original shape they cannot be classified definitively as either *opus sectile*, veneer, or paving. For example, Candler collected a large piece of alabaster with a curved shape and beveled edge (cat. no. 291) "from the Great Theater at Ephesus" on September 6, 1867, that is likely to be wall veneer. Archaeological excavations at Ephesus have uncovered a *scaenae frons* richly adorned with imported marbles that included several varieties of alabaster.[21]

Ambiguities between decorative stone, building materials, and architecture are evident in a handful of fragments in the collection. Such materials were classified as "indeterminate architecture," and are included in this study because they were treated at some point in their history as paving or veneer, having one surface sawn, waxed, or polished.[22] Of special interest is cat. no. 603, whose label records that Candler took it from the Great Pyramid of Cheops on October 4, 1867. During his trip to Greece, Kelsey gathered architectural fragments from important monuments. Cat. no. 495 comes from the temple of Athena Alea at Tegea; cat. no. 596 is an ancient white limestone "block used in rebuilding of s. wall of Olympeium" in Athens. Similarly, cat. no. 595 came directly from the excavations of the Baths of Diocletian, from the "bottom of trench for foundations of new building," in October 1900.

Secure identification is also complicated by the fact that over one hundred fragments have been polished and covered with a thin layer of wax, which has since yellowed. The Kelsey accession book records that cat. no. 272, a sample bought from F. Romano on March 9, 1901, was cut with a copper wheel and then polished. Done to show texture and color, identifying features such as the fluting on columns were usually paid special care and left intact. Kelsey bought a number of fragments from Romano, who may have been responsible for selling Kelsey all the fragments treated in such a manner.

Of additional interest are the paving and veneer gathered en masse, put into bags, and assigned a single museum inventory number. On February 8, 1893, a bag of paving and veneer (cat. no. 701) was taken from Timgad, Algeria, as part of a large purchase Kelsey made in Tunisia from French authorities in Carthage and dealers in Tunis.[23] The thirteen pieces of cat. no. 701 include seven possible fragments of *breccia di Settebasi*, four of *africano*, and one each of *alabastro* and *portasanta*. Furthermore, Giuseppe De Criscio acquired many pieces of decorative stone from Pozzuoli and its vicinity. Cat. no. 704 is composed of assorted lithotypes, while cat. no. 702 includes thirty-eight pieces of *porfido verde* paving or veneer, *opus sectile*, and molding strips. Such groups likely belong to the same assemblage, but unfortunately, precise find locations were not recorded.

Our very broad definition of paving and veneer also includes non-decorative slabs used for more utilitarian purposes, such as pavement in courtyards, sidewalks, steps, thresholds, windowsills, and pedestals. Other fragments had more specific functions, such as the wedged-shaped cat. no. 700, which would have been fitted to a corner. A label on cat. no. 326, an alabaster fragment, records that it was taken from the "corner of [a] wall" on the Palatine Hill in Rome. Fifteen fragments are preserved in such a way that they have been tentatively labeled ancient paving or veneer. These are pieces that have no preserved edges, surfaces that are not flat or parallel, or that are simply too small to be identified with certainty. As with all the fragments, it

19. Fant, Russell, and Barker 2013, 199, 201.
20. S. Barker, personal communication.
21. Heberdey 1904; Heberdey 1905; Öztürk 2005, 8–10.
22. Cat. nos. 155, 296, 495, 503, 595, 596, 603, 611.
23. Pedley 2012, 51.

is difficult to tell if they were collected from original or secondary contexts. While reuse of veneer and paving was extensive in antiquity, it is almost impossible to tell if marble veneer, for example, was cut from an old block or reused as secondhand veneer.

The collection contains fifty-two pieces of *opus sectile*, identified as having a geometric, curvilinear, or figural shape. Scholars have adopted the term to refer to the Roman technique wherein marble is cut and inlaid to fit a specific design for mural or floor decoration. Pliny seems to be describing this technique when he says, "Slabs of marble not only cover entire walls but are also engraved with patterns and decorated with twisting lines that represent objects and animals."[24] Appreciation for *opus sectile* in the Roman world stemmed from both a love of exotic stones and an appreciation of the technical difficulty of cutting thin pieces of stone into intricate and ornate shapes. The most common decorations were simple geometric shapes in repeating patterns. Well represented within the collection are the triangles (9), rectangles (4), lozenges (4), segmented circles (4), and hexagons (1) cut from favorite polychrome marbles that formed parts of larger decorative ensembles (figs. 2.6–9).[25] Cat. nos. 157 (*giallo antico*) and 420 (*cipollino rosso*) are large semicircular *opus sectile* that recall a common motif in geometrically designed floors or walls where segmented disks formed a focal point or part of a repetitive pattern (fig. 2.9c and d).[26] In the homes of the Roman elite, interiors were decorated with elaborate scenes from mythological stories and natural landscapes. Individual features were carefully cut out of different materials and fitted together. Details on the interior could be engraved after the inlays were set in place. Curvilinear ornamentation was more challenging to execute than geometric shapes. Placement on walls allowed for better viewing of the variegated colors and intricate craftsmanship that made them desirable, but *opus sectile* was just as often positioned on floors.

In addition to *opus sectile*, the technique of intarsia was used in mural decoration. While an *opus sectile* design is achieved by juxtaposing separate pieces directly over a foundation layer of mortar or plaster, *opus interrasile* uses the champlevé technique, in which colorful marbles are inserted into stone recesses to create an elaborate composition.[27] Examples of intarsia date exclusively from the mid-first to second centuries, with *opus sectile* coming into favor during the Late Roman and Byzantine periods.[28]

Scholars have interpreted the *interraso marmore* Pliny refers to as engraved pieces set into a hollowed-out background matrix, and have coined the term *opus interrasile*.[29] While Pliny's language is by no means definitive, the relatively rare examples of engraved marble approximating his descriptions as "paintings in stone" are suggestive. Dancing figures composed of separate pieces of stone have been uncovered in the House of the Colored Capital at Pompeii, the nymphaeum of Nero's Domus Transitoria in Rome, and the Claudian nymphaeum at Baiae, Punta Epitaffio.[30] Botanical motifs of the kind seen in the Kelsey collection are preserved in Hadrian's Villa, an imperial domus on the Janiculum, and the summer residence of Domitian at Castel Gandofoldo.[31] Vegetal intarsia depicted flowers, intertwining plant stalks, or acanthus leaves; a slate setting was preferred, or, in special cases, obsidian. Cat. no. 449 preserves the more complex intarsiated decoration wherein cells were carved out to serve as supports for thinner decorative elements (fig. 2.10). It has the typical gray slate setting with recesses for the inlay of *rosso antico* in a floral design. Two *rosso antico* pieces remain in the channels.

Based on similarities in shape, color, and lithotype, five different sets of figural *opus sectile* or intarsia have been identified in the Kelsey collection. With the exception of cat. no. 449, which was collected by Kelsey in Rome, provenance for the *opus sectile* is not known. Candler's acquisitions deserve special attention because he appears to have collected them as groups. One such group is composed of two fine pieces of *rosso antico* (cat. nos. 464 and 465) cut into S-shapes that would have fit into a larger composition (fig. 2.11).

24. Pliny, *Natural History* 35.2, trans. Healy 1991.
25. Triangles: cat. nos. 9, 69, 110, 112, 113, 152, 344, 490, 528, 540?. Rectangles: cat. nos. 466, 517, 540? 546, 699. Lozenges: cat. nos. 118, 120, 475, 582. Segmented circles: cat. nos. 153, 157, 308, 420. Hexagon: cat. no. 583.
26. Guidobaldi and Salvatori 1988.
27. Bonanni 1998, 259–78.
28. Dunbabin 1999, 263–64.
29. Pliny, *Natural History* 35.2.
30. Bonanni 1998, 262–64.
31. Dubois-Pélerin 2007, 106–12; Violante 2013; Filippi 2005b; Sapelli Ragni 2009, 104; De Nuccio and Ungaro 2002, cat. nos. 195–98.

Figure 2.6. *Opus sectile* triangles. Cat. nos. (a) 9, (b) 69, (c) 110, (d) 112, (e) 113, (f) 152, (g) 344, (h) 490, (i) 529.

Figure 2.7. *Opus sectile* rectangles. Cat. nos. (a) 466, (b) 517, (c) 546, (d) 699.

Another group collected by Candler is comprised of six well-preserved *giallo antico* fragments with scalloped edges (cat. nos. 142, 143, 145, 146, 147, 148; fig. 2.12a–f). Their flowing shapes indicate they were part of an ensemble depicting the natural world. Lines imitating vegetal stalks were incised on the interior surfaces of the pieces to enhance the motif. A circular wedge of *giallo antico* (cat. no. 145), only 0.42 cm thick, and a thin fragment of a rectangle (cat. no. 150) may also have been part of this group (fig. 2.12g–h).

Illustrated together in figure 2.13 are four separate sets of light-colored stones cut into vegetal and floral shapes; they bear stylistic similarities to one another and are well preserved. The only piece collected by Van Deman is a large white marble acanthus leaf (cat. no. 590; fig. 2.13a), while the others were all acquired by Candler. Nearly identical to Van Deman's acanthus leaf are three matching white marble leaves with incised lines representing foliage veins (cat. nos. 568, 569, 570; fig. 2.13b–d). Another set of four stylistically similar pieces may have originally belonged to the same decorative composition (fig. 2.13e–h): Two fragments made from a creamy white limestone, possibly *palombino*, are leaves with incisions that preserve traces of pink paint (cat. nos. 573, 574; fig. 2.13e–f). Cat. no. 572 (fig. 2.13g) is a delicate fragment of white marble in an S-shape of tapering width, much thinner (0.44 cm) than the others. The most elaborate is a quadrangular piece (cat. no. 575, fig. 2.13h) with engravings along the border and interior. In the middle is a stylized lotus flower similar to an adaptation of an easternizing motif seen in Roman Third Style wall painting. The last group consists of four pieces of a creamy white limestone (*palombino*?) depicting a flower bud, the curling tips of leaves, and the body of a leaf (cat. nos. 576, 577, 578, 579; fig. 2.13i–l).

Given the relative rarity of well-preserved vegetal *opus sectile* and intarsia, that the Kelsey pieces do not preserve their background setting and have no contextual information to help determine a chronology, attribution as either *opus sectile* or intarsia is not possible. The shapes and technical execution of the leaves are most similar to the vegetal intarsia found in Hadrian's Villa and the Roman villa of Nonii Arrii on Lake Garda, where leaves of *giallo antico* and creamy white

Figure 2.8. *Opus sectile* lozenges and hexagon.
Lozenges: cat. nos. (a) 118, (b) 120, (c) 475, (d) 582; hexagon: cat. no. (e) 583.

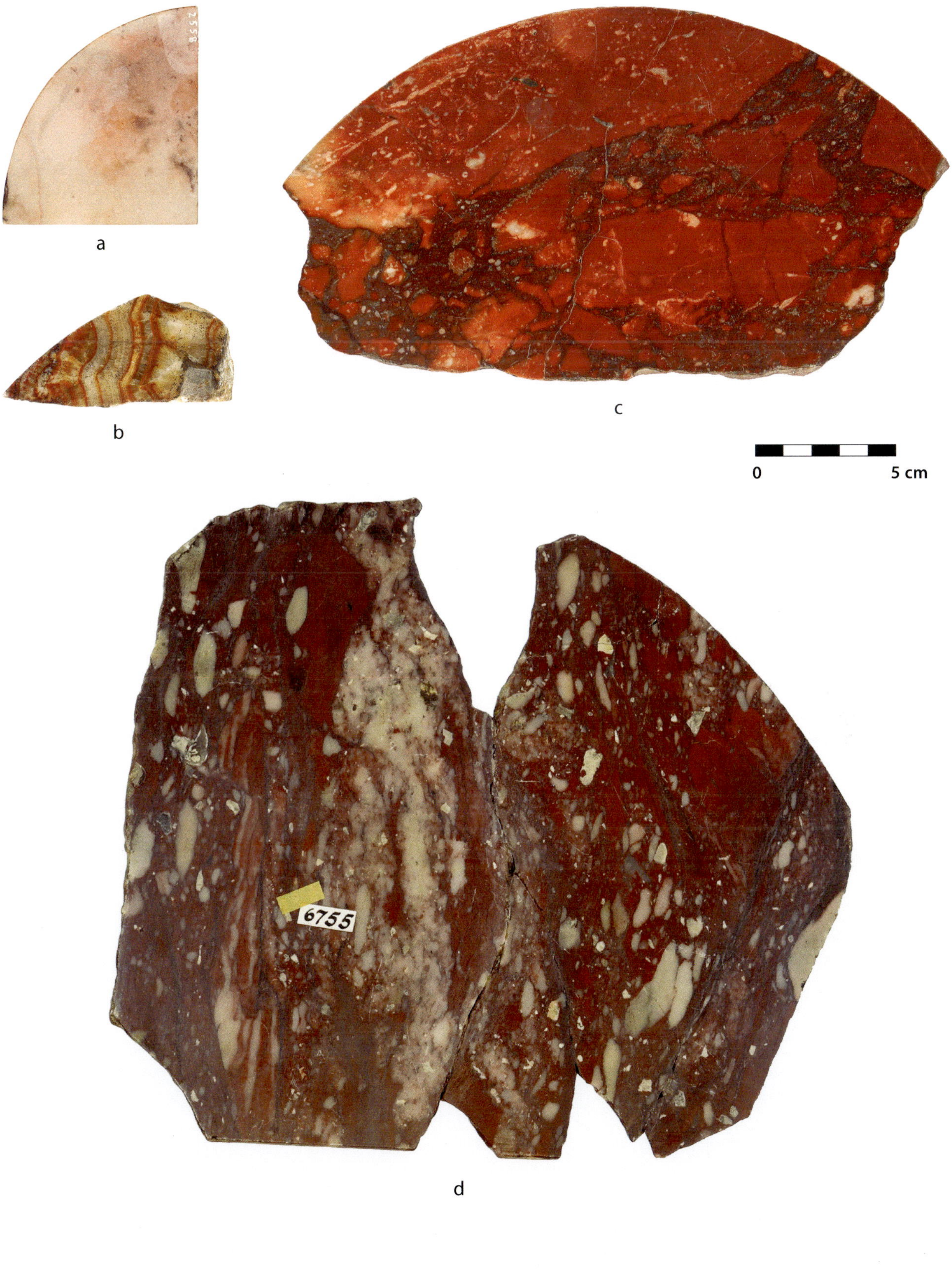

Figure 2.9. *Opus sectile* circular elements.
Cat. nos. (a) 153, (b) 308, (c) 157, (d) 420.

Figure 2.10. *Opus interrasile.* Cat. no. 449.

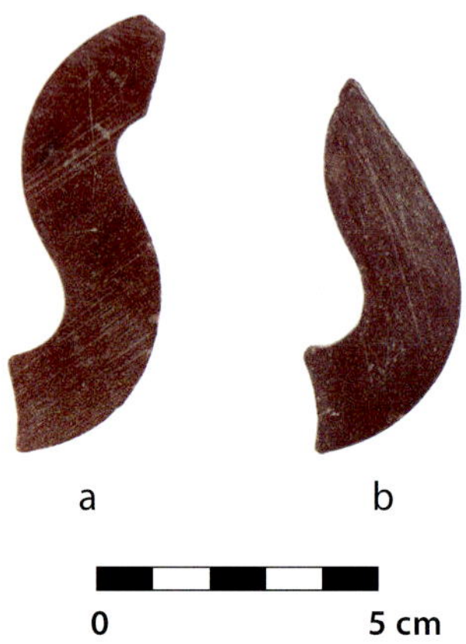

Figure 2.11. S-shaped *opus sectile*. Cat. nos. (a) 464 and (b) 465.

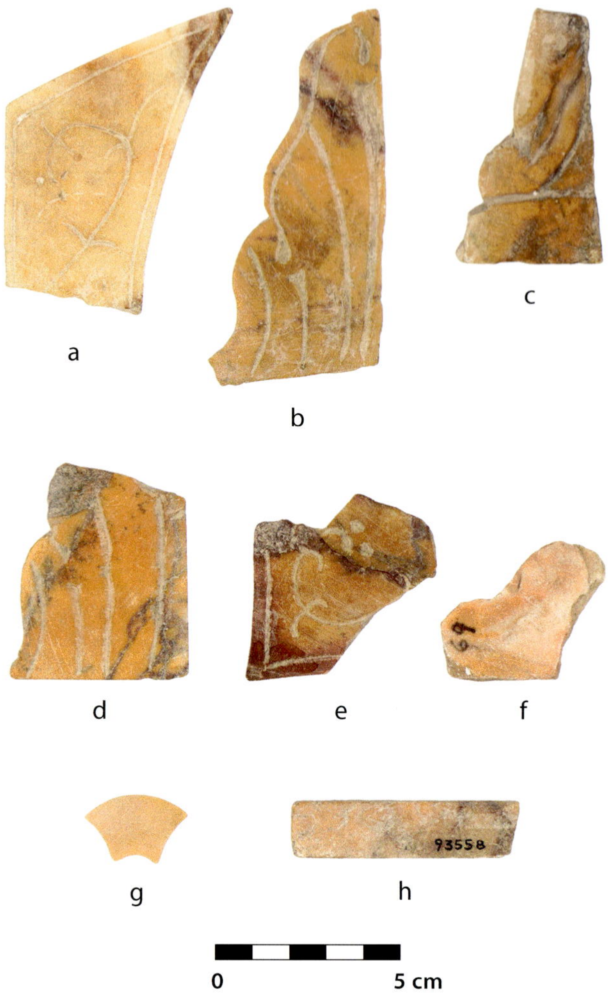

Figure 2.12. *Giallo antico opus sectile*. Cat. nos. (a) 142, (b) 143, (c) 144, (d) 146, (e) 147, (f) 148, (g) 145, (h) 150.

palombino predominate.[32] Likewise, the leaves may have been set into a Corinthian pilaster capital. Examples are preserved from the imperial villa of Trajan at Arcinazzo as well as the Horti Lamiani in Rome, where a *rosso antico* pilaster capital held multi-colored acanthus leaves (fig. 2.14).[33]

The extraordinarily well-preserved *opus sectile* floral scrolls in the Porta Marina in Ostia offer additional comparanda for imagining the original arrangement of the Kelsey fragments (figs. 2.15–16).[34] The *antae* pilasters that frame the exedra are composed of similar individual-cut pieces of leaves, vine tendrils, and flowers of a creamy white *palombino* set against a dark green background. The engraving on cat. no. 575 (fig. 2.13h) resembles in miniature version the volute capitals in the floral pilasters.[35] Decorative schemes found in situ from Roman villas in Italy preserve similar designs, but were executed with a range of materials using combinations of stone and glass. The S-shaped *rosso antico* pieces illustrated in figure 2.11 could have

32. Sacchi 2015, 141–43, fig. 21; Violante 2013; De Nuccio and Ungaro 2002, 471–74, fig. 2.
33. Mari 2004, 6; De Nuccio and Ungaro 2002, 418–19, cat. nos. 129–32.
34. Becatti 1969; Guidobaldi 2000a; Kiilerich 2014.
35. This is well illustrated in Guidobaldi 2000a, 258, fig. 7. For an example of a drawing of this motif on a pavement in an Early Christian baptistery in Piacenza, now lost, see Guidobaldi 2009, 387, fig. 22.

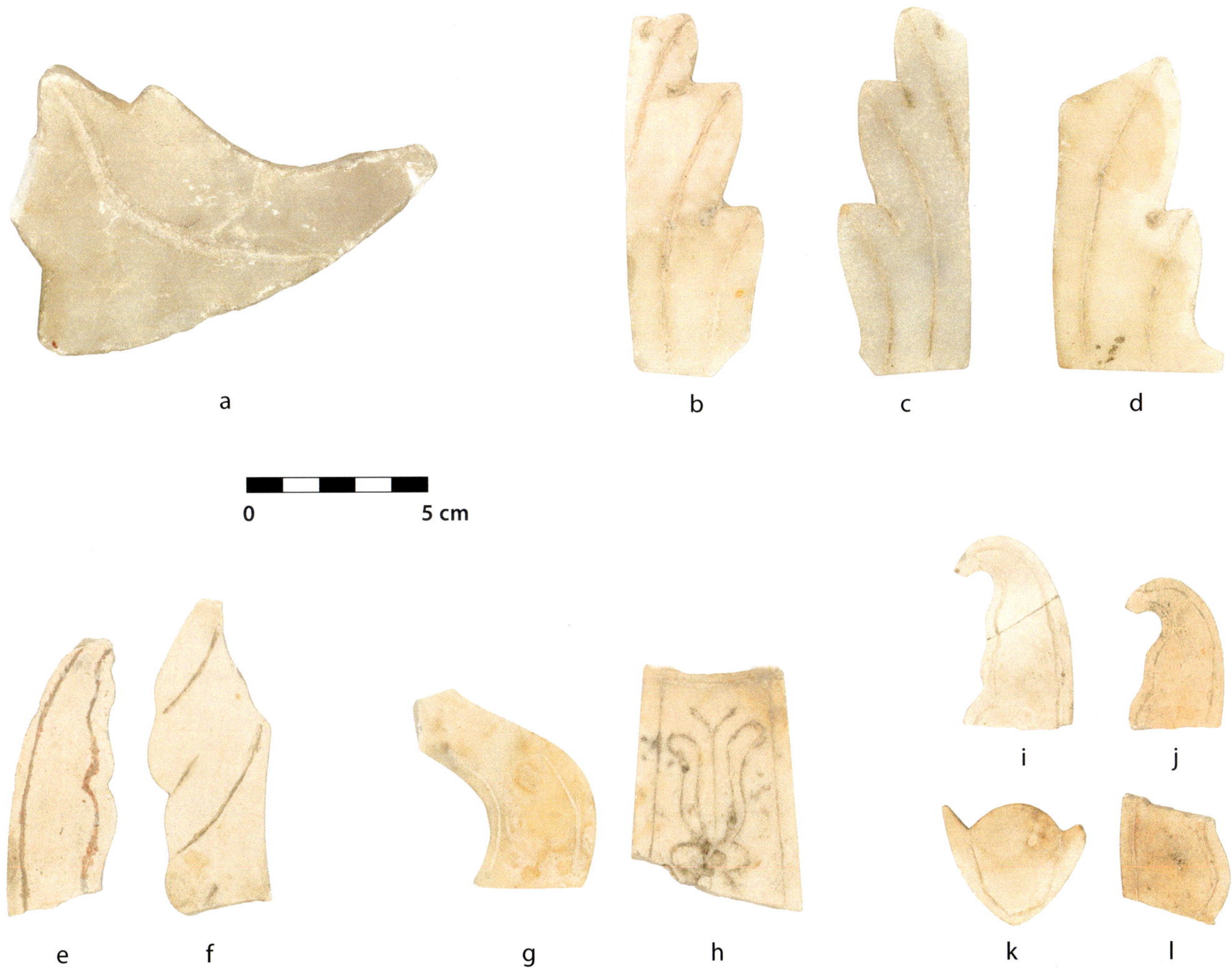

Figure 2.13. Vegetal *opus sectile*.
Cat. nos. (a) 590, (b) 568, (c) 569, (d) 570, (e) 573, (f) 574, (g) 572, (h) 575, (i) 576, (j) 577, (k) 578, (l) 579.

originally belonged to a larger composition, such as the glass *opus sectile* panel with vine tendrils or volute scrolls from the villa of Lucius Verus in Acqua Traversa.[36] In the *stibadium* of a Late Roman villa at Faragola, richly colored panels framed by pilasters bear vegetal motifs, creating a complex combination of geometric and figural patterns.[37]

The intrinsic value of these techniques was linked to the use of expensive imported stone and an astounding attention paid to craftsmanship. Each individual piece had to be ground and then abraded to achieve a close fit in its foundation. The beveling preserved on the sides of cat. nos. 464, 465, and 577 (figs. 2.11a–b and 2.13j) allowed for easy insertion into their compositions. Impossibly thin slabs of marble were cut by means of a saw, the markings of which are preserved on cat. nos. 420, 514, and 527. Cat. no. 145 (fig. 2.12g), which is 0.42 cm thick and 1.6 cm wide, shows the extraordinary fineness and exacting proportions to which these miniscule pieces were carved.

36. Saguì 1998; Saguì, Santopadre, and Verità 2012.
37. Volpe, De Felice, and Turchiano 2005; Turchiano 2008.

Figure 2.14. Pilaster capital from the Horti Lamiani. Capitoline Museum. Photo by Leah E. Long.

Figure 2.15. Detail of volute capital from Porta Marina, Ostia. Museo Nazionale dell'Alto Medioevo, Rome. Photo by Leah E. Long.

Figure 2.16. *Opus sectile* decoration from Porta Marina, Ostia. Museo Nazionale dell'Alto Medioevo, Rome. Photo by Leah E. Long.

Architectural Revetment

The Kelsey collection has a number of exquisite pieces that display how marble revetment could take on an architectural character. Marble veneer cut to imitate three-dimensional architecture was a structurally superficial decoration, distinguished from plain revetment as having a profile or relief. This kind of surface ornament was intended to reference canonical classical architecture and was usually applied to the inner façades of Roman buildings. Roman writers did not have a specific term to describe these faux architectural façades, lumping them into the category of *crustae*. In his disapproval for the latest fashion in wall painting, Vitruvius refers to the original marble architectural veneer that preceded the trend:

> *Ex eo antiqui, qui initia expolitionibus instutuerunt, imitati sunt primum crustarum marmorearum varietates et conlocationes. Postea ingressi sunt, ut etiam aedificiorum figuras, columnarum et fastigiorum eminentes proiecturas ...*
>
> Hence those of the ancients who first used polished coats of plastering, originally imitated the variety and arrangement of inlaid marbles. Afterwards the variety was extended to the cornices, and the yellow and red frames of pannels, from which they proceeded to the representations of buildings, columns, and the projections of roofs ... (*On Architecture* 7.5.1–2, trans. J. Gwilt).

Holes for clamps and dowels that were used to affix panels to walls are seen on many Roman buildings and attest to the ubiquity of decorative marble façades. Such patterns seen on brick skeleton walls in the Domus Flavia, Horti Sallustiani, and Hadrian's Villa suggest elaborate orthogonal revetment panels.[38] Illusionistic revetment was employed in both public and private contexts, adorning buildings as diverse as fora, amphitheaters, nymphaea, libraries, temples, houses, and apartments, yet the most impressive were undoubtedly reserved for imperial palaces, villas, and public buildings in and around Rome. Roman baths in particular received sumptuous ornamentation. Imperial *thermae* in Rome as well as in the provinces were enveloped by high, marble-veneered walls, which were in turn complimented by richly adorned *aediculae*. So popular was the decoration that First and Third Style wall paintings developed as convincing substitutes for real marble masonry or revetment in Roman houses, succeeding if they fooled the viewer into believing that painted walls were made from precious marbles.[39]

Because revetment was expensive, marble slabs could be combined with stucco or plastered wall paintings for economy. The array of designs preserved in the archaeological record indicates that customers had a great deal of choice in the arrangement of veneer, from the simple confinement of decoration to the lower parts of walls to complex arrangements of orthogonal revetment panels, pilasters, and entablatures. At Ephesus, the so-called Marble Hall in the second-century CE apartment block known as the Terrace Houses presents an extraordinary example of the conspicuous consumption of imported marbles, surpassed only by imperial villas like the Horti Lamiani in Rome. Thin panels were shaped like fluted pilasters, capitals, and moldings; piers, *antae*, and capitals, usually placed at doorways and openings to rooms, framed the succession of rows of marble revetment (fig. 2.17).

Attractive in situ architectural revetment is also preserved in Ostia at the Forum Baths, House of the Augustali, and House of Cupid and Psyche, as well as in luxury villas on the Bay of Naples.[40] At Oplontis, fragmentary dados with moldings remain attached to the walls in rooms 60, 64, 65, and 74 in Villa A, and fine, colorful veneer also decorated the lower zone in a similar manner at the Villa of the Papyri at Herculaneum.[41] Strikingly, far more examples are extant in imitation wall paintings than in actual marble. Even painted décor could be combined with real marble trimmings.[42] The corpus of evidence allows for decorative zones to be reconstructed. At the lowest level was the dado or socle course, which was composed of rectilinear arrangements of marble panels framed by horizontal fillets and moldings. Above the dado are the pilasters and orthostats that made up the middle zone and the focal point of the wall. In the Marble Hall at the Terrace Houses at Ephesus, the highest level contains square panels surrounded by *emblemata*, inset

38. Bruto and Vannicola 1990.
39. Clarke 2014, 356–57.
40. Bruto and Vannicola 1990.
41. Barker and Fant 2019a; Guidobaldi and Esposito 2009, 356–57.
42. McAlpine 2016; Barker and Fant 2019a.

Figure 2.17. Reconstructed Marble Hall at the Terrace Houses, Ephesus, Turkey. Photo by Sinan İlhan.

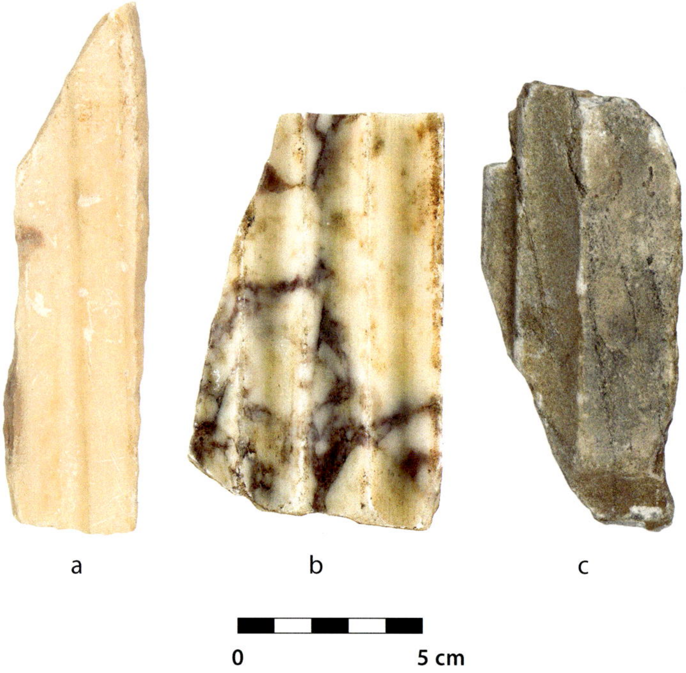

Figure 2.18. Architectural revetment, pilaster fragments. Cat. nos. (a) 176, (b) 178, (c) 195.

Figure 2.19. Architectural revetment, pilaster capital. Cat. no. 456.

circular disks or square panels of exotic colored stone (fig. 2.17).

The collection of architectural revetment in the Kelsey Museum is remarkable for its diversity. Nearly every element described above is represented along with a wide range of stone types. Classification of architectural revetment has previously been somewhat arbitrary, as it oftentimes falls between architecture and surface ornament. For morphological identification, I follow Boldrighini's proposals and terminology.[43]

Pilasters

There are three fluted pilaster fragments in the Kelsey collection (fig. 2.18). Cat. nos. 176 and 178 are fragments of *pavonazzetto* pilasters with shallow fluting meant to resemble a three-dimensional column. Cat. no. 178 is a typical example with its thin walls and polished surface, whereas cat. no. 195, a *cipollino* pilaster fragment from Palestrina, has thick and unpolished fluting. Decorative schemes within the pilaster zone have been reconstructed in the Marble Hall in the Terrace Houses at Ephesus as well the imperial baths at Sagalassos, and help to visualize the original placement of the Kelsey fragments (fig. 2.17).[44] The fluted pilaster fragments in the Kelsey would have had bases below and capitals above, and been surrounded by adjacent molded panels.

While in Rome, Kelsey acquired a fragment of a *rosso antico* pilaster capital (cat. no. 456; fig. 2.19). This fragment preserves the lower right portion of the capital and is decorated with a five-petal rosette with a central button. The rosette is tethered to a branch on the right and surrounded by scrolling vines and vegetation. The motif calls to mind the *lyrakapitelle*, where the calyx of the acanthus plant has vine tendrils with budding rosettes.[45] *Rosso antico* was especially favored for such Corinthianizing capitals, which capped vertical pilaster columns.[46] The back surface has been roughly chiseled to allow for better adherence to the wall (see fig. 2.39d, below).

Dados and Orthostats

Six fragments in the collection come from the dado or orthostat course (fig. 2.20), identified as such by having a molding in the form of simple raised edging. On either course, the molding served as a decoration to frame the outer portion on the top, side, or bottom of a panel. Cat. no. 183, made of *pavonazzetto*, and cat. no. 562, of Carrara marble, have simple, beveled moldings. Cat. no. 109, of *africano*, cat. no. 132, of *giallo antico*, and cat. no. 524, of *greco scritto*, have similar moldings with two grooved incisions, while cat. no. 567, of white marble, has three. This piece is notable as one of Candler's curios collected from the Basilica of St. John in Ephesus, Turkey, although publications on the excavations there do not record the unearthing of moldings.[47]

Fillets

Cat. nos. 48, 52 (both porphyry), 252 (*portasanta*), 385 (*verde antico*), 521 (*greco scritto*), 586 (white marble), and 646 (beige marble) are fillets, or, in Italian parlance, *listelli* (fig. 2.21), narrow molding strips that have a conventional height of 0.5 to 3 cm and tend to be square in section.[48] They come in a variety of profile shapes, as seen in the rounded edges of cat. no. 521, the thick triangle of cat. no. 646, and the concave profile of cat. no. 52. Cat no. 48, with its segmented curves, can be classified a *listello rettilinei con apicature* (rectilinear fillet with apexes).[49] In tectonic revetment, fillets marked a transitional zone and divided the dado, pilaster, and *emblemata* courses into registers. In *opus sectile* figural decoration they framed the figure within a border, and in geometric *opus sectile* formed rectangular outlines.[50] The beveling on cat. nos. 252 and 586 helped lock the compositions into place.

43. Boldrighini 2007.
44. İlhan 2014, 803; Waelkens et al. 2002, 377.
45. Gans 1992, 88–107.
46. A similar *rosso antico* pilaster capital belongs to the Museum of Fine Arts in Boston (1987.187), having been excavated at the Roman villa of Montegrotto Terme, Via Neroniana. For others from Ostia and the Capitoline Museum, see De Nuccio and Ungaro 2002, cat. nos. 125, 126, 127.
47. Keil 1951.
48. Boldrighini 2007, 55.
49. Boldrighini 2007, fig. 98.
50. Barker and Fant 2019a; Guidobaldi and Salvatori 1994.

Figure 2.20. Fragments of architectural revetment from the dado or orthostat course. Cat. nos. (a) 109, (b) 132, (c) 183, (d) 524, (e) 562, (f) 567.

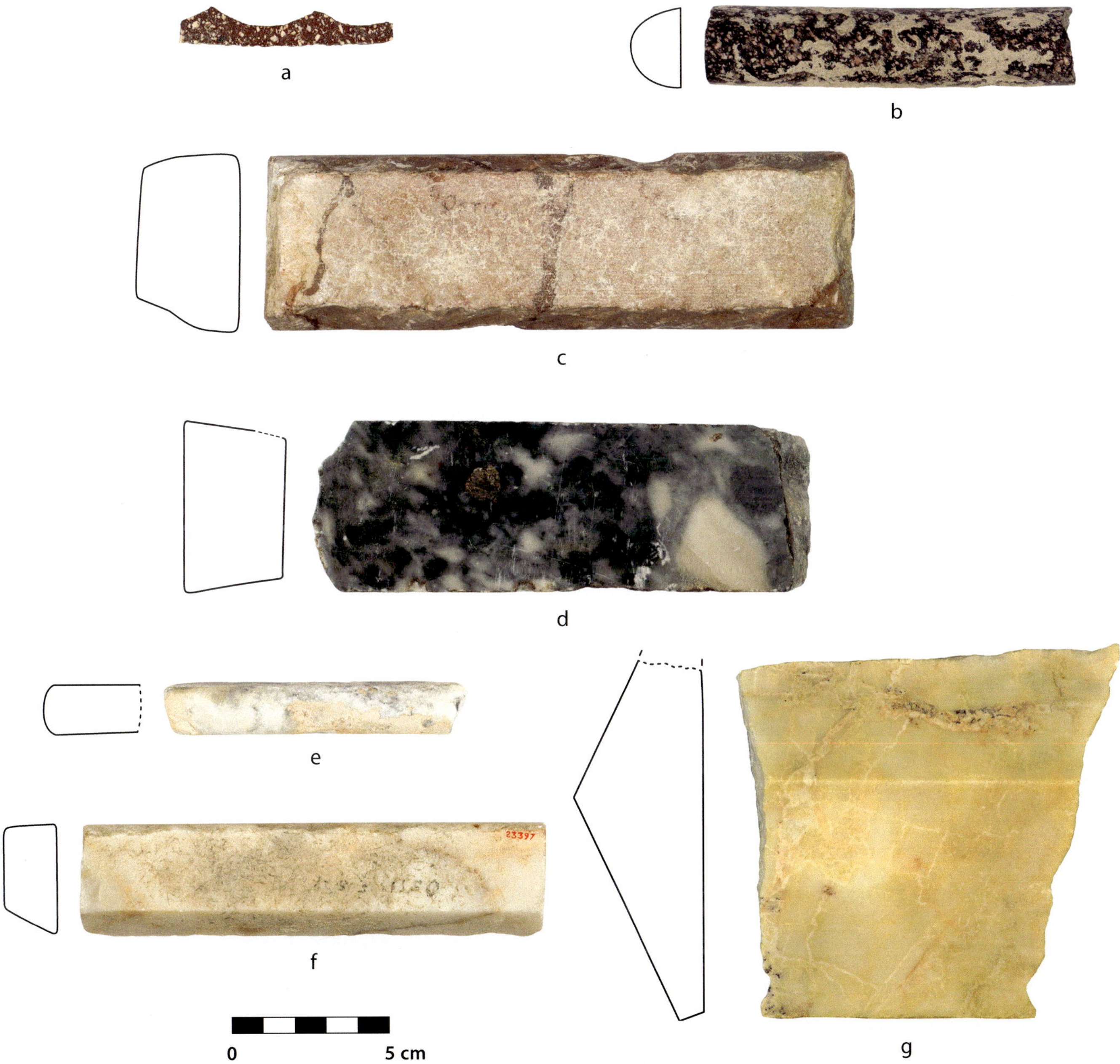

Figure 2.21. Architectural revetment. Fillets, cat. nos. (a) 48, (b) 52, (c) 252, (d) 385, (e) 521, (f) 586, (g) 646.

Figure 2.22. Fragment of architectural revetment. *Tondino*, cat. no. 486.

Tondini

Similar to the fillets are the so-called *tondini*, circular moldings used in wall decoration. Cat. no. 486 (*breccia gialla*), the single fragment of this type in the Kelsey collection (fig. 2.22), has a three-quarter convex semicircular profile. Its posterior edge has been left unworked, indicating that it was designed to protrude from its setting. As seen in the Domus Transitoria in Rome and in the House of the Augustali in Ostia, *tondini* were placed perpendicular to orthostats to serve as another option for partitioning.[51]

Moldings and Cornices

The twenty-four pieces of moldings and cornices represent the largest corpus of architectural revetment in the collection. In general, such materials have not been subject to systematic cataloguing and analysis, and in many cases have been discarded, heaped up in piles at the site of excavation, or shelved in museum depots without proper documentation. Dell'Acqua's 2014 study on fragments excavated at Brescia in northern Italy represents an attempt to interpret this class of materials independently, although a number of studies publish moldings with other categories of finds.[52] Used as framing devices within features of wall revetment or floor paving, they could also form separations in walls, offsetting doors and windows as well as decorative niches and *aediculae*. Some notable examples of in situ finds are from the *aediculae* of the interior cella walls of the Temple of Apollo Sosianus and on the bottom of the sculptural niches in the courtyard of the Horrea Epagathiana at Ostia.[53]

Several observations about provenance of the moldings and cornices can be made. De Criscio's cornices (cat. nos. 154, 468, 469, 470, 471, 472, 473) come from Pozzuoli and its vicinity. Kelsey collected fragments from in and around Rome, including at the Baths of Diocletian (cat. nos. 217, 223, 409, 535), 38 Via Firenze (cat. no. 451), the villas near Palestrina (cat. no. 584), and Roviano (cat. no. 452). Cat. no. 494 was picked up from the Hill of Kronos at Olympia, Greece. Three identical cornices of white marble (cat. nos. 487, 488, 489) were gathered as surface finds during the 1924 excavation at Pisidian Antioch in Turkey. Despite the fact that the collectors gathered them at different times, several moldings have identical forms. While their dimensions differ slightly, the *rosso antico* cornices collected by Kelsey from near Roviano (cat. no. 452), 38 Via Firenze (cat. no. 451), and Rome (cat. nos. 447, 448) have identical profiles. Cat. no. 462, collected by Van Deman from an unknown location, has the same profile and dimensions as cat. no. 452. Matching cornices gathered by different collectors add credence to hypotheses about a specialized production center in Rome and the circulation of pattern books.[54] The predominance of *rosso antico* is striking. The red marble, which could take a high polish, was the favored polychrome source for decorative molding, and its abundant use is attested elsewhere.[55]

While moldings in Roman architecture could assume a horizontal, vertical, or curved shape, those used as architectural revetment embellished a flat, two-dimensional surface. Embellishment could take on various forms, either with smoothed, geometric profiles, or, more frequently, with decorative motifs such as continuous dentils or egg and dart. The moldings here are all of the smooth type and consist of either a single curve or composite, complex elements. Seventeen distinct morphologies were identified, and were classified according to the ordering of the molded elements and sizes of the pieces.

Single-curved moldings in the collection include cat. nos. 223 and 254 (both *portasanta*), which are *torus* moldings with identical profiles. The *torus* molding has one curve that is a convex semicircle of rounder and larger proportions in comparison to other types of moldings (fig. 2.23).

The varieties of compound or composite moldings in the collection are numerous. Cat. no. 584 (fig. 2.24a) is an ogee or *cyma reversa*, or *gola diritta*, molding type, where the upper portion is convex, so that a *cavetto* (concave quarter-round) is above an *ovolo* (a convex quarter-circle). Cat. no. 535 (fig. 2.24b) has an *ovolo*, set off by *listelli*. The molding sequences of several

51. Bruto and Vannicola 1990, 368–71; Boldrighini 2007, 57, fig. 119.
52. Dell'Acqua too adopts Boldrighini's terminology of *incorniciatura* and *modanture*, and here the Italian will be used since it is the original language for describing architectural molding. Dell'Acqua 2014; Boldrighini 2007, 55, 73–74.
53. For the Apollo Temple, see Viscogliosi 1996, 111–12, 164–71; De Nuccio and Ungaro 2002, 449–50; For Ostia, see Pensabene 1994, 375.
54. Pensabene 1994, 375.
55. See above, nn. 46, 53.

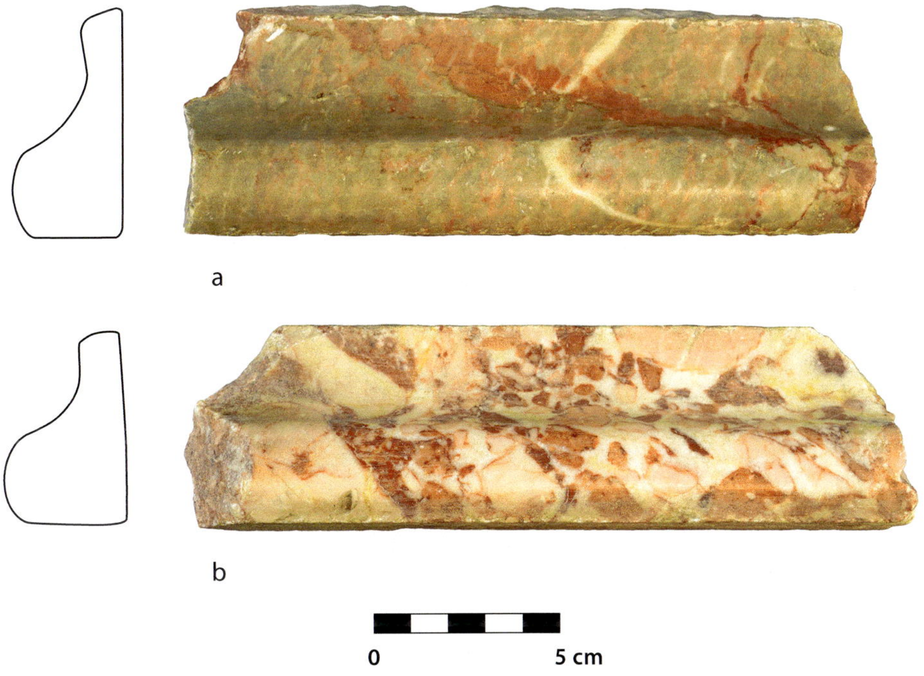

Figure 2.23. Single-curved moldings. Cat. nos. (a) 223 and (b) 254.

fragments are particularly unusual; they are most likely not ancient. Cat. no. 217 (fig. 2.24c) has a large, dominating astragal, whose half-round surface is surrounded by two flat planes. From the top to the bottom, it has the following smooth molding: a shallow *cavetto*, astragal, and a *gola rovescia* (a reversal of the *gola diritta*). Cat. no. 409 (fig. 2.24d) has a very unusual recurving profile, as if to create a channel. Without the recess of the channel, the shape of would be a *torus*; above the hollowed-out *torus* is a *listello* and a *cavetto*.

Cornices are also composite moldings. Cornice moldings have long, rectilinear shapes with a smooth or ornamental decoration and a profile imitating a cornice block from the traditional architecture of the entablature. The eighteen cornice moldings in the Kelsey collection all have a smooth decoration. In general, their profiles are characterized by a succession of rounded elements framed by *listelli*. The twelve distinct morphological types are illustrated in figures 2.25–28.

The matching cornices cat. nos. 487, 488, and 489 from Pisidian Antioch (type 1) have a profile with a *listello*, *scotia*, *listello*, *ovolo*, and *listello* (fig. 2.25). The source for the white marble was presumably Dokimeion, located nearly 70 km distant; import of marble to the city from that locale is attested on a seated statue of Zeus signed by a Menandros from Dokimeion. The cornices likely come from the imperial cult complex where the University of Michigan focused its excavations in 1924, one year prior to accessioning the cornices into the museum. The few examples of architectural white marbles reported by excavation director David Robinson all come from the area of the imperial cult.[56]

The only other white marble cornice collected by Kelsey is cat. no. 494 (type 3; fig. 2.27a), which has a *torus*, *listello*, *gola rovescia*, *listello*, and *scotia* profile. The remaining cornices acquired by Kelsey are of *rosso antico*. Type 2 cornices (cat. nos. 447, 448, 451, 452, 462; fig. 2.26) are characterized by an *ovolo liscio*, *listello*, *gola rovescia*, and *listello*. Cat. no. 452 (collected by Kelsey) and cat. no. 462 (collected by Van Deman) are nearly exact matches, while cat. nos. 447, 448, and 451 display minor variations in size and cuttings. On cat. no. 451, carvers skillfully confined a vein of white marble to the back so that only pure red *rosso antico* was visible. Cat.

56. Robinson 1926.

Figure 2.24. Compound or composite moldings. Cat. nos. (a) 584, (b) 535, (c) 217, (d) 409.

Figure 2.25. Identical type 1 cornices collected by Kelsey in Pisidian Antioch, Turkey. Cat. nos. (a) 487, (b) 488, (c) 489.

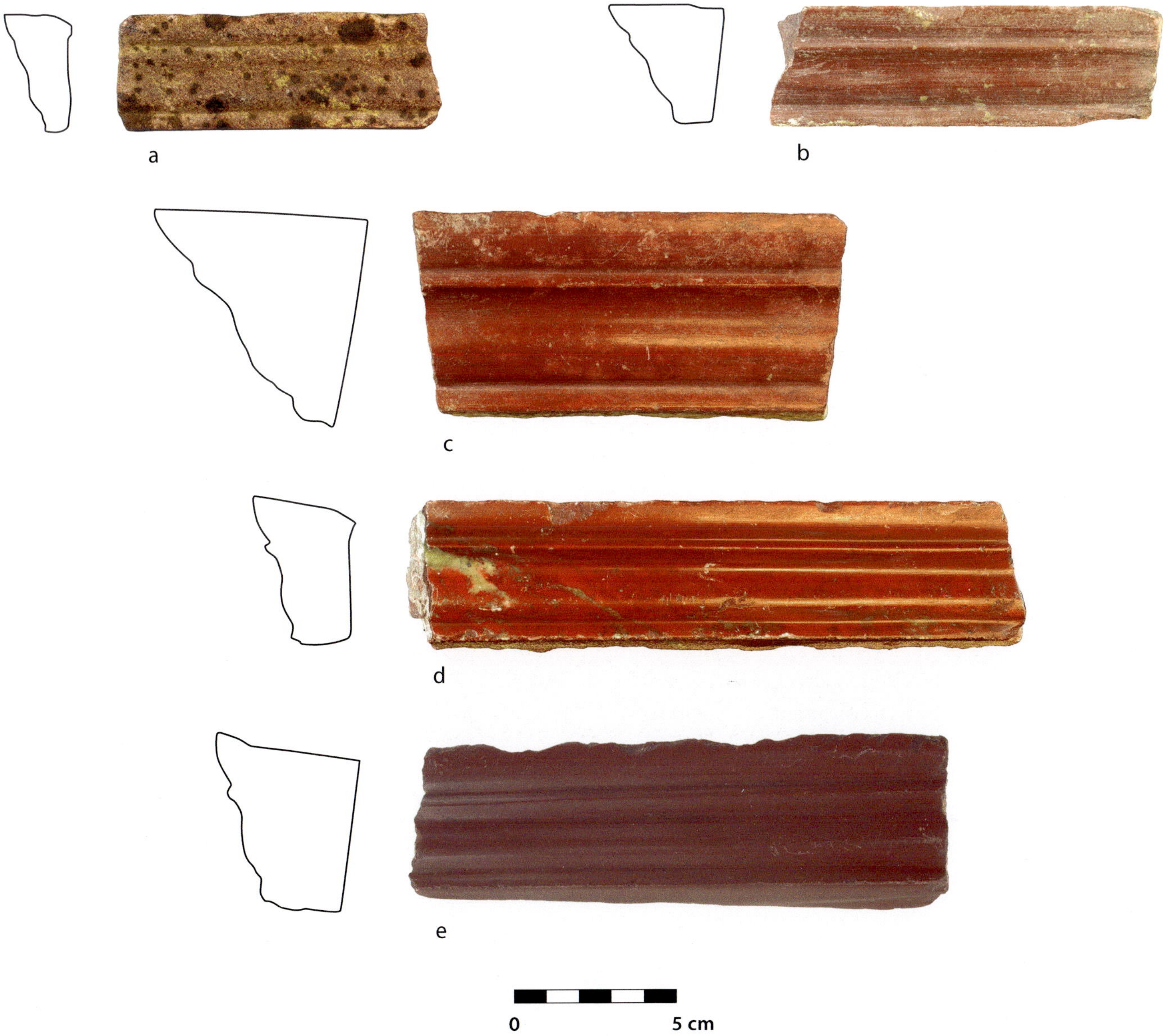

Figure 2.26. Type 2 *rosso antico* cornices. Cat. nos. (a) 447, (b) 448, (c) 451, (d) 452, (e) 462.

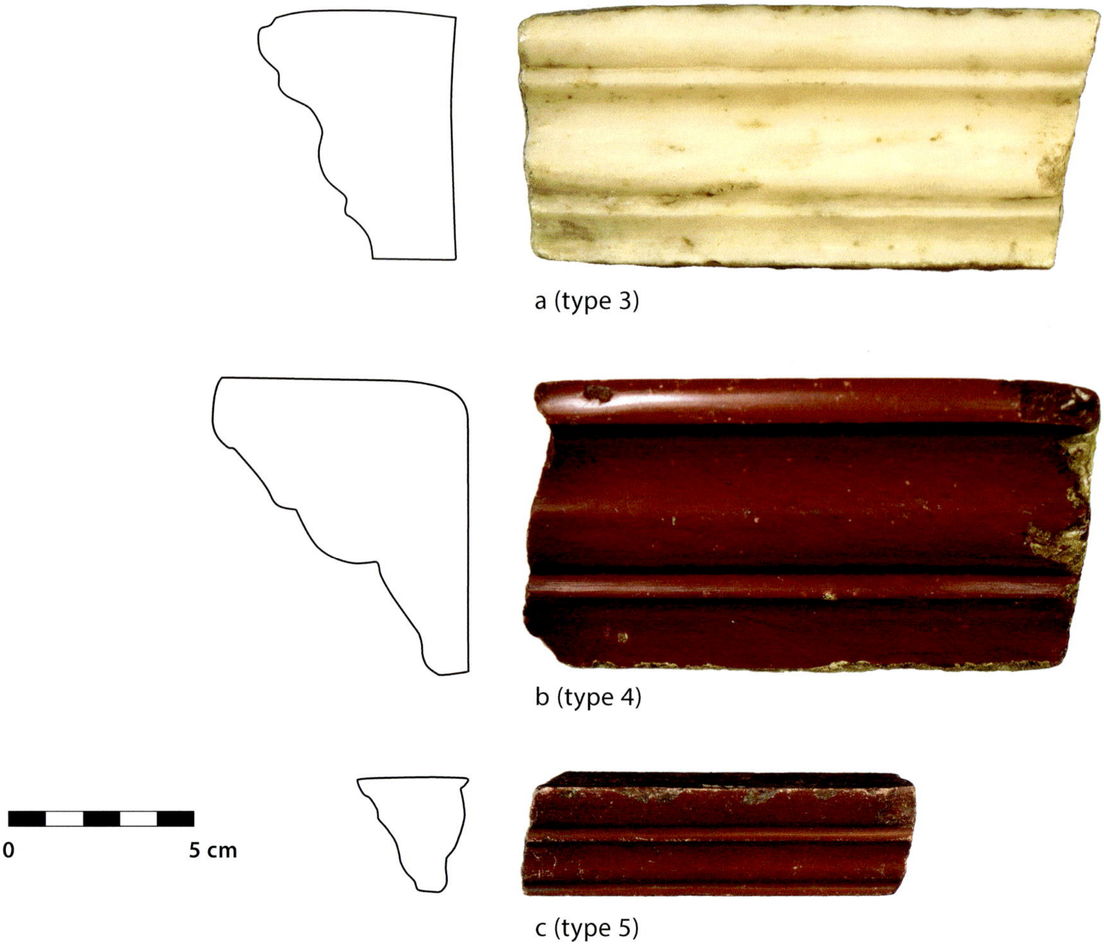

Figure 2.27. Other cornices collected by Kelsey and Van Deman. Cat. nos. (a) 494 (type 3), (b) 453 (type 4), (c) 458 (type 5).

no. 453 (type 4; fig. 2.27b) has a *listello, gola rovescia, listello*, and *scotia*, and is similar to one of the most commonly reported types (type M) in Dell'Acqua's study.[57] Cat. no. 458 (type 5; fig. 2.27c), the smallest of Kelsey's cornices, has a double profile, with an *ovolo, listello, gola rovescia*, and a *listello* on one side and a *gola diritta* on the other.

De Criscio's cornices (fig. 2.28) are each a distinct type, but of smaller proportions, and are all *rosso antico* with the exception of cat. no. 154, which is *giallo antico*. Cat. no. 468 (type 6), with its *listello, gola rovescia, listello*, and *scotia*, is similar to type 4, but smaller and with a less pronounced curvature. Cat. no. 470 (type 7) is characterized by a *listello, gola diritta, listello*, and *torus*; cat. no. 154 (type 8) has a *torus, listello, gola rovescia*, and *listello*; cat. no. 469 (type 9) is a simple *gola rovescia*; cat. no. 471 (type 10), a *torus, ovolo, tondini*, and *ovolo*; cat. no. 472 (type 11), a *torus, gola rovescia*, and *gola rovescia*; cat. no. 473 (type 12), an *ovolo, listello, gola rovescia*, and *listello*.

It is important to note that similar typologies to those identified within this collection are also attested elsewhere in the archaeological record. Cornices have been excavated from the Nonii Arrii villa, an imperial domus on the Janiculum in Rome, and a domus at Aix-en-Provence in ancient Gallia Narbonensis.[58] From these examples, it seems carvers had a finite but flexible set of combinations to which they adhered.

57. Dell'Acqua 2014, 456.
58. Sacchi 2015, 136; Filippi 2005a; Digelmann 2003, 235–37.

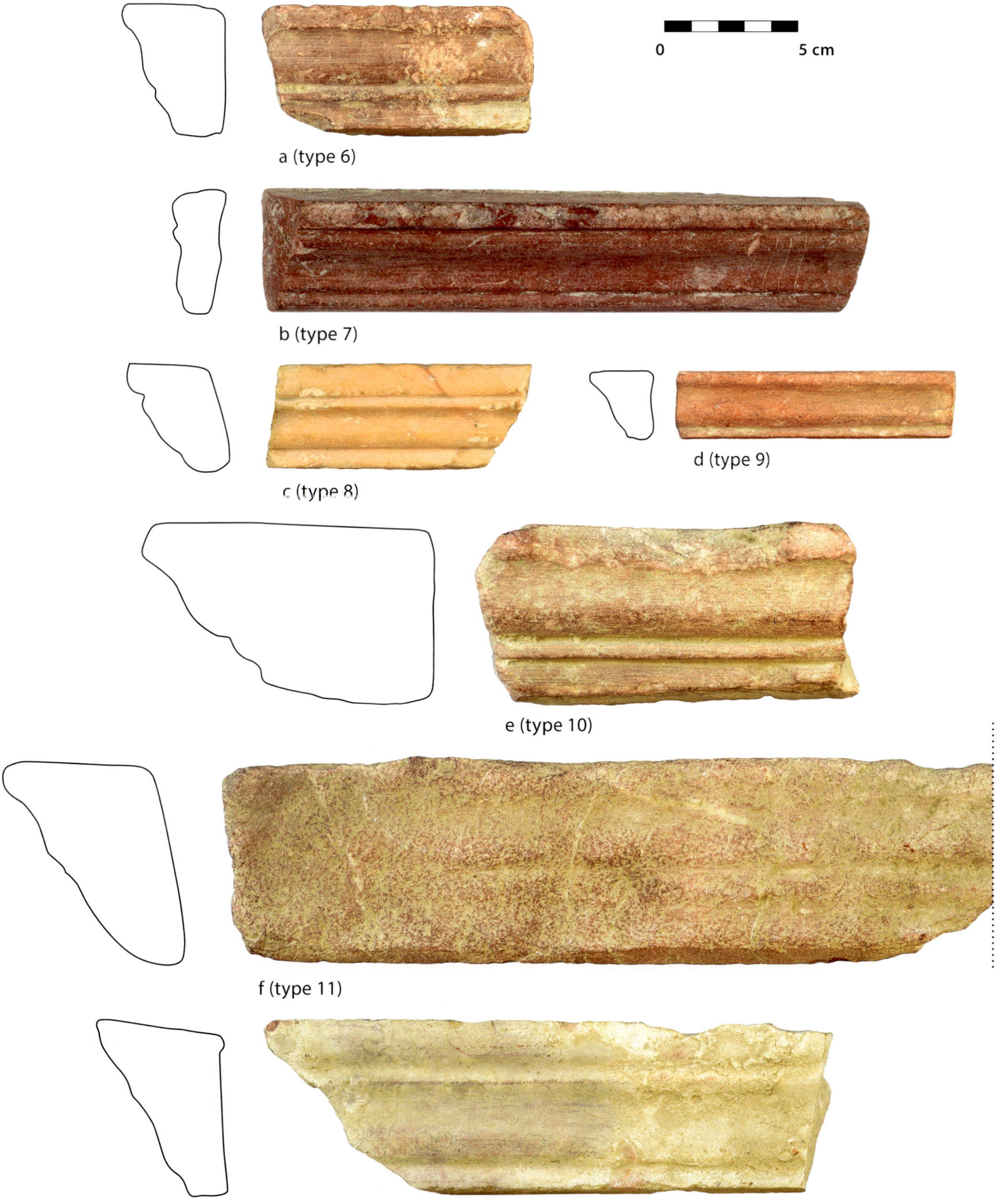

Figure 2.28. Cornices collected by De Criscio. Cat. nos. (a) 468, (b) 470, (c) 154, (d) 469, (e) 471, (f) 472, (g) 473.

Interestingly, a remarkable continuity and conservatism in design has endured in interior decoration to this day, evident in stores such as Home Depot and Restoration Hardware, which sell crown moldings that offer "a unique classical look."[59]

Columns

The Kelsey Museum's collection of Roman marbles has twenty-three column fragments, many of which were recut into paving or veneer, hence their inclusion in this study. As elsewhere, few have a recorded archaeological context. F.W. Kelsey's objects come primarily from Rome and Ostia. On January 21 and 23, 1901, four *giallo antico* fragments (cat. nos. 121, 123, 124, 126) were collected from 38 Via Firenze in Rome. Kelsey acquired cat. nos. 213, 119, and 338 from the Baths of Diocletian with the help of Boni and Romano.

The dressings of the columns in the collection are characteristic of classical architecture (figs. 2.29–32). Shafts are unfluted, smoothed (cat. nos. 70, 97, 216, 226, 250, 322, 336, 338, 367, 415, 550), or are fluted (cat. nos. 121, 122, 225, 491), sometimes with finely cut fillets (cat. nos. 119, 123, 156). Four (cat. nos. 124, 126, 213, 373) are *rudentes* columns, named by Vitruvius for having rope infill between the flutes on the lower third of the column shaft. The *rudentes*, usually an inch in width, are separated into twenty-four channels and have a convex profile, raised above shallow grooves and offset by thin molding strips.

Polychrome columns were a hallmark of Roman construction. Roman builders did not follow the Greek tradition of creating columns out of stacked drums of white marble. Instead, they preferred colored columns hewn from single blocks of stone imported at great distances from across the Roman Empire. To achieve the desired aesthetic effect, shafts were complemented with capitals and bases of contrasting colors, and orders were sent to various quarries. The most prized columns were the colored monoliths of 30, 40, or 50 Roman feet[60] reserved for lavish public building projects such as the Pantheon and the Baths of Caracalla. Favored materials for monolithic columns are well represented in the Kelsey collection: *marmor Numidicum* from Chemtou, Tunisia; *marmor Luculleum* from Teos, Turkey; *marmor Chium* from Chios, Greece; and *breccia di Settebasi* from Skyros, Greece.

In some instances, where enough of a column's circumference is preserved, it is possible to reconstruct its original diameter. To extrapolate the diameter in full, the extant perimeter of the object was traced on paper and a caliper used to connect the known edges of the column together. Table 2.1 presents the diameters for seventeen out of the twenty-two column fragments, while figures 2.30–32 show their profiles and reconstructed diameters in illustrated form.

Table 2.1. Reconstructed measurements of column diameters.

Cat. No.	*Diameter (cm)*
367	34.5
415	36.0
336	36.0
250	36.5
322	38.0
550	41.0
226	41.5
156	43.0
124	44.0
338	45.5
119	52.0
373	54.5
213	56.5
216	62.0
126	65.5
122	83.0
97	117.0

What are now small fragments stored in museum drawers were once parts of monolithic columns of varying sizes. The reconstructed diameters of the column shafts tend to fall in multiples of 40, 50, or 60 cm, while cat. no. 97 has the largest diameter in the collection with a measurement of 1.17 m. The diameter of a column was significant for the rules that Roman architects followed when designing classical orders. The mathematical ratio of the lower diameter of the column shaft was tied to intercolumniation, the height

59. "Moulding & Millwork," "Moulding Styles," The Home Depot website, https://www.homedepot.com.
60. One Roman foot is generally understood to be 29.6 cm, although in practice there was likely a spread around this mean.

Figure 2.29. Unfluted, smooth column shafts.
Cat. nos. (a) 70, (b) 97, (c) 216, (d) 226, (e) 250, (f) 322, (g) 336, (h) 338, (i) 367, (j) 415, (k) 550.

Figure 2.30. Fluted column shafts. Cat. nos. (a) 121, (b) 122, (c) 225 (with reconstructed profile), (d) 491; with finely cut fillets: (e) 119 (with reconstructed profile), (f) 123, (g) 156 (with reconstructed profile). Drawings by Lorene Sterner.

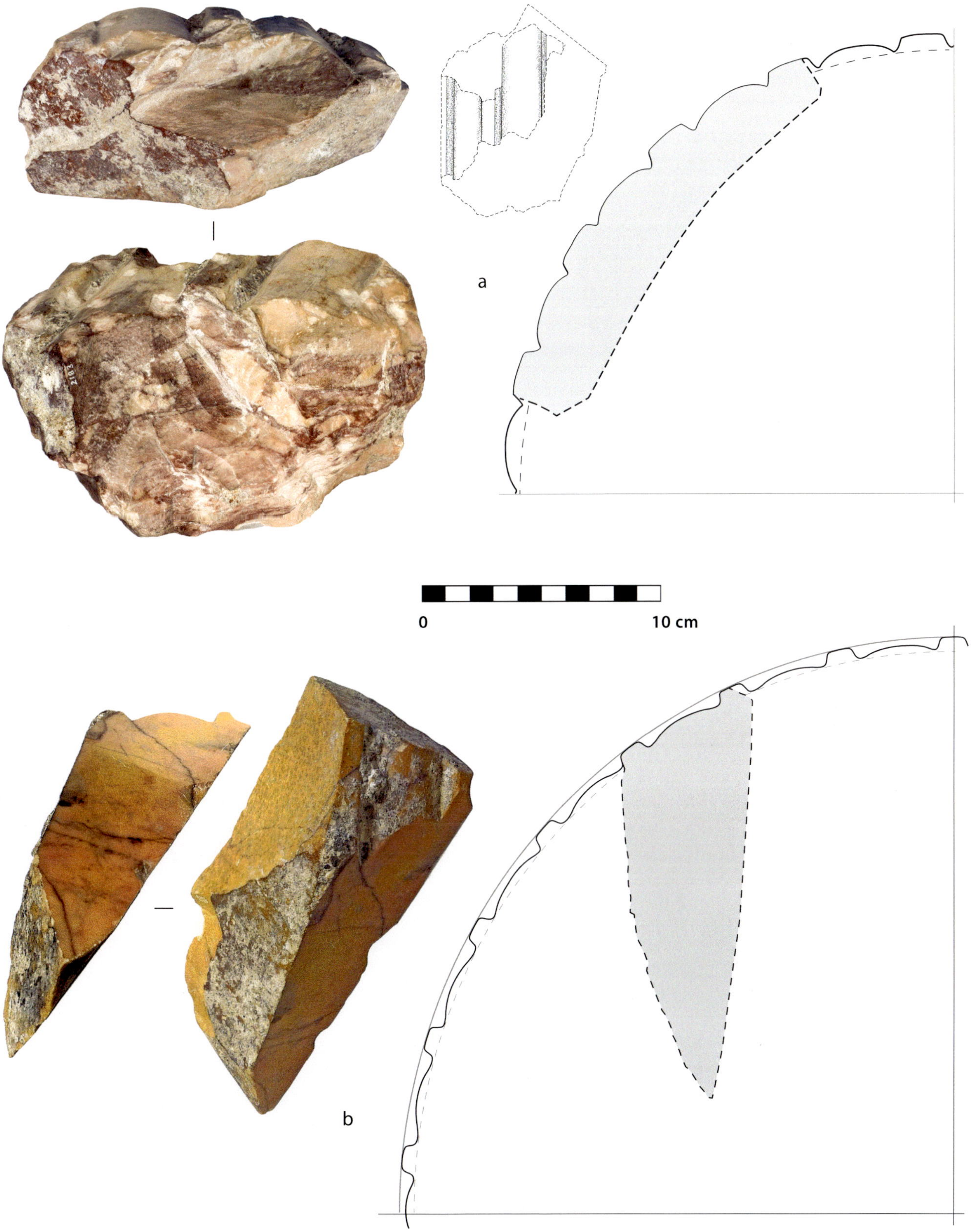

Figure 2.31. *Rudentes* columns. Cat. nos. (a) 124 (with three-quarter view and reconstructed profile), (b) 126 (with reconstructed profile). Drawings by Lorene Sterner.

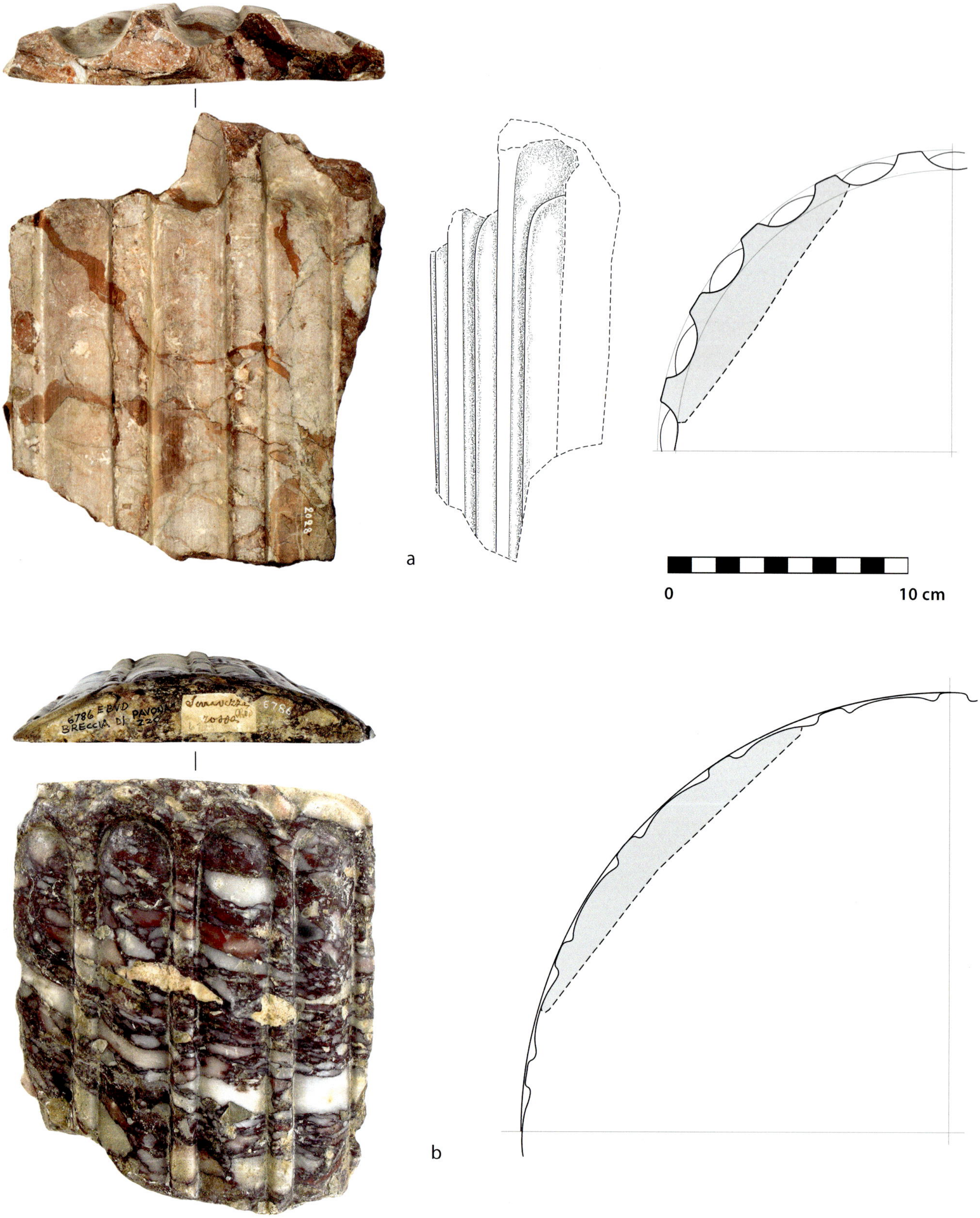

Figure 2.32. *Rudentes* columns, *cont*. Cat. nos. (a) 213 (with three-quarter view and reconstructed profile), (b) 373 (with reconstructed profile). Drawings by Lorene Sterner.

of the column (including the base and capital), the height of the shaft, base, and capital, individually, as well as other elements.[61] Monolithic columns generally had a slenderness ratio (height to diameter) of 8 to 1.[62] It is impossible to know whether the columns were ever associated with the rules prescribed by the architect Vitruvius for the production of huge monoliths and modular construction, or even the degree to which Roman architects themselves followed these rubrics in practice. Along with a great diversity in design, stockpiling materials for future use, reuse, and recycling resulted in columns having variable proportions. While reconstructing the original height of the column fragments based on Vitruvius' system of column diameter is difficult, comparanda from Roman buildings may be illuminative. The interior monolithic columns of the Pantheon as well as the columns in the portico of Octavia in Rome adhered strictly to the 8 to 1 ratio.[63] With diameters of 1.105 and 1.1 m, respectively, the heights of column shafts measure 8.87 and 8.78 m. With cat. no. 97's comparable diameter of 1.17 m, the shaft of the fragment may have originally been over 9 m (9.36 m) high. At 26.64 Roman feet, cat. no. 97 was similar in size to column shafts in imperial monuments such as those in the Forum of Nerva (29.75 m) and the Library of Hadrian at Athens (24 m).

Stone Vessels, Labra, and Candelabrum

The stone vessels, *labra*, and single candelabrum are included in the Kelsey decorative stone collection because of their exotic materials and ornamental function, and because they are typical of the small furnishings in outdoor decoration.[64] Lush gardens were an important part of the homes and villas of the Roman elite as an ostentatious display of the owners' wealth. Gardens were embellished with sculptures of mythological subjects linked to the rural world, fountains, pillars, and basins.

Cat. nos. 20 and 21, both made of granite from the Aswan region, and cat. no. 545, of *bardiglio*, are fragments of Roman *labra* (fig. 2.33). Large tubs for holding water, *labra* were frequently installed in public and private baths for washing and were also used as fountains, ornamental ponds in private residences, tanks for sacred ablutions, and funerary urns.[65] While they served a practical need for hygiene, Roman *labra* made of precious and beautiful stones were also monumental, decorative objects. Cat. no. 545 bears a typical shape with its flat bottom and curved walls ending in an everted, flat lip. Its good state of preservation allows for its measurements to be reconstructed as 57.5 cm in diameter and 15 cm in depth, indicating it was a shallow, small to medium-sized bowl.

Fragments with thinner walls were labeled more generically as stone vessels (fig. 2.34). Cat. no. 12, a rim fragment of a krater whose extrapolated diameter is 36 cm, was made from the *granito della colonna* from Wâdi Umm Shegilat in the Eastern Desert of Egypt. Cat. no. 53 is a porphyry vessel with thin walls. Cat. no. 591 is a rim fragment from a small vessel, originally 18 cm in diameter, and bears a leaf bud decoration. There is a striking prevalence of Egyptian materials among the marbles in the Kelsey collection, and it is representative of the Egyptian tradition, dating to Predynastic times, of using colored stone for canopic jars and drinking cups.[66] Funerary urns made of Egyptian alabaster, purple porphyry, and Aswan granite were fashionable among the Roman elite, in part because Egyptian stones were laden with symbolism of distant landscapes and were deemed to have magical and regenerative properties.[67]

Unique in the collection is a Roman candelabrum acquired by Kelsey in Rome. Cat. no. 504 preserves a twisting double helix column shaft, 11.3 cm in diameter and 21 cm in height, whose fillets turn into acanthus leaves (fig. 2.35a). Roman candelabra of this type are rare, but a few extant examples of similar pieces shed light on the original form of the Kelsey piece (fig. 2.35b).[68] A square base was surmounted by a

61. Wilson Jones 2000, 147–51.
62. This ratio was endorsed by Roman surveyor's formulas recorded in a medieval codex in Munich. Wilson Jones 2000, 148 n. 45.
63. Wilson Jones 2000, 148.
64. Sacchi 2015, 144–45.
65. Ambrogi 2005.
66. Von Bissing 1940; Aston 1994.
67. Perna 2015.
68. Comparanda for such Roman candelabra include one sold at Sotheby's as N08603, Lot 50, under the Egyptian, Classical, and Western Asiatic on December 10, 2009, which is now on loan to the Metropolitan Museum of Art. For other related examples, see Berger 1990,

Figure 2.33. Fragments of Roman *labra.* Cat. nos. (a) 20, (b) 21, and (c) 545.

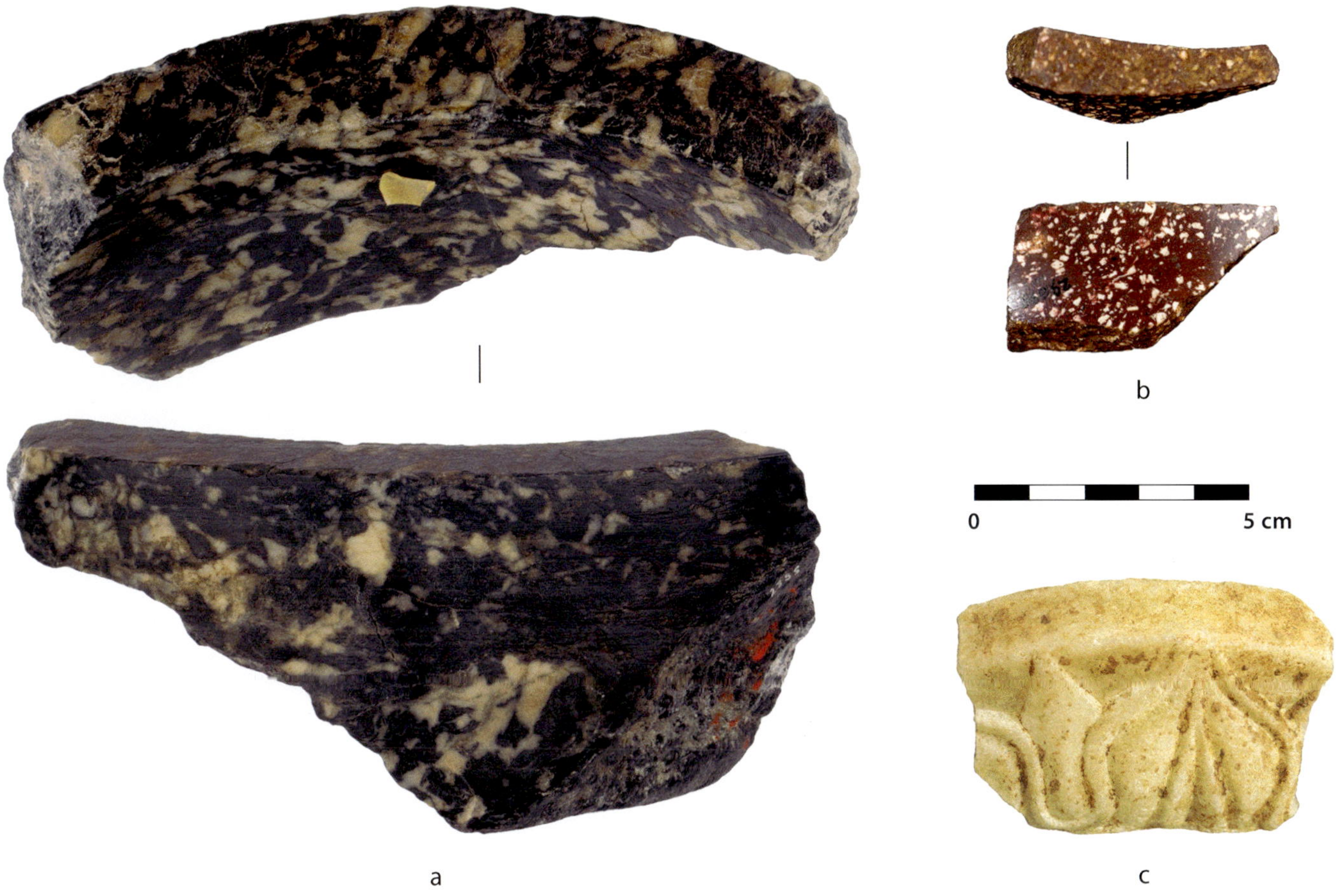

Figure 2.34. Fragments of stone vessels. Cat. nos. (a) 12, (b) 53, and (c) 591.

helical shaft with large calyx leaves of the acanthus, which culminated in entwining stalks. A large double pinecone or artichoke crowned the shaft. Pinecones appeared as decorative elements in funerary art on urns and *cippi*, and also served as symbols of fertility and wealth. Set among gardens, parks, and shrines, candelabra were luxurious, ornamental household furniture. In the Hellenistic period, the shaft held a candle or lamp for illumination, but the Romans embellished the form further. Most of the examples of candelabra with twisting shafts ending in acanthus leaves have been found in and outside Rome, such as at the Villa of Quintus Voconius Pollio. Their elaborate ornamentation inspired the twisting Solomonic columns found in Renaissance and Baroque architecture.

Sculpture and Sarcophagi

While sculpture was rejected as part of our definition of decorative stone, sculptures reworked for secondary purposes are included. The three fragments that can be classified in this manner are distinctive for their recycling and the exotic stone from which they are made (fig. 2.36). Cat. no. 1 preserves a knee or thigh, and is an example of basanite from Wâdi Hammâmât. Cat. no. 2, also made of basanite, preserves three fingers holding an object. On both, a single side was cut to create a flattened surface so the object could be reused as paving or veneer. Cat. no. 298 is the lower part of a small alabaster bust, and has a hole on the bottom filled with the remains of an iron dowel. It is interesting to note that Kelsey acquired these materials from

509–16; Vermeule, Cahn, and Hadley 1977, 52; Breuer and Gatier 2001, no. 7, figs. 41–42.

the same dealers in Rome who also sold him the sculptures, now identified as the Hartwig-Kelsey fragments from the Templum Gentis Flaviae.[69] The encrusted drapery on the front of the bust indicates its exposure was a result of reuse. A single sarcophagus fragment of Carraran marble was also included as part of the collection. Cat. no. 560 is identifiable by its width and shape as the wall from the body of a sarcophagus. It was obtained by Kelsey at the Baths of Diocletian in October 1900.

Ancient Reuse and Repairs

Builders in the Roman world recycled materials on a massive scale. The widespread practice of demolition, salvage, and reuse made sound economic sense as it saved on the costs of quarrying, shaping, and transporting raw materials.[70] The decorative marbles, granites, and porphyrys in the collection were quite valuable and not immune from these measures. Around eleven of the fragments in the collection show signs of ancient reworking, reuse, or repair, with most redeployed as cladding. Marble stripping is well attested in the archaeological record, and can be seen in numerous buildings throughout the Roman world. The removal of marble slabs had a clear economic benefit in that it generated a great deal of reusable material.[71]

Originally monolithic or fluted columns, cat. nos. 119, 121, 122, 123, 124, 225, and 491 (see fig. 2.30) were reprocessed as veneer or paving. The shape and size of columns made them well suited to being sawed down into thin slabs. Cutting down a column to obtain the desired dimension could take place in any direction. The exterior portion of column fragment cat. no. 124 (fig. 2.31a) was shaved off so that the *rudentes* fluting was preserved, while elsewhere the width of a flute was used as a guideline for cutting a cross section. Mortar preserved in the flutes of cat. nos. 122, 123, and 124 (figs. 2.30b, 2.30f, 2.31a, 2.37a) show that they were placed into a bedding so that the original architectural function was hidden from view. *Opus sectile*, paving, and veneer could also be recycled and redeployed in new decorative programs. Two pieces in the Kelsey

a

b

Figure 2.35. (a) Candelabrum fragment cat. no. 504. (b) An example of a fully preserved candelabrum from Sotheby's Sale N08603. Photograph courtesy of Sotheby's, Inc. © 2017.

collection were likely refurbished to be fit into a new floor or wall design. The lozenge shape of cat. no. 506 may have been cut at a later date, while saw marks on the straight edge of the curved fragment cat. no. 308 (fig. 2.9b), as well as traces of mortar, indicate it may have originally been *opus sectile*. While recut materials can be recognized as recycled from architecture or sculpture, it is almost impossible to identify the

69. Gazda and Haeckl 1996, 8–9.
70. Barker 2010.
71. Barker 2012.

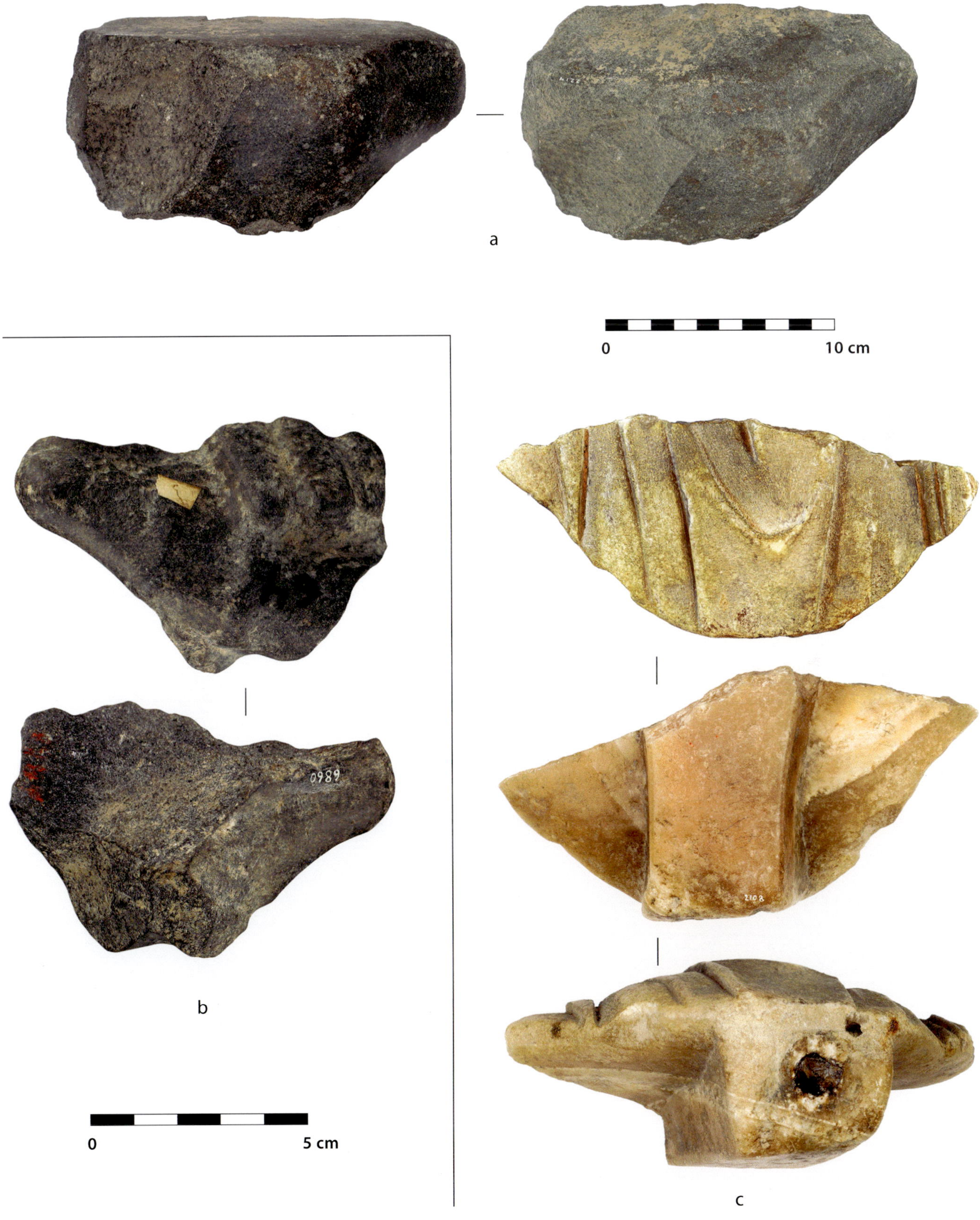

Figure 2.36. Fragments of recut sculpture. Cat. nos. (a) 1, (b) 2, (c) 298.

a

b

0 5 cm

Figure 2.37. Ancient repairs seen on (a) cat. no. 124, a recut column showing cement, and (b) cat. no. 545, a repaired *labrum* showing a metal dowel.

internal fragments of secondhand materials whose elements have been erased.

Veneer and paving were particularly susceptible to cracking and fracturing and, if damaged, could simply be repaired. Mortar on the edges of a finely cut *rosso antico* paving or veneer (cat. no. 461) and mortar on the rough breaks on *africano* paving or veneer (cat. no. 103) indicate they were broken and glued together using mortar as an adhesive. Although it is always difficult to determine whether traces of mortar were used for fixing breaks or as the preparatory layer for setting stone, the application of mortar to internal edges and absence on the back suggests mending. Elsewhere, a large metal dowel inserted into cat. no. 545 (fig. 2.37b), a *labrum* made of *bardiglio*, appears to have been part of an attempt to repair a break and restore the object to its original state. The restoration was ultimately unsuccessful, and the vessel later broke again in the same spot.

Ancient Tools

The collection is a trove for understanding Greco-Roman stoneworking practices. That Kelsey himself donated many of the pieces bearing vestiges of ancient craftsmanship indicates he intentionally selected pieces with tool marks. The collection represents the types of tools used in the last stages of finishing and fitting marble floor paving and wall veneers.[72]

The heavy single-point pick marks preserved on a number of fragments reflect the roughing-out process that took place in the quarries themselves (fig. 2.38).[73] The scars from straight-on blows seen on many of these fragments came from holding the quarry pick vertically. The strike patterns created a series of regular circular indentations that provided a rough surface for fitting revetment into place.

Marks from multiple different tool types left on a single object can be found in examples throughout the collection. Stone carving has always been a conservative craft with a limited repertoire of tools, yet these could be used in tandem with one another in multiple ways to achieve different effects and produce a wide range of products. A piece of *portasanta* paving or veneer (cat. no. 209) from the Baths of Diocletian and a piece of white marble paving or veneer (cat. no. 505) from Eleusis, Greece, bear markings of the pick along with the point chisel, while a *pavonazzetto* paving or veneer (cat. no. 169) preserves both point chisel and saw marks. The point chisel made shallow, short grooves and was used to flatten the surface of the revetment (cat. nos. 91, 92, 93, 129, 138, 282, and 456; fig. 2.39). A small piece of a white limestone (cat. no. 596) collected from the south wall of the Olympeium in Athens during Kelsey's 1901 trip bears vestiges of both quarry pick and claw chisel (fig. 2.40). The back of a piece of *portasanta* paving or veneer (cat. no. 162) that Kelsey brought from Rome has point chisel markings, while its edge preserves marks of both the point and claw chisel. The use of the round chisel and small pick can be seen on cat. no. 553.

The most commonly attested tool is the saw. Twenty-seven fragments bear traces of a metal blade as it moved back and forth, cutting the stone by means of a mixture of abrasives and water (fig. 2.41).[74] The saw was perhaps the most economical tool for cutting thin panels out of marble blocks, and at the same time exposed the visual properties of the stone, ensuring its suitability for decoration. Saws were regularly used for flooring or revetment in the Roman period, and are archaeologically attested as tool marks, installations, and representational images at Ostia in Italy; Dokimeion, Ephesus, and Hierapolis in Turkey; and Gerasa in Jordan.[75]

The marble fragments also display familiar features from Roman architecture used for securing decorative stone revetment in place. Many preserve traces of mortar on the back, indicating they were attached to a floor or wall in a mortar bedding.[76] Circular holes drilled into the sides of a *portasanta* paving or veneer (cat. nos. 478, 479) and a cornice molding (cat. no. 468) indicate another method used to anchor the

72. Rockwell 1993.
73. Quarry pick marks: cat. nos. 87, 105, 170, 171, 172, 190, 208, 209, 210, 228, 232, 233, 244, 246, 248, 251, 265, 270, 296, 340, 341, 453, 505, 515, 553, 596, 640, 692.
74. Saw marks: cat. nos. 22, 75, 125, 130, 163, 169, 208, 215, 254, 259, 275, 284, 290, 302, 308, 311, 351, 378, 427, 432, 434, 442, 527, 531, 641, 685, 691, 701.
75. Röder 1971, 303–11; Bruno 2002b, 188–89; Ritti, Grewe, and Kessener 2007.
76. Traces of mortar: cat. nos. 89, 103, 120, 122, 123, 128, 164, 167, 184, 188, 204, 209, 234, 294, 308, 311, 391, 397, 399, 445, 461, 531, 542, 595, 600, 644.

slabs to a wall (see fig. 2.5). The sockets originally held metal clips or pins; one end of the clip was engaged in the hole and the other was embedded in a layer of mortar. The *portasanta* fragments were likely part of the same wall decoration, having identical measurements in thickness and drill holes (1.1 cm). Their relatively heavy weight (cat. no. 478: 4.08 kg; cat. no. 479: 1.5 kg) shows that Roman builders needed to rely on several techniques to keep bulky pieces of veneer in position.[77] A metal pin is preserved on the back of cat. no. 487, a cornice molding from Pisidian Antioch. Cat. nos. 91 and 298 have rust residue or staining, associated with the iron pins during their original use for wall revetment mounting.

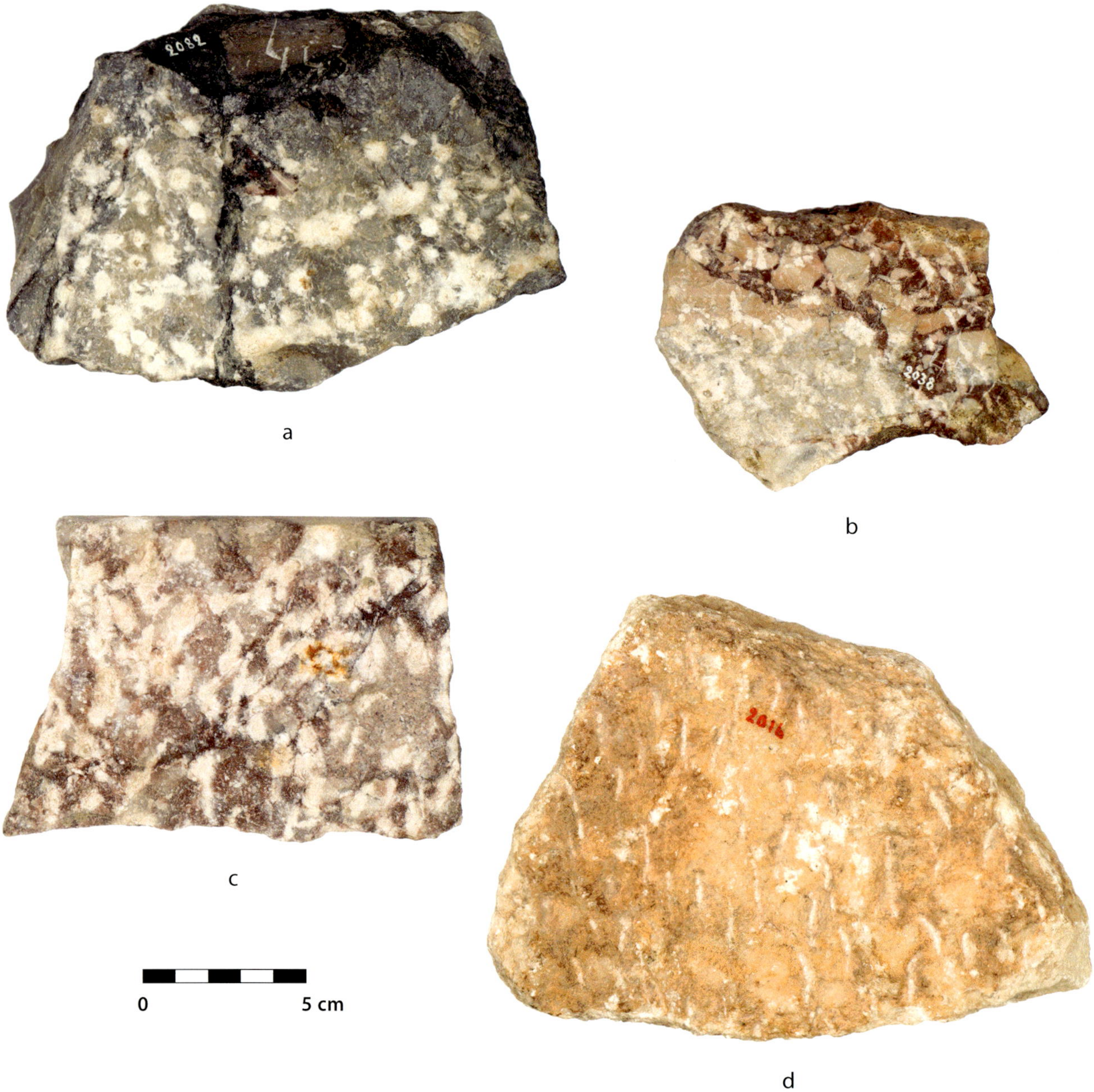

Figure 2.38. Examples of quarry pick marks. Cat. nos. (a) 105, (b) 244, (c) 265, (d) 515.

77. For extant examples, see İlhan 2014, 827, pl. 437. Although Ball (2002) discusses these techniques in detail, he does not mention the use of metal clips.

Figure 2.39. Examples of point chisel marks. Cat. nos. (a) 138, (b) 129, (c) 209, (d) 456.

Figure 2.40. Example of claw chisel marks. Cat. no. 596.

Figure 2.41. Examples of saw marks. Cat. nos. (a) 254, (b) 434, (c) 691.

Chapter 3

PETROGRAPHIC, STABLE OXYGEN AND CARBON ISOTOPIC ANALYSES

Leah E. Long and Fabrizio Antonelli

This chapter presents the archaeometric study of eleven fragments in the collection of decorative stone at the Kelsey Museum of Archaeology. The sampled pieces come from floor or wall decoration, such as paving or veneer, architectural revetment, or architectural components from ancient civic buildings in Greece and Italy, as well as from a single sarcophagus. With ancient stone studies on the rise in recent decades, classical archaeologists and art historians have worked in tandem with geologists, chemists, and physicists to establish a scientific index of the main marbles exploited in antiquity. By experimenting with a variety of analytical techniques, scientists have made advances in matching white marble artifacts to Greco-Roman quarry sources, and a multi-method approach is now standard. Ratios of stable isotopes of carbon and oxygen have reliably been combined with petrographic analysis to determine the source for ancient white marbles.[1] Such methods are currently the most widely used and reliable combination, which is the reason for their adoption in the current study. Oxygen and carbon isotopic analyses were carried out in the Stable Isotope Lab in the Earth and Environmental Sciences Department at the University of Michigan, and petrographic analysis in the Laboratory for the Analysis of Ancient Materials (LAMA) at the Iuav University of Venice, where final interpretation was completed by Fabrizio Antonelli.[2]

Museum curators are increasingly engaging in collaborative efforts to undertake geochemical analyses on the collections in their museums. Without laboratory work, the origins and production histories of objects remain unexplored and new data cannot be added to existing studies. The analysis of materials in the Kelsey Museum represents a unique opportunity to examine decorative stones related to early twentieth-century archaeological fieldwork. Because Francis Kelsey and Esther Van Deman were interested in amassing building materials from ongoing excavations and ancient sites for teaching purposes, the decorative stone collection is ripe for studying materials with known provenance. That several were taken from famous locales, such as the Baths of Diocletian in Rome and Olympia in Greece, adds to their scientific value.

While the lion's share of the approximately 700 pieces of decorative marble in the Kelsey Museum collection are polychrome, a significant number, around fifty, are white. These pieces pose a special challenge since the quarry origins of white marbles are not securely identifiable by the naked eye. Often the museum archives provide identifications (for example, a label "from Mt. Pentelicon, Greece"), making scientific analysis desirable. A good example is cat. no. 492, a piece labeled as "bit of Carrara marble" and recorded as having been collected at Sparta, Greece. From what is known of the trade in white marble, however, it seems unlikely that Italian Luna marble was exported to Greece, which was itself a source of many good marbles. The pieces selected for study all have a source attributed to them, but it is not always clear who was responsible for determining that information (table 3.1). Perhaps Kelsey himself or the museum curators made the assignments; in one case, the archives show that the marble collector Orlandi determined that a fragment from the Baths of Diocletian was Carraran (cat. no. 560).

It was not feasible to examine all fifty pieces of white marble in the Kelsey collection. The eleven

1. Herz 1987; Herz 1988; Moens, De Paepe, and Waelkens 1992; Gorgoni et al. 2002; Antonelli and Lazzarini 2015 and references therein.
2. Support for the scientific analysis at both institutions was provided by a faculty research grant from Virginia Commonwealth University in Qatar.

Table 3.1. Sampled objects with information about original use, provenance, and marble identification from Kelsey Museum accession books. All pieces scaled to 20%.

Cat. No. 492
Original Use: Paving or veneer
Site/City: Sparta, Greece
Context: Theater
Accession Book: Carrara

Cat. No. 493
Original Use: Paving or veneer
Site/City: Sparta, Greece
Context: Theater
Accession Book: Parian?

Cat. No. 494
Original Use: Cornice molding
Site/City: Olympia, Greece
Context: Mount Kronos
Accession Book: Pentelic

Cat. No. 495
Original Use: Architectural fragment
Site/City: Tegea, Greece
Context: Temple of Athena Alea
Accession Book: Carrara

Cat. No. 496
Original Use: Paving or veneer?
Site/City: Pherae, Greece
Context: Acropolis
Accession Book: Hymettian?

Cat. No. 497
Original Use: Paving or veneer
Site/City: Patras, Greece
Context: Odeum
Accession Book: Pentelic

Cat. No. 498
Original Use: Paving or veneer
Site/City: Rome, Italy
Context: Baths of Diocletian
Accession Book: Pentelic

Cat. No. 499
Original Use: Paving or veneer
Site/City: Rome, Italy
Context: Baths of Diocletian
Accession Book: Pentelic

Cat. No. 510
Original Use: Paving
Site/City: Rome, Italy
Context: 38 Via Firenze
Accession Book: Parian

Cat. No. 560
Original Use: Sarcophagus
Site/City: Rome, Italy
Context: Baths of Diocletian
Accession Book: Carrara

Cat. No. 562
Original Use: Molded dado / orthostat
Site/City: Rome, Italy
Context: Tunnel under Quirinal
Accession Book: Pentelic

0 10 cm

pieces that we did test were chosen with the hope that they would yield the most fruitful evidence. These pieces were selected primarily because each one had a white color and an archaeological context, ideally, both a known city and a specific building. All eleven were acquired by Francis Kelsey, and unfortunately, he did not record exact findspots beyond general site or building. This limits our interpretation somewhat, however, many other museums are beset by the same problem and yet scientific analyses of their materials have contributed important new information to the field of marble sourcing.

The fragments fall into two groups: those from civic buildings in the city of Rome (five items) and those collected from various sites across southern Greece during Kelsey's 1901 trip (six items). Table 3.1 illustrates the objects and provides identification of their original use, archaeological context, as well as the supposed quarry source indicated in the accession materials. With such criteria, it was hoped new light could be shed on provenance identification, architectural design of specific buildings, and the marble trade in general, and that archaeometric results be considered together with previous studies.

In many cases, isotopic or petrographic analyses alone cannot definitively distinguish white marble sources from one another, or from several other major sources of white marble traded in antiquity. Without performing the full battery of tests on our samples, our goal was not to pinpoint the exact quarries from which the materials derived, but rather to identify or suggest one or several possible regions as a marble's source. Given that these eleven pieces were collected from diverse locations around the Mediterranean, we felt that the combination of isotopic and petrographic analysis should be sufficient for determining potential source areas.

Sampling and Analytical Procedures

Kelsey Museum curator of conservation Suzanne Davis prepared the samples for destructive analysis in accordance with museum guidelines. A single small piece no larger than a sugar cube was taken from the back or an otherwise hidden part of each artifact. Photographs of the object were taken before and after sampling to document the area of intervention. The small chunks were then ground down into the powder required for analysis of the stable carbon and oxygen isotopes.

All eleven samples were analyzed in the Stable Isotope Lab in the Earth and Environmental Sciences Department at the University of Michigan under the direction of Kyger Lohmann and Lora Wingate. To measure the carbon and oxygen isotopic composition of marble, the carbon and oxygen were first chemically separated from the sample as CO_2. Carbonate samples weighing a minimum of 10 micrograms were placed in stainless steel boats and were roasted at 200°C *in vacuo* for one hour to remove volatile contaminants and water. Samples were then placed in individual borosilicate reaction vessels and reacted at 77°±1°C with four drops of anhydrous phosphoric acid for eight minutes in a Finnigan MAT Kiel IV preparation device coupled directly to the inlet of a Finnigan MAT 253 triple collector isotope ratio mass spectrometer. Data are reported in ‰ notation (parts per thousand, or per mil) relative to the Pee Dee Belemnite standard (VPDB). Precision and accuracy of data were monitored through daily analysis of a variety of powdered carbonate standards. At least four standards were reacted and analyzed daily. Measured precision is maintained at better than 0.1‰ for both carbon and oxygen isotope compositions. The results of the isotopic analysis were compared to numerical values available with the latest databank, which includes more than one thousand samples from the main quarries throughout the Mediterranean.[3]

In the case of the five objects from the civic buildings of Rome, thin sections were also prepared and studied under the optical microscope (at LAMA) in order to determine the specific petrographic characteristics of the marbles. The main petrographic features studied include microstructure, carbonate crystal boundaries, maximum grain size of the largest crystal of calcite/dolomite expressed in millimeters (MGS), and the presence and relative quantity of accessory

3. Antonelli and Lazzarini 2015.

minerals. The samples were also compared with the most recent published data.[4]

The six samples from southern Greece were not subject to petrographic analysis due to financial constraints, but above all because the most updated petrographic database to date does not include all the minor white marble sources from the region. To provide at least some approximate indication of MGS of these fragments, examination was done with a magnifying loupe hand lens, a light source, and a millimeter scale.

Results and Discussion

The results of petrographic and isotopic analyses of the marble artifacts from the decorative stone collection in Kelsey Museum of Archaeology are listed in table 3.2, with the isotopic ratios of δ^{13}C and δ^{18}O shown in the diagrams of figure 3.1.

As for the objects from Rome, the combination of the minero-petrographic characteristics and isotopic signatures of the marbles clearly show that the two pieces of paving or veneer from the Baths of Diocletian (cat. nos. 498, 499) are both of Pentelic, and that the sarcophagus fragment from the Baths of Diocletian (cat. no. 560) and molded dado (cat. no. 562) from a tunnel under the Quirinal Hill are of Carraran. The isotopic values of cat. nos. 499 and 498, ranging from −7.59 to −3.97‰ in oxygen, show the wide δ^{18}O spread characteristic of Pentelic marble signatures.[5] The petrographic study revealed typical features of fine-grained pure calcitic marbles (fig. 3.2b–e). Distinguishing

Table 3.2. Petrographic characteristics and isotopic composition of the sampled pieces.

Cat. No.	Original Use	Accessory Minerals									MGS (mm)	Calcite Crystal Boundaries	Texture	Fabric	δ^{18}O (VPDB)	δ^{13}C (VPDB)	Origin
		K-mica	Serpentine	Quartz	Plagioclase	Graphite	Pyrite	Ore Minerals	Apatite	Dolomite*							
510	Paving	±	±	±						+++	2.25	Sutured ± curved	HE	Mosaic with some strained crystals	−3.07	3.41	Cape Vathy, Thasos
499	Paving or veneer	++		++	±	±	±				1.15	Curved ± embayed	HE	Foliated	−7.59	2.57	Pentelic
560	Sarcophagus	±		++		+					0.65	Curved/straight	HO	Mosaic with rare triple points	−1.62	1.98	Carrara
562	Molded dado/orthostat	±		±		±		+	±		0.75	Curved/embayed	HE	Mosaic	−1.38	2.09	Carrara
498	Paving or veneer	+		±		±	+				0.95	Curved ± embayed	HE	Slightly foliated	−3.97	2.64	Pentelic
492	Paving or veneer										1.10				−0.42	2.57	Doliana
493	Paving or veneer										1.10				−4.05	2.66	Doliana
495	Architectural										1.00				0.26	2.50	Doliana
497	Paving or veneer										1.30				−1.47	3.34	Hymettus or Mani?
494	Cornice molding										0.40				−6.08	2.52	Doliana or Pentelic?
496	Paving or veneer?										1.80				0.09	2.03	Doliana?

Legend
± traces
\+ present
++ average abundance
+++ very abundant

* The presence of dolomite was detected by means of XRD analysis.

4. Antonelli and Lazzarini 2015 and references therein.
5. Pike 2009.

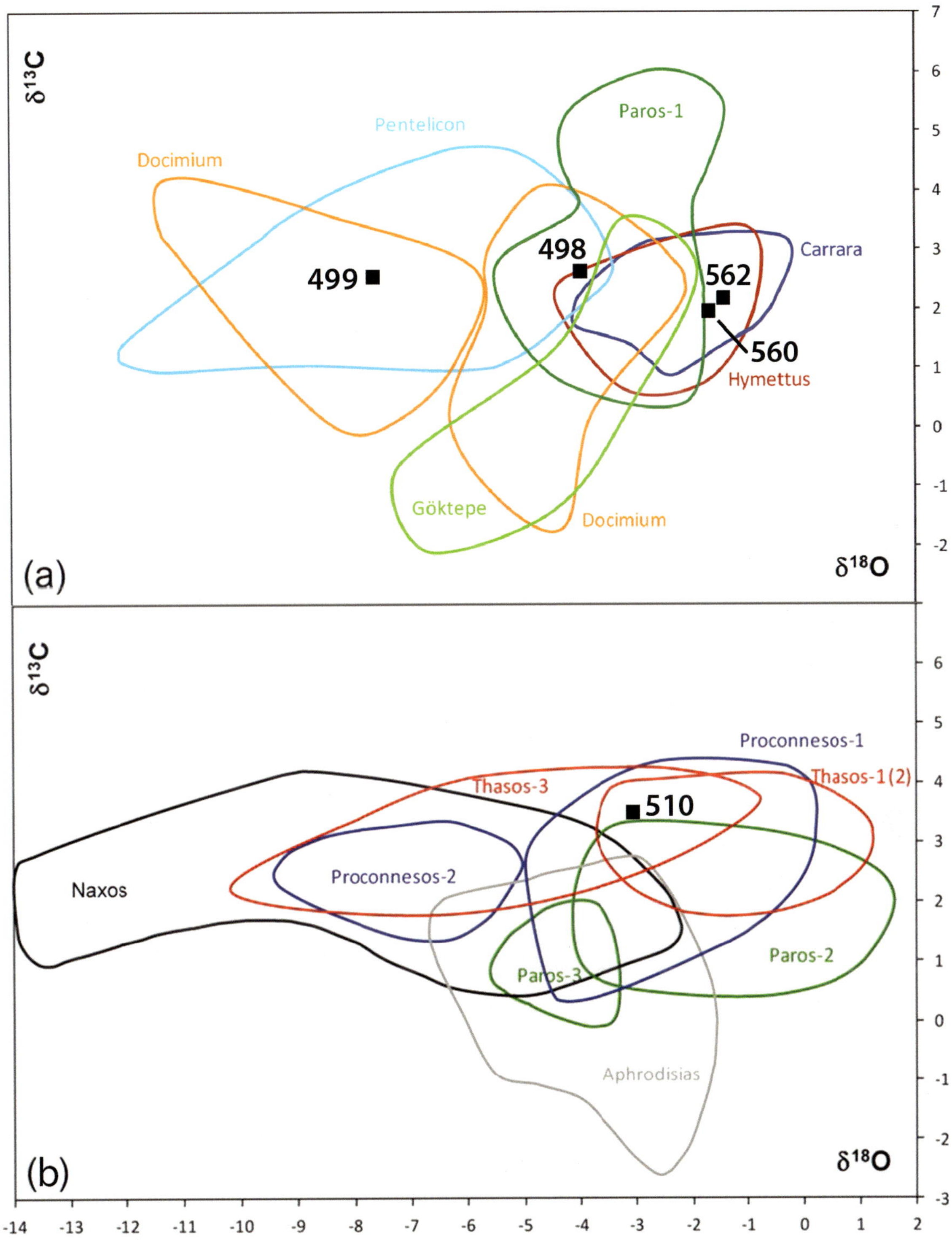

(a) Fine-grained marbles:

Paros-1 = Paros Stefani

(b) medium- to course-grained marbles:

Paros-2 = Lakkoi
Paros-3 = Karavos
Proconnesos-1 = Saraylar
Proconnesos-2 = Çamlik
Thasos-1(2) = Aliki
Thasos-3 = Cape Vathy

Figure 3.1. The isotopic composition of the marble samples compared with the reference databank of Antonelli and Lazzarini 2015 for (a) fine-grained marbles and (b) medium- to course-grained marbles.

petrographic features of the Italian Carrara and Greek Pentelic marbles are mainly the boundary shapes of the carbonate crystals and the different heteroblastic fabrics, with the latter being lineated and often stressed. Quartz, graphite, K-mica, apatite, and pyrite are found in the Carraran and Pentelic samples, and are accessory minerals typically present in these marbles. Carraran is also associated with variable quantities of ore minerals, as seen in cat. no. 562. Pentelic sample cat. no. 499 contains some plagioclase.

It is not surprising that two of the pieces collected in Rome were identified as *marmor Lunense*. As seen in cat. no. 560, Carraran marble was the mainstay of sarcophagus production in Italy.[6] The use of the marble for decorative revetment, such as cat. no. 562, has been observed in an exedra of the Baths of Trajan in Rome, in the temple of Venus Genetrix, as well as in private villas outside of Rome in Oplontis.[7] Beginning in the first century BCE, *marmor Lunense* was the preferred decorative marble for public and private buildings in Rome. Convenient proximity and imperial involvement made large-scale exploitation possible.[8]

Importation of the expensive Pentelic marble from Greece to Rome had occurred since the Republican period when Roman generals built their manubial temples with it.[9] By the early Imperial period, Roman emperors made extensive use of Pentelic marble for their public buildings.[10] Although the Attic marble was a symbol of prestige, distribution networks from the mid-first century CE onward gave rise to a huge increase in its availability. While there are no specific contexts for cat. nos. 498 and 499 within the huge 13-hectare district of the Baths of Diocletian, such as were reconstructed in the exotic marbles decorating the walls of the *natatio*,[11] the use of Pentelic marble in objects as utilitarian as marble slabs shows that no expense was spared for the imperial building project.

The petrographic features of cat. no. 510, a coarse-grained dolomitic marble, suggest a source from Cape Vathy, Thasos, Greece (fig. 3.2a). Its measured isotopic ratios confirm a Thasian provenance (fig. 3.1b).

Identifying the source of fragments from Greece without petrographic analysis is problematic. The fact that some of the sampled pieces fall outside the Roman period and the absence of exact findspots further limits interpretation. Moreover, a large number of minor varieties of white marbles can be found in the Peloponnese, and the probable use of local marbles should also be taken into account. The isotopic ratios of $\delta^{13}C$ and $\delta^{18}O$ are shown plotted against some major and minor white Greek marbles in figure 3.3.

Although the results are inconclusive, in several cases we have high confidence in the identifications because excavations corroborate the scientific results. For example, cat. no. 495 is an architectural fragment collected from the fourth-century BCE temple of Athena Alea in Tegea, Arcadia. The excavators believed the locally available Doliana marble was used throughout the entire superstructure of the Late Classical temple, as well as in more limited amounts in the Archaic-Classical temple and nearby associated buildings.[12] Our results also suggest Doliana as the probable origin although the oxygen levels fall just outside the normal range for that source (fig. 3.3).[13] The isotopic signature and MGS of the paving or veneer from the theater at Sparta (cat. nos. 492, 493) also suggest a Doliana marble, yet a Mani origin for cat. no. 492 cannot be excluded.[14] Investigations at Sparta have led the excavators to understand that the primary building material of the theater was white marble from several different local Laconian sources.[15] Ancient white marble quarries have been found in the region of Laconia, and while they are not included in the database here and have not been the focus of scientific analysis, they should also be considered as possible sources.[16] Identification of the remaining pieces—from the odeum of Patras (cat. no. 497), a cornice molding found on Mt. Kronos in Olympia (cat. no. 494), and paving or veneer(?) from the

6. Koch and Sichtermann 1982.
7. De Fine Licht 1974 30–31; Attanasio, Brilli, Bruno, et al. 2012, 337; Barker and Fant 2019b.
8. Dolci 1980.
9. Bernard 2010.
10. Bruno et al. 2002; Attanasio, Brilli, Bruno et al. 2012.
11. Beste, Arnold, and Wolf 2015, 200–08.
12. Østby 2014a; Østby 2014b, 57, 60, 305–09, 311, 314.
13. Antonelli and Lazzarini 2015.
14. Herz 1987; Cooper 1988; Attanasio, Brilli, and Ogle 2006, 108–14.
15. Waywell and Wilkes 1995; Waywell, Wilkes, and Walker 1998; Waywell and Wilkes 1999.
16. Kokkorou-Alevras et al. 2009.

a

b

c

d

e

(a) Cat. no. 510: Thasian marble from Cape Vathy showing a heteroblastic fabric made of dolomite crystals having sutured boundaries.

(b) Cat. no. 499: Pentelic marble showing a foliated heteroblastic fabric made of calcite crystals having mainly curved boundaries.

(c) Cat. no. 498: same features as cat. no. 499.

(d) Cat. no. 560: Carrara marble showing a homeoblastic mosaic with triple points made of calcite crystals having curved/straight boundaries.

(e) Cat. no. 562: Carrara marble showing a heteroblastic mosaic fabric made of calcite crystals having curved/embayed boundaries.

Figure 3.2. Photomicrographs of thin sections.

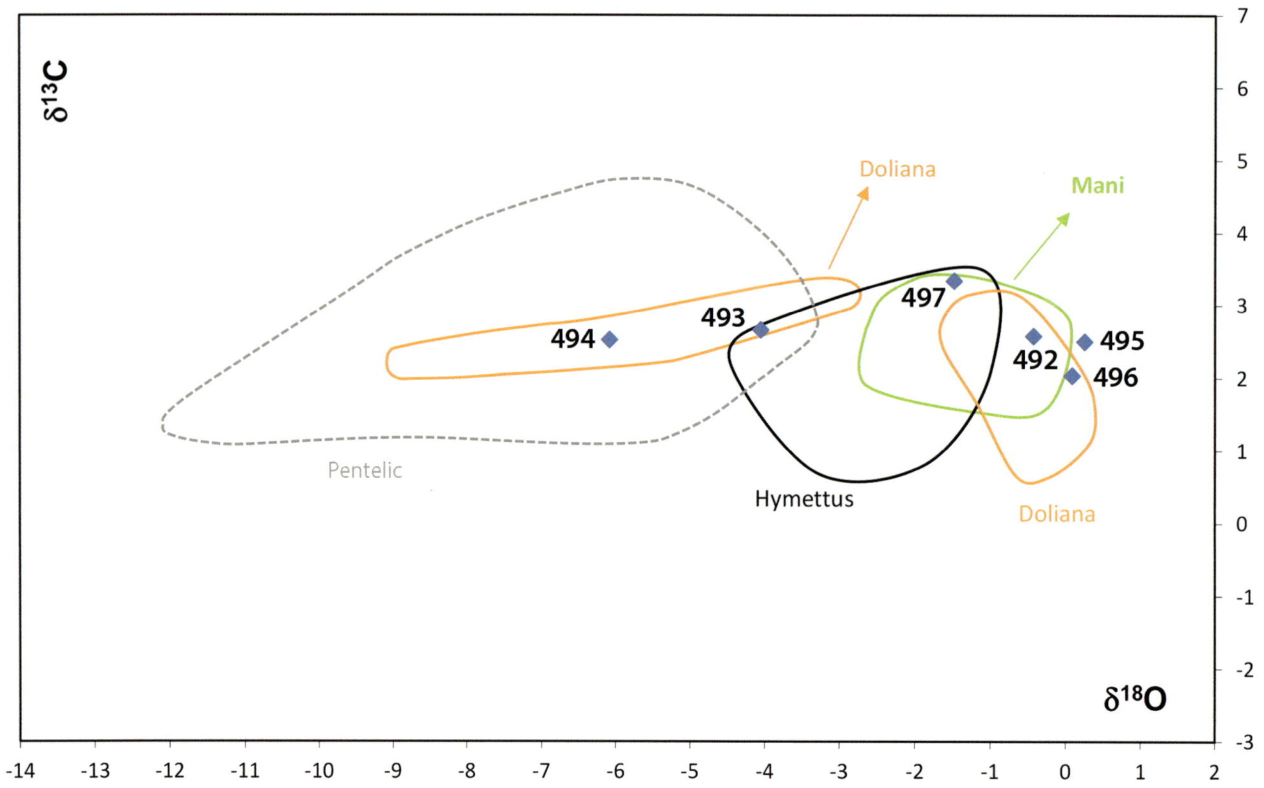

Figure 3.3. The isotopic composition of the marble samples from Greece compared with the isotopic signatures of major white sources in Greece.

acropolis at Pherae (cat. no. 496)—are shown in figure 3.3. Because these buildings have not been subject to architectural study, it was not possible to make further hypotheses about the provenance of the pieces.

Conclusions

The results of scientific analysis of the marbles in the Kelsey Museum collection tallies with general trends for the Roman marble trade. In Rome, the use of Luna, Pentelic, and Thasian marble for architectural surface decoration corresponds with numerous archaeometric studies. While Rome always relied on her favored, nearby source of Lunense marble for adornment, the import of prestigious marbles from Greece also played a major role. With regard to the results from the Greek sites, which are admittedly inconclusive, results indicate that abundant local sources of good-quality marbles supplied local building projects. Unfortunately, given that the archaeological findspots and chronological horizons of the pieces are not entirely secure, theories about production and workshops cannot be proposed.

The identifications in the museum archives show the well-known pitfalls of assigning provenance to white marbles by eye. This is especially true of the samples from Greece, which were all misidentified. For example, while the accession books labeled the marble of cat. no. 495 from the temple at Tegea as Carraran, its measured isotopic signature falls well outside of data points for marble from Carrara. If the piece was indeed taken from the Late Classical temple, the quarries at Luna would not have been opened for another three centuries. Yet in the case of the pieces gathered in Rome, the listed sources usually match the scientific results, showing that experts who made these assignments at the turn of the twentieth century either had great familiarity with Rome's ancient marbles or were simply lucky guessers.

Chapter 4

LITHOTYPE PROFILES

J. Clayton Fant, with contributions by Leah E. Long

All lithotypes given a specific identification in the catalogue are listed here. Those identified speculatively ("possibly ...") or generically (e.g., white marble, green limestone, etc.) are excluded. North American lithotypes and Mediterranean ones which had no ancient use are not treated here, but additional information is presented in the catalogue entries themselves. Full references for citations are found in the bibliography at the end of the volume. The order of entries is by the most familiar name, in most cases the Italian artisanal name.

Africano — Cat. Nos. 75–117, 701

Also known as: *m. Luculleum*
Source: Sığacık (ancient Teos), western Turkey

Tectonic breccia or conglomerate with assorted limestone clasts, fossils, fractures filled with calcite or quartz.

The source was unknown before Ballance (1966) found quarries at Sığacık (ancient Teos) near Seferihisar, west of Izmir. Subsidiary quarries have since been found nearby at Beyler (Benler) (Türk et al. 1988) and three other locations; a newly discovered inscription refers to the sites as *lapidicinae Luculleiae* (Adak and Kadıoğlu 2017). At Sığacık and Beyler the dominant lithotype is a uniform dark gray, and this seems to be what Pliny (*Natural History* 36.49) speaks of despite the fact that *Luculleum* is the multicolored variety. Dated quarry inscriptions attest to imperial involvement at the source and also suggest that it was abandoned in the last decades of the second century CE; the listing of *Luculleum* in Diocletian's Price Edict with one of the highest costs probably reflects the scarcity value of reusable pieces.

Pliny's location of the source as Chios has usually been taken as a paleographical mistake, but recent work shows that the same geological formation extends across the straits to Chios (Lazzarini 2007, 125).

Africano was one of the first polychrome marbles imported to Rome and was tightly wound up with the political rivalries of the late Republic. The marble was clearly one of the spoils of Lucullus' Asian campaigns. In the Empire it was one of the three most prized marbles, along with Phrygian *pavonazzetto* and Numidian *giallo antico* (q.v.).

Africano was used in the Basilica Aemilia and most subsequent imperial buildings. Broken trunks of original 40-foot columns from the *Templum Pacis* are visible near the tourist entrance to the Roman Forum. JCF

References

Corsi 1845, 99; Gnoli 1988, 147–51; Borghini 1989, 133–34; Fant 1989; Dworakowska 1990; Price 2007, 136–37; Adak and Kadıoğlu 2017

Alabastro

Cat. Nos. 280–335

Also known as: *alabastro cotognino, alabastro Egiziano, l. alabastrites,* alabaster
Source: Nile Valley, Egypt

This calcitic alabaster (travertine, in proper geological terminology; Harrell 1990) is found in many places in the Nile Valley. It has been quarried and used from Predynastic times to the present. ("Oriental," in traditional collecting terminology, may simply point to Egypt and is also used by geologists to distinguish calcite alabaster from gypsum; Price 2007, 45 and under "Further Information" in the alabaster entries of her online Corsi Collection, OUMNH). The most important quarry, at least in pre-Roman times, was at Hatnub, some 15 km from Amarna. The stone's advantages, beyond the aesthetic, are abundance and easy workability. Alabaster, being a freshwater sedimentary rock, is far from rare and today we are not much better at rigorous provenancing since it has not attracted the same research effort as marble and granite.[1] But, given the ubiquity of it in Egyptian culture and given the quarries' easy access to the river, there is no reason to doubt that most of the alabaster exported to the western Mediterranean was from Egypt. One exception may be an alabaster, visually indistinguishable from its Egyptian counterparts, found at ancient Thyatira (modern Akhisar) in Turkey; however, its orbit beyond local use remains unclear (Çolak and Lazzarini 2002).

Pliny (*Natural History* 36.59–60) begins his history of Egyptian alabaster in Italy by quoting Nepos on the amazement caused by Lentulus Spinther's public display of large amphora-shaped vessels. But only five years later (Spinther was consul in 57 BCE), Nepos saw 32-foot columns (monolithic, surely—otherwise the anecdote lacks punch). The volume of imports in the late Republic was sufficient to allow supplies to trickle down to Pompeii, where both the rich and the aspiring could use Egyptian alabaster, among others, in floors and represent it in painted form on walls from the period of the Second Style onward (Barker and Perna 2018). Pliny's next item, four small columns in the theater of Cornelius Balbus, has already made the transition into Roman Egypt. Since production from one quarry or another was probably continuous, no particular effort was perhaps necessary on the Romans' part to keep the supplies coming. JCF & LEL

Note

1. Except for Egypt itself, where Klemm and Klemm (1993, ch. 4) have surveyed eight quarries and succeeded in distinguishing them by chemical composition (p. 222).

References

Harrell 1990; Aston, Harrell, and Shaw 2000, 59–60; Çolak and Lazzarini 2002; Klemm and Klemm 2008, ch. 4, esp. 147–66; Barker and Perna 2018; Barker and Perna 2024

Alabastro fiorito

Cat. Nos. 289–294

Source: Pamukkale (ancient Hierapolis), western Turkey

This decade's spate of fieldwork on ancient quarries in Turkey has revealed important new information on the alabasters of ancient Hierapolis. Strabo first mentioned the "stone of Hierapolis" (*Geographica* 9.5.16) in his list of new stones that were being brought to Rome in the first century CE. Paul the Silentiary records that the stone was used in an ambo of the Hagia Sophia in Constantinople, attesting to its continuing popularity in the Byzantine period. Nineteenth-century Italian stonemasons similarly took poetic license in describing alabaster's appearance according to the patterns displayed, giving rise to the terms most commonly used today.

Calcitic alabasters were quarried at many sites throughout the ancient Mediterranean, with the most famous hailing from the Nile Valley in Egypt, but also Italy, Algeria, Tunisia, and elsewhere in Turkey. Today much confusion arises in differentiating these beautiful stones from one another and in locating their original source, in large part because archaeometric studies are rare, but also because single quarries can produce multiple varieties. The Italian mission's recent archaeological surveys of ancient quarries has helped to clarify the archaeological and geological setting of Hierapolitan alabasters as well as provide thorough geochemical and petrographic analysis. At Hierapolis, the stone was found in travertine quarries, with three main geological varieties. One is yellowish-brown in color with a more porous and earthy appearance, while the others are more compact, most readily distinguished by linear or undulating bands, with colors varying from white to reddish and yellow. The latter type, which was the most sought after in antiquity, is onyx-like with vertical bands of white, brown, and rusty red veins. Depending upon the direction of the cut along the grain, the stone can change its appearance to be an *alabastro fiorito* (with inclusions that create a flowering effect), *alabastro listato* (veins in less densely packed bands), *alabastro a ghiaccione* (white ice), or *alabastro a rosa*. Strabo was likely referring to the *fiorito* and *listato* types as the colored stones brought to Rome during the reign of Augustus (Scardozzi 2016, 142 n. 12).

Until recently, the quarries near the Turkish village of Gölemezli, also in the Lykos River Valley and not far from the city near ancient Tripolis, were considered the main source for Strabo's stone (Bruno 2002a). Systematic investigation within the territory of Hierapolis has uncovered evidence for ancient alabaster exploitation both in the immediate vicinity of the city and at 3 km distance from its urban core. In some locations, quarrymen followed the travertine bands into fissure-ridges where the stone was deposited, resulting in dramatic narrow and deep excavation trenches with vertical faces that could be 2–10 m narrow, 5–20 m deep, and over 100 m long. Used in civic monuments at home and in the region, it may have been exported as a prestige marble across the Mediterranean. Since the Hierapolis *fiorito* may be confused with other alabasters from Gebel Oust (Tunisia) and Bou Hanifa (Algeria), only archaeometric analyses can determine its distribution (Brilli et al. 2018).

Seen in Rome as early as the Late Republican period, the *fiorito* and *listato* varieties were popular for floor and wall cladding, small columns, vases, urns, and busts of statues. Examples presumed to be from Hierapolis are in surface decoration of the Horti Lamiani, the Domus Flavia in Rome, and in the private houses of Pompeii and Herculaneum. Closer to home, in Istanbul, lavish displays can be seen in the Hagia Sophia, in imperial sarcophagi for Theodora, wife of Justinian, and the wife of Emperor Anastasios II, both of which were once housed in the Church of the Holy Apostles in Constantinople. LEL

References

Bruno 2002a; Çolak and Lazzarini 2002; Scardozzi 2012; Scardozzi 2016; Barker and Perna 2018; Brilli et al. 2018; Scardozzi 2019; Barker and Perna 2024

Alabastro a ghiaccione

Cat. Nos. 300–301

Also known as: ice alabaster

Sources: Pamukkale (ancient Hierapolis), western Turkey; La Batteria quarries near Monte Circeo, Lazio, Italy

The artisanal term refers to calcitic alabaster with large areas of continuous color ranging from light gray to dark beige, often with wavy combinations of color, of exceptional translucency even for alabaster. A sharp visual discrimination from Egyptian *alabastro cotognino* is not always possible, but chemically they are distinct in their strontium and lead isotope ratios (Barker and Perna 2018). Possible sources for *alabastro a ghiaccione* are the La Batteria quarries near Monte Circeo, Italy, and Pamukkale (ancient Hierapolis) in western Turkey. The fourteen thresholds in *alabastro a ghiaccione* at Villa A ("of Poppaea") at Oplontis, part of the original construction of ca. 50 BCE, are most likely from La Batteria. Work currently under way is likely to shed considerable light on the provenience of this alabaster. JCF

References

Bruno 1998; Bruno 2002a; Scardozzi 2012; Perna and Barker 2017; Barker and Perna 2018; Barker, Perna, et al. 2018; Barker and Fant 2019a; Barker and Perna 2024

Alabastro di Palombara, alabastro palombino

Cat. No. 309

Subvarieties of *alabastro a tartaruga* (q.v.)

Alabastro a pecorella

Cat. No. 312

Source: Aïn Tekbalet and Bou Hanifia, western Algeria

The North African *alabastro a pecorella* stands out as the most richly colored alabaster quarried in antiquity. The patchy patterns of red, white, and to a lesser extent brown and yellow were thought to resemble sheep's wool and gave this banded travertine an unusual three-dimensional appearance. These colorful dendritic markings are in fact metal oxide minerals deposited by hot springs. Research, scant until recently, has shown the Romans acquired varieties at Aïn Tekbalet and Bou Hanifia near Oran, in western Algeria. Not all sources were distinctive, though, and the resemblance of the horizontal banding seen at the Aïn Tekbalet quarry to *alabastro fiorito* from Turkey causes further consternation. Likewise, Gebel Oust in Tunisia may have produced *pecorella* (Lazzarini 2012, 437).

Since so many quarries produced multiple varieties of alabasters, it is possible to see why pinpointing a source remains elusive, and why ongoing studies on ancient alabasters are so important. Smaller in supply than its North African counterparts, *alabastro a pecorella* nonetheless enjoyed distribution throughout the central Mediterranean, especially to Herculaneum and Rome, where it was a favorite among Roman emperors. The Mauretanian kings' close relationship to Roman emperors was undoubtedly connected to the stone's appearance in the capital buildings of Rome in the early first century CE. Yet its most extensive use was in the local North African cities of Cuicul, Thamugadi, Hippo Regius, Carthage, and Utica, in public and private buildings alike. LEL

References

Lazzarini 2009, 462; Herrmann, van den Hoek, and Tykot 2012; Herrmann and van den Hoek 2014; Barker and Perna 2018; Barker and Perna 2024

Alabastro a tartaruga

Cat. Nos. 309, 310?, 315–316

Also known as: *Alabastro tartarugato*
Source: Jano di Montaione, Tuscany, Italy

Alabastro a tartaruga (or *tartarugato*) is a compact banded calcitic travertine with a rich brown ground and blobs and veins of the same or lighter brown, occasionally white. Artisans of the Italian tradition compared it to the shell of the tortoise (*tartaruga*). Now traced to a very small quarry near the village of Jano di Montaione in Tuscany, Italy (Lazzarini, Villa, and Visonà 2006), and chemically distinguishable from other alabasters for its high strontium content (Lazzarini et al. 2012), it occurs as one of three local varieties, including *alabastro cinerino*, named for its blue-gray layers the color of ash mixed with white layers (these are also known as *alabastro di Palombara*, cat. no. 309 and possibly 310, or *alabastro palombino*, cat. no. 311). It was also used in Florentine tables in the *pietre dure* tradition around 1700 (Betti and Brovadan 2012, 265). Cf. OUMNH no. 358, with some uncertainty about the source. The alabaster of Castelnuovo dell'Abate (*alabastro di Montalcino*) is also quite similar. Gnoli (1988, 228) reported seeing fragments of small columns and floor tiles; one of the latter is documented at Herculaneum and Pompeii in a Third Style context (Barker and Perna 2018, 56–58 and personal communication). JCF

References

Mielsch 1985, 39, nos. 83–88; Gnoli 1988, 228; Borghini 1989, 151; Lazzarini, Villa, and Visonà 2006; Price 2007, 52; Betti and Brovadan 2012, 265; Lazzarini et al. 2012; Barker and Perna 2018; MUSNAF nos. 1705, 1716, 1752, 1878; OUMNH nos. 354–358; Barker and Perna 2024

Astracane / Castracane

Cat. Nos. 660–661

Subvariety of *Lumachella* (q.v.)

Basalto

Cat. Nos. 593–594

Also known as: *selce di Roma*, basalt
Source: Alban Hills south of Rome

Dark gray to black extrusive volcanic rock abundant in the old lava flows south of Rome on the north slopes of the Alban Hills. This basalt was used extensively for road paving in Rome for many centuries right into the modern era. JCF

References

DeLaine 1997, 87–88; Jackson and Marra 2006

Basanite

Cat. Nos. 1–3

Also known as: *basalto verde, basalto verde antico, grovacca, l. basanites, bekhen*-stone, *pietra bekhen*
Source: Wâdi Hammâmât, Eastern Desert of Egypt

Commonly but incorrectly called *basalto*, *bekhen*-stone is a sedimentary sandstone/siltstone called graywacke that has been exposed to metamorphic pressures, hence a metagraywacke. Its color ranges from nearly black to gray to olive shades.

Quarried in several places in Wâdi Hammâmât in the Eastern Desert of Egypt, about 80 km east of Koptos/Qift. Exploitation began in the Predynastic period (Goyon 1957, 12) and extended to the reign of Maximinus Thrax in 238 CE. In the late Augustan period, activity in the Wâdi Hammâmât started anew, and evidence from the workers' huts points to continuous occupation, a change from the episodic expeditions of earlier millennia (Brun 2003, 193). An agent of P. Iuventius Rufus, a military tribune who was in charge of Egypt's mineral resources, was at work in Wâdi Hammâmât with a cohort of soldiers, a secretary, and an architect, assigned apparently to build the quarry village. Later evidence shows a large and specialized organization of quarry crews and military guards (Kayser 1993, 114–15, nos. 1, 4, 12). Epaphr(oditos), a quarry contractor known also at Mons Porphyrites, Mons Claudianus, and Wâdi Fawakhir, is attested in the early Hadrianic period,[1] and ostraca from the rooms of the settlement mention quarrymen (*sklerourgoi*), picks (*skapheia*), and wedges (*spheës*), and a neat list scratched into a flat stone catalogues some twenty blocks by dimensions (Kayser 1993, nos. 43, 40, 20).

Roman sculptural use of the *bekhen*-stone shows that it retained its connection with rulers and gods, from a bust of Julius Caesar (Belli Pasqua 1995, no. 1), three images of Augustus, two of Livia, one each of Tiberius, Germanicus, a generic prince, and Agrippina the Younger (Belli Pasqua 1995, nos. 4–4, 7–8, 10, 11, 9, 12; Talamo 2000). *Bekhen*-stone was also useful for the creation of faux-Egyptian artifacts on demand, as at Hadrian's Villa. The Romans were also happy to use it, as well as Egyptian porphyry and Aswan granite, for secular luxury objects such as basins in baths, like the well-known example in the Cortile Ottagonale of the Vatican Museums. JCF

Note

1. Kayser 1993, no. 19, *I. Ko.Ko.* 54 & 55; at Mons Porphyrites: *I. Pan* 21 = *IGRR* I 1256 = *SEG* XIII 601; at Mons Claudianus: *I. Pan* 42 = *SEG* XV 863 = *OGIS* II 678 = *IGRR* I 1255, *I. Pan* 42; at Wâdi Fawakhir: Guéraud 1942, no. 30. The architectural elements here bearing his name may have been intended for a shrine similar to the ones he built at Claudianus and Porphyrites (Kayser 1993, 126).
   - *I. Ko.Ko.* = A. Bernand, *De Koptos à Kosseir*. Leiden, 1972.
   - *I. Pan* = A. Bernand, *Pan du désert*. Leiden, 1977.
   - *IGRR* = R. Cagnat et al., *Inscriptiones Graecae ad res Romanas pertinentes*. Paris, 1906–27.
   - *OGIS* = W. Dittenberger, *Orientis graeci inscriptiones selectae*. Leipzig, 1903–05.
   - *SEG* = *Supplementum Epigraphicum Graecum*.

References

Corsi 1828, 199; Couyat and Montet 1912; Gnoli 1988, 93–97; Borghini 1989, 195; Kayser 1993; Klemm and Klemm 1993,

Basanite (cont.)

365–66; Ambrogi 1995, 193–94; Belli Pasqua 1995; Harrell, Brown, and Lazzarini 2002; Brun 2003

Bigio antico

Cat. Nos. 530?, 538?, 550?, 551?, 553

Also known as: *m. Lesbium* (ancient names of other varieties of *bigio antico* are unknown)

Sources: Moria, island of Lesbos, Greece; Belevi near Ephesus, Sığacık (ancient Teos), and Göktepe, all in western Turkey; Mani Peninsula, mainland Greece; island of Chios, island of Rhodes

Bigio antico, "ancient gray," is the Italian artisanal name for gray marble with medium to coarse crystals in varying gray hues. Corsi noted that varieties with spots entirely white or entirely black were called *bigio brecciato* while ones with fossils were *bigio lumachellato* (Corsi 1845, 106). Neither the artisanal tradition nor Pliny's treatment was consistent in distinguishing marble and limestone (Pliny, *Natural History* 36.135; Lazzarini 2007, 97). Gray marbles are widespread, and sources and artifacts have become only recently subjects of sustained study. All five attributions in the Kelsey collections are qualified or secondary.

Gray marbles, apart from the *bardiglio* of Luna, were not widely used in Roman Italy before the last third of the first century CE, and then chiefly in private buildings. Many sources are mentioned in the literature in the Aegean and on the coast of Asia Minor (Chios, Miletus, Halicarnassus, Rhodes), but are not verified. Those of Lesbos, Sığacık (Teos), and Cape Matapan (Taenarum) on the Mani Peninsula of mainland Greece stand out. The Mani varieties drew literary attention, partly from the presence of *rosso antico* (q.v.) and *bigio morato* (see the profile on *nero antico*), not to mention an entrance to the underworld. The quarries of Moria on the island of Lesbos were used largely for architecture (large columns of the east wing at Oplontis Villa A: Barker and Fant 2019a and b, visual identification). Large sources of *bigio antico* for sculpture are now identified at Belevi (near Ephesus), already well known for its white marble deposits (Attanasio, Bruno, Prochaska, and Yavuz 2017). The Teos quarries seem to have continued to export gray marble long after the splashier but less abundant *africano* (q.v.) had been worked out (Fant 1989; Brilli et al. 2019). JCF & LEL

References

Corsi 1845, 94, 106–07; Gnoli 1988, 179-80; Lazzarini 2007, 97; Price 2007, 76; Attanasio, Bruno, Prochaska, and Yavuz 2017; Brilli, Giustini, and Kadıoğlu 2019

Botticino

Cat. Nos. 615–617

Source: Botticino, Lombardy, Italy

A buff-colored limestone with round algae patches and tan to brown stylolites. The quarries at Botticino near Brescia in northern Italy are still active.

Used only locally in the Roman period but widely in Italy from the Renaissance onward and today even at big box home centers in North America. JCF

Reference

Price 2007, 169

Breccia di Aleppo

Cat. Nos. 336–337?

Source: Kariès, island of Chios, Greece

Groundmass of brick red, angular clasts of gray and yellow, sometimes with marine fossils, appearance quite variable.

Gnoli explains that this breccia, which does not come from Aleppo in Syria, got its artisanal name from a slip of the pen by an eighteenth-century French mineralogist who wrote "Alep" instead of "Alet," the source of a somewhat similar breccia near Aix-en-Provence. This was enshrined by Faustino Corsi's 1845 catalogue. Its actual source remained unknown until 2001 when Lazzarini, following up a conjecture by Gnoli, identified it at the village of Kariès, about 6 km inland from the town of Chios, on the island of that name. This is close to the well-known quarries of Latomi, source of *portasanta*, with which it has geological affinities.

Breccia di Aleppo was quite popular in Italy through the Julio-Claudian period but pretty well vanishes after the Flavian period. It was not used in imperial projects but is relatively common in private contexts, including luxurious ones like the insert paving of the long portico flanking the pool of the East Wing of Villa A at Oplontis. JCF

References

Corsi 1845, 141–42; Gnoli 1971, 202ff.; Gnoli 1988, 237; Borghini 1989, 161; Lazzarini 2007, 137–49; Barker and Fant 2019a

Breccia cenerina — Cat. Nos. 475, 476?

Source: unknown

Breccia of marble clasts, gray to white, in a calcareous ground of purple hues. Long considered a variation of *breccia di Seravezza* (q.v.), it is now cogently argued on geological grounds to be unrelated to the Apuan Alps and so remains of unknown origin (Barteletti and Amorfini 2003). It was used for small objects, especially floor tiles. Although it is sometimes said to be entirely post-Roman, Pensabene points to its use in Rome during the Julio-Claudian period (2013, 448). It can be a prominent element in *opus sectile* floors in the Vesuvian cities, for instance in room 5 (triclinium) of the House of the Stags at Herculaneum. JCF

References

Gnoli 1971, 208; Borghini 1989, 165; Bartelletti and Amorfini 2003; Pensabene 2013, 448

Breccia corallina — Cat. Nos. 230?, 338–339, 340?, 341–343, 344?, 345–348, 362?

Also known as: *m. Sa(n)garium*
Sources: Bilecik, Vezirhan Province, Turkey; Karaburun Peninsula, Turkey

Limestone clasts of varying size in a color recalling coral but ranging from orange to salmon pink or light purple matrix.

Quarried throughout Turkey, most famously in the area of the village of Bilecik in the province of Vezirhan but also in the Karaburun Peninsula as recent surveys have discovered (Bruno, Elçi, et al. 2012). This was probably the marble that Pliny the Younger had in mind during his governorship of Bithynia when he proposed to the emperor Trajan (*Epistulae* 10.41–42) digging a canal to connect the river Sangarius (modern Sakarya) to Lake Sophon (modern Sapanca) near Nicomedia, giving quick access to the Sea of Marmara (the river curves around to empty into the Black Sea north of the Bosphorus). Lazzarini's recent study of the marbles and quarries of the region (2002b) reassigns the name *m. Sa(n)garium* from *occhio di pavone* to *breccia corallina*.

Despite Pliny's interest, *breccia corallina* is essentially absent from imperial building projects (before the Severan Atrium Vestae; Gnoli 1988, 238–40) but otherwise popular in Italy as columns. It is present, in very small quantities, in the fragmentary bar claddings of Pompeii and Herculaneum, suggesting a modest use in buildings of the early and mid-Julio-Claudian period (Fant, Russell, and Barker 2013, 188). Its most prominent use in Campania is as the dominant elements (socle and panels) of the luxurious rooms facing the pool of the East Wing of Villa A at Oplontis. Here the cement is often of a striking purplish hue. JCF

References

Mielsch 1985, 45–46, nos. 201–50, pls. 7–8; Gnoli 1988, 238–40; Borghini 1989, 166; Lazzarini 2002b; Bruno, Elçi, et al. 2012; Fant, Russell, and Barker 2013, 188; Barker and Fant 2019a

Breccia gialla

Cat. No. 486

A subvariety of *giallo di Siena*
Source: Montagnola Senese near Siena, Italy

Weakly metamorphosed limestone with fine crystals, colors similar to Numidian *giallo antico* but often with paler yellows and large white areas. *Breccia gialla* is known to have been quarried in the hills of the Montagnola Senese near Siena, Italy, from the Renaissance and is still commercially available (Price 2007, 92). *Breccia gialla* and the variety called *breccia dorata*, which the Kelsey *tondino* cat. no. 486 closely matches, were prospected by the Romans and used as early as the Severan period (Bruno and Lazzarini 1999, 79–80). Since they are interbedded with *giallo di Siena* the Romans must have been familiar with it too. *Breccia gialla* has been pressed into service as a post-antique substitute marble for Numidian *giallo antico*, notably in a few of the discs of the Pantheon's pavement, which was relaid in 1868–70, mostly with ancient marble recovered in the excavation of the Marmorata (Fant 2001, 188; Maischberger 1997, 166). JCF

References

Maischberger 1997, 166; Bruno and Lazzarini 1999; Fant 2001; Price 2007, 92–93

Breccia di Hereke

Cat. No. 618

Source: Hereke, northwest Turkey

The breccia and puddingstone of Hereke are two facies of the same formation. The clasts of the puddingstone are rounded instead of angular, and larger. Hereke is near Gebze in northwest Turkey, just west of ancient Nicomedia. The stone is very common at Constantinople/Istanbul, being used from the early Byzantine period to the Ottoman, and it appears also in St. Mark's Basilica in Venice. JCF

References

Gnoli 1971, 248, note to p. 72; Lazzarini 2012, 75

Breccia medicea

Cat. Nos. 349, 350, 351

Subvariety of *breccia di Seravezza* (q.v.)

Breccia rossa Appenninica

Cat. Nos. 477, 478, 479?, 480

Source: Coregna, La Spezia, Italy

Limestone conglomerate with clasts of varying shades in a groundmass containing iron ore (Price 2007, 126).

Used locally and at Rome (a small quarry block in the Antiquarium del Celio, a column at Ostia in the Domus fuori Porta Marina (Becatti 1969), in the Fourth Style salon 5 of the House of the Stags, Herculaneum, which gives a *terminus ante quem* of the early/middle of the first century CE. Quarries at Coregna (La Spezia), Italy, reopened in the seventeenth century and were worked to exhaustion until the 1970s. JCF

References

Borghini 1989, 189; Bruno and Lazzarini 1999; Price 2007, 126

Breccia di Seravezza

Cat. Nos. 349–351, 353?, 373?, 483?, 484

Source: Apuan Alps, northern Italy

Breccia of one or more kinds of marble clasts in a calcareous ground, colored red, yellow, green, and sometimes violet (*breccia di Seravezza*); it can sometimes pass for Phrygian *pavonazzetto*. A notable variety is *breccia medicea* (cat. nos. 349, 350, 351)

with strong markings. Quarried in the Apuan Alps in Tuscany, Italy. JCF

References

Lazzarini and Turi 1999; Bartelletti and Amorfini 2003; Price 2007, 131; Pensabene 2013, 447–48

Breccia di Settebasi Cat. Nos. 174?, 352–363, 364?, 365?, 366, 367?, 368–378, 379?, 701?

Also known as: *m. Scyrium*(?)
Source: Renes and nearby inlets, island of Skyros, Greece

This colorful breccia from the island of Skyros in the Sporades archipelago of the northern Aegean draws its name from the suburban villa so named which served as a virtual quarry for spoil in the post-antique period. The marble was not exploited by the Greeks, who preferred the island's white marble. The ancient name may have been *marmor Scyrium*, but textual support is weak. The Roman *m. Scyreticum* in Diocletian's Price Edict seems to refer to the white variety, the ancient name of the polychrome varieties remaining unknown. It was being imported to Rome in at least small quantities by 50 BCE and was in wide use by Augustan times.

Strabo (*Geographica* 9.5.16) compares it to stones of Carystos, Dokimeion, and Hierapolis and notes that at Rome polychrome marbles were used in public and private construction, making white marble less valuable (he may be referring specifically to the white marble quarries of Skyros itself). Although extant remains at Rome do not support Strabo's assertion about columns, Corsi (1845, 149–50) recorded six columns plus two of *semesanto* (a variety with much smaller clasts). Despite a hint of imperial interest in the quarries (*sub cura Sergi Leg(ionis) XV* inscribed on a block at the Marmorata [Bruzza 1870, 152–53], possibly recording a single expedition of this centurion), *breccia di Settebasi* was not used in imperial constructions until the Villa del Casale at Piazza Armerina in the late period. JCF

References

Belli 1842, 57–60; Corsi 1845, 198–201; Bruzza 1870; Dubois 1908, 129–30; Dworakowska 1972, 7 nn. 1–4; 10 nn. 14–22; Dworakowska 1975; Dworakowska 1983, 23, 25, 102; Mielsch 1985, 47, pls. 9–10, nos. 273–312; Gnoli 1988, 232–35; Borghini 1989, 192–93; Lazzarini and Turi 1999; Lazzarini and Cancelliere 2000; Lazzarini 2007, 160–81; Price 2007, 129

Breccia traccagnina Cat. No. 485?

Source: unknown

Limestone clasts of a range of colors in a calcareous matrix, sometimes called a conglomerate. Named for Traccagnino, a *commedia dell'arte* character better known as Arlecchino (Harlequin), who dressed in patchwork. The term is subjectively applied to a number of stones found in some notable Roman churches; its use in antiquity is not certain. Color varies greatly, and its origins remain unknown. JCF

References

Corsi 1845, 146; Mielsch 1985, nos. 320–29; Gnoli 1988, 250–51; Borghini 1989; Price 2007, 148

Breccia verde d'Egitto Cat. No. 3?

Also known as: *hexatontolithos, hecatontolithos, centopietre*
Source: Wâdi Hammâmât, Eastern Desert of Egypt

Polychrome metaconglomerate of well-rounded pebbles and cobbles in a green matrix of chlorite. For the development of the Wâdi Hammâmât quarries, see above, at *basanite*.

This striking stone, called "sixty-stone stone" (Constantinos Rhodios cited by Gnoli 1988, 118) and "one hundred-stone stone" (Pliny, *Natural History* 37, 167), was exploited for table supports, vases, and bowls, but chiefly for columns of middling dimen-

Breccia verde d'Egitto (cont.)

sions. The period of first imports is not clear since most extant pieces are reused, but Gnoli reasonably suggested that it paralleled the Augustan revival of *bekhen*-stone production. There was no quarrying after antiquity until the 1980s, when the Egyptian Geological Survey and two Italian-Egyptian companies, Marmonil and Petrobel, began a small-scale operation. JCF

References

Gnoli 1988, 117–21; Borghini 1989, 195; Harrell, Brown, and Lazzarini 2002; Klemm and Klemm 2008, 309–12

Broccatello

Cat. Nos. 395–397, 398?, 399–404

Also known as: *broccatello di Spagna, jaspi de la Cinta de Tortosa*
Source: vicinity of Tortosa (ancient Dertosa), northeastern Spain

A biomicrorudite limestone tinted with iron oxides and notable for a variety of fossils (mostly rudists) often with recrystallized calcite infills (it is not a jasper, despite the local Catalan name). Quarried at a number of sites around the city of Dertosa (modern Tortosa) on the northeast coast of Spain, particularly Barranc de la Llet, La Cinta, and Els Valencians.

Abundantly used at the Roman capital Tarraco (modern Tarragona) of the province of Hither Spain and in surrounding villas from the Augustan period, *broccatello* was also its most important marble export to Rome and Italy, almost exclusively in the form of *crustae* for paving and *opus sectile* compositions. Locally, it was also favored for inscriptions as well as occasional columns and sculpture. It had another period of popularity in the late Baroque period. JCF

References

Corsi 1845, 115–16; Mielsch 1985, 43–44, pls. 5–6, nos. 165–75; Gnoli 1988, 210–11; Borghini 1989, 198; Falcone and Lazzarini 1998; Mayer and Rodà 1999; Antonelli 2002, 273–74; Price 2007, 164–65; Gutiérrez Garcia-Moreno 2009, 229–45

Broccatellone

Cat. Nos. 405?, 406?, 407?, 408

Source: Çakmakli, Manisa, Turkey

A breccia with large and small limestone clasts in a fine pink matrix, bivalve fossils. Quarried at various sites in western Turkey. JCF

References

Mielsch 1985, 46, pl. 8, nos. 251–54; Gnoli 1988, 240; Borghini 1989, 199; Lazzarini 2002b; Price 2007, 111

Cipollino mandolato

Cat. No. 184

Also known as: *cipollino dei Pirenei, campan (vert), griotte*
Source: Haute Pyrenees and Campan, France

A lightly metamorphosed limestone with granular calcite nodules and fossils in a green, pink, or red calcitic matrix (Price 2007, 134).

Quarries in the French Pyrennes were worked in the first century CE but more extensively in later centuries. Used more in Gaul than Italy. There is also a red-violet version with the same network patterning. JCF

References

Mielsch 1985, pl. 17, nos. 596, 599; Borghini 1989, 204; Antonelli et al. 2002; Price 2007, 134

Cipollino rosso

Cat. Nos. 409–420

Also known as: *rosso brecciato, m. Carium, m. Iassense*
Source: Kıyıkışlacık (ancient Iasos), western Turkey

The marble that the Romans called *m. Carium* or *m. Iassense,* after its geographical location in Iasos, Caria, came in three distinct varieties. The most common, *cipollino rosso,* was a red, gray, and white marble with swirling veins like a sliced red onion, and in the nineteenth century was also referred to as *africanone* (big *africano*). The less abundant *rosso brecciato* was a brecciated red and white marble, sometimes with gray and purple inclusions. The rarest form was a monochrome red marble, used as a substitute for the pure red *rosso antico* from the Mani Peninsula in Greece, and mostly for small-scale statuary.

The quarries were located in three different areas of exploitation just to the northeast of ancient Iasos (Andreoli et al. 2002), near modern-day Kıyıkışlacık on Turkey's west coast. Famed polychrome marbles were located in the territories of several coastal cities in Asia Minor of moderate size and administrative importance, and their proximity to modest but well-equipped harbors was sufficient to facilitate the export of local materials. Quarry remains show the marble was exploited as early as the Hellenistic period, but it was in the Late Antique and Early Byzantine periods that *m. Carium* reached its zenith in popularity.

Writers like Constantine the Rhodian and Paul the Silentiary praised the stone for its beauty, such as in the latter's description of Justinian's Hagia Sophia in Constantinople: "You may see the bright green stone of Laconia and the glittering marble with wavy veins in the deep gullies of the Iasian peaks, exhibiting slanting streak of blood-red and livid white."[1] The great builder also had the stone installed in his Holy Apostles Church in Constantinople as well as in his crown jewel of San Vitale at Ravenna. Part of *m. Carium*'s appeal to Christians was the red color's symbolic association with the flesh of martyrs. Reuse in medieval and Renaissance architecture occurred in both the Western and Islamic worlds, where it was redeployed as columns and mihrabs in the Ummayad Mosque in Damascus, Dome of the Rock in Jerusalem, and St. Mark's Basilica in Venice. LEL

Note

1. Paul the Silentary, *Description of the Hagia Sophia*, lines 617–46, quoted in Mango 1972.

References

Monna and Pensabene 1977, 109–13; Mielsch 1985, 46, pl. 8, nos. 257–59; Gnoli 1988, 244–45; Borghini 1989, 289; Andreoli et al. 2002; Lazzarini 2009, 463

Cipollino verde

Cat. Nos. 185–203 (196 and 197 are *cipollino grigio*)

Also known as: *m. Carystium, m. Styrium, Karystia lithos*
Source: Karystos and Styra, Euboea, Greece

Folded green calcitic marble with strata rich in chlorite. The reference of the Italian artisanal name to an onion alludes to the layers, both in appearance and in tendency to separate from one another. Abundant along 60 km of mountains on the southeast coast of the island of Euboea with five chief sites of quarrying around Styra and the port of Karystos/Carystus, the greenish color shifts somewhat toward gray near the southern tip of the island. The gray variety, called by Italian artisans *cipollino grigio* or *bigio*, was mostly avoided by Roman quarrymen.

Although it quickly became one of the most widely used polychrome marbles in the Roman world, its entry to Rome was late. Pliny, quoting Nepos (*Natural History* 36.48), tells us that Caesar's *praefectus fabrum* Mamurra, who should be credited with opening the Luna quarries on Caesar's behalf (Fant 1988), was the first to have no columns in his whole house but monolithic (*solidas*) ones of Carystian and Lunense. The date is likely to be in the 40s BCE after the Gallic campaigns, and the absence of Carystian from the rich marble floor of the house under the Ludus Magnus, not earlier than 50 BCE, is consistent with this (Guidobaldi and Salvatori 1988, 174). Whether Mamurra was further involved in organizing the import of Carystian, as he must have been with Lunense, is a matter of conjecture. An elegy in the *corpus Tibullianum* (3.3.13–14) already uses Carystian columns as a token of luxury,

Cipollino verde (cont.)

and two generations later Strabo (*Geographica* 10.446) gave a description of the quarries in a matter-of-fact tone that suggests that the product was already well known.

Abundant epigraphical evidence for imperial officials in Euboea was first collected by Dubois (1908). Quarry blocks found at the imperial *statio marmorum* at Portus show that Carystian was inventoried here alongside marbles both under clear imperial control and of uncertain ownership (Baccini Leotardi 1979, 1989; Pensabene 1994; Pensabene and Bruno 1998b). A new and subtle examination by Russell (2013, 45–49) suggests that imperial involvement here and elsewhere was less direct and more open to both private enterprise and informal ventures by officials on the side than suspected. For Carystian marble, this tallies well with the sheer extent of the quarry fields: extensive and continuous control would be a daunting task. The diffusion of Carystian geographically is the widest of the "imperial" marbles (Sutherland 2013), and it continued into the Byzantine period.

From its earliest uses in Augustan architecture, Carystian marble was prominent alongside Lucullan, Numidian, and Phrygian. The pavements of the Basilica Aemilia, Forum of Augustus, and Temple of Concord are good examples. Carystian also lends itself well to the production of monolithic columns (those still in the quarries at Kylindri grace several book covers). The tonal range of the marble made it particularly at home in bath buildings, as Martial (*Epigrams* 9.75.7) observed. JCF

References

Corsi 1845, 97; Dubois 1908, 115–25; Papageorghakis 1964; Baccini Leotardi 1979; Lambraki 1980; Fant 1988; Guidobaldi and Salvatori 1988; Baccini Leotardi 1989; Pensabene 1994; Vanhove 1996; Borghini 1997, 202–03; Pensabene and Bruno 1998a; Lazzarini 2007; Price 2007, 174–75; Chidiroglou 2009; Russell 2013, 45–49; Sutherland 2013

Diaspro, jasper

Cat. Nos. 269–279

Also known as: *diaspro di Corsica, diaspro di Sicilia,* jasper
Sources: San Pietro di Rostino, Corsica; western Sicily

True jasper is flint formed of silica of chemical or biological origin, precipitated in the deep sea. It can form strata, veins, and nodules. Intensity of color and patterning are essentially what distinguishes flint from the more valuable jasper, but the colors depend on impurities and appear in a very wide range. Jaspers are commonly divided into "hard" and "soft," although chemically they are unrelated in origin. Most of the jaspers in the Kelsey collections are labeled soft jaspers from Sicily; they are dolomitic limestones and easily worked. JCF

References

Borghini 1989, 209; Evangelista and Lazzarini 1998; Price 2007, 246–47; Lazzarini 2019

Dokimeion white marble

Discussed under *Pavonazzetto*

Doliana

Cat. Nos. 491?, 492, 493, 494?, 495, 496?

Also known as: *marmo di Doliana, marmo di Tegea,* white marble
Source: Ano Doliana, Peloponnese, Greece

The white marbles from Ano Doliana, some 20 km south of Tripolis in the central Peloponnese of Greece, bear some visual resemblance to Pentelic marble but are often somewhat darker, less translucent, and free of the micaceous and chlorite veining characteristic of the latter. Its uses seem to have been largely local, which is not surprising given both the requisite overland transport and the stiff competition from more famous and logistically sited Athenian and island marbles. Its use in the temple of Athena Alea at Tegea, argued by recent excavators (Østby 2014), is logical, while claimed uses at Mantineia, Olympia, and Bassai await confirmation (Palagia 2002). The marble was not exported; it found its way into the Kelsey collections because Francis Kelsey's Greek trip of spring 1901 included Tegea and Sparta in its itinerary (see Pedley 2012, 67–74). JCF

References

Palagia 2002; Attanasio, Brilli, and Ogle 2006, 105ff.; Østby 2014a and b

Fior di pesco

Cat. Nos. 418?, 421–422, 423?, 424–429

Also known as: *fior di persico, m. Chalcidicum*
Source: Eretria, Euboea, Greece

Marmor Chalcidicum was the name given by the Romans to the quarry near Eretria south of ancient Chalkis, the main town on the island of Euboea in Greece. The quarries were neither the largest of antiquity nor was the stone diffused widely; nevertheless, its color set it apart and it should be considered one of the more important polychrome marbles of the Roman era. Red, brown, white, and violet colors predominate in varying ways. The red and white colors can diffuse into delicate shades of pink, or thin streaks of red and violet can criss-cross through veins of white marble. At other times it is a breccia with white clasts set in a cherry-colored matrix.

The stone was popular in the Baroque period in Rome, when it became known as *fior di pesco* ("peach blossom"), and most important churches have at least a single piece. While the monuments in Rome with original contexts were dismantled, elsewhere in Italy and the central Mediterranean it was preferred for columns and surface adornment. The *fior di pesco* columns in the Severan theater at Lepcis Magna and in the forum in Syracuse were unusual both for their quantity and use in public buildings and were likely the result of a specific commission. LEL

References

Corsi 1845, 100–01; Gnoli 1971, 157–59; Mielsch 1985, 57, pl. 16, nos. 548–60; Gnoli 1988, 184–86; Borghini 1989, 212; Antonelli et al. 2002; Lazzarini 2002a; Lazzarini et al. 2002; Lazzarini 2007, 205–09; Price 2007, 128; Schmid 2009

Gabbro eufotide

Cat. Nos. 4–7

Source: Wâdi Maghrabîya, Eastern Desert of Egypt

Mettagabbro, mottled white to light gray, dark green, medium- to coarse-grained.

The quarries were identified only recently (Harrell, Brown, and Lazzarini 1999), a dozen small sites in Wâdi Maghrabîya in the Eastern Desert of Egypt and a tributary, located north of Wâdi Hammâmât and roughly equidistant between the roads leading from the Nile to the quarrying areas of Mons Claudianus (ending at Safaga) and to Wâdi Hammâmât (ending at Quseir). This is also the source of *granito verde plasmato*, not represented in the Kelsey collections. Pottery found was not diagnostic for a period more specific than first and second centuries CE, but uses in Italy point back to an introduction possibly as early as the late Augustan period.

This stone was used primarily in *opus sectile* paving but also for small columns, moldings, and table supports. It appears in Campania in pavements at the House of Fabius Rufus at Pompeii and is a prominent element in triclinium 5 of the House of the Stags at Herculaneum, and a pavement from the Villa of the Papyri now in room CXL–CXLI of the Naples Archaeological Museum.

Price notes that the name *gabbro eufotide* is used in contemporary literature primarily to denote gabbro quarries in Liguria and Tuscany in Italy. JCF

References

Gnoli 1971, 132; Mielsch 1985, pl. 24, nos. 815–16; Gnoli 1988, 158–59; Borghini 1989, 210; Dolci and Nista 1992, 89; Brown and Harrell 1995; Pensabene and Bruno 1998b, 12, nos. 65–66; Harrell, Brown, and Lazzarini 1999; Lazzarini 2002a, 233; Price 2007, 196

Giallo antico

Cat. Nos. 118–160, 255?

Also known as: *m. Numidicum*, Numidian marble
Source: Chemtou (ancient Simitthus), Tunisia

A brecciated fine limestone from Chemtou, ancient Simitthus, in the Roman province of Africa, modern Tunisia. Pliny (*Natural History* 5.2) caustically remarked that Numidia was notable for nothing except Numidian marble and wild beasts, meaning lions, whose coats fall into the same tawny color range as the marble. A lion sculpted from Numidian in the Sala degli Animali of the Vatican Museums shows that Pliny was not alone in appreciating the visual pun. Modern quarrying on a modest scale east of the ancient site in the same ridge makes this a rare ancient marble commercially available today.[1]

The marble is inconsistent in color and patterning, ranging from a plain pale yellow to a darker yellow-orange ground with dense brecciation and purple infills to a yellow ground with larger calcite clasts resembling mortadella. A walk through the quarries reveals an even wider range of colors, down to a consistent dark red like liver, and vividly reminds us that Roman supervisors were well briefed as to which varieties to exploit and which to shun.

The name Numidian first appears in the first half of the second century BCE in a gloss on *pavimenta Poenica* which the lexicographer Festus attributes to the Elder Cato (*On the Meaning of Words*; ed. Lindsay 1913, 282). Pliny (*Natural History* 36.49), however, asserts that the first instance of imported Numidian marble he found was in the house of M. Aemilius Lepidus, colleague of Q. Lutatius Catulus and consul in 78 BCE (Pliny adds this to identify the Lepidus, not to date the event), who installed a threshold block of this marble. It is notable that Numidicum, while consistent with Roman preference for geographical designations in naming marbles, is unusual in that no variant forms or names are on record.

The marble had been imported to Rome for as long as a century before Numidia became a Roman province in 46 BCE. The quarries seem to have been a project of the Numidian kings and a symbol of their monarchy (Rakob 1993, 1994). Caesar annexed Numidia as Africa Nova after the battle of Thapsus, one of the victories celebrated in the triple triumph of 46 (Appian, *The Civil Wars* 2.101). This may explain why the site of Caesar's improvised funeral pyre and spontaneous cult was marked with a 20-foot monolithic column of Numidian marble (Fant 1993, 147).

Under Augustus, Numidian marble was given the prominent role in imperial projects which it enjoyed to the end of the western empire, for example, in the paving of the Basilica Aemilia and the porticos of the Forum of Augustus. In the Trajanic-Hadrianic Pantheon, Numidian was paired with Phrygian for the columns of the niches of the Pantheon as well as the pavement (Claridge 2010, 69, 178, 230). Simmithus became a Roman colony under Augustus, and there are indications of a Roman presence in the same period. Blocks in the quarries with inventory numbers and names of imperial freedmen and procurators are good evidence for intense imperial involvement, if not perhaps for outright imperial ownership. This picture is reinforced by the very wide diffusion of Numidian marble around the Mediterranean. Trade was largely limited to wealthy coastal cities, not just because of transport costs but also because its high initial cost left smaller cities out of the bidding.

A heavily built square structure near the quarries, first claimed as an *ergastulum* for *damnati ad metallum* (prisoners condemned to work in the quarries), has recently been more convincingly reinterpreted as a workshop (Mackensen 2005). This makes sense of the fact that the strong closures are mounted inside, not outside, the building and were thus intended to protect the valuable artifacts being produced, and the producers, rather than to confine prisoners (who in fact have proved very difficult to locate archaeologically at any of the major quarries).

Activity as attested by quarry inscriptions continued throughout the second century CE and possibly quite a bit later, as an inscription (if correctly restored) names an *officina* (workshop) founded by a man describing himself as *agens in rebus* (Dubois 1908, no. 113 = *CIL* VIII 14,600). Numidian is listed in Diocletian's Edict on Maximum Prices as the second most expensive (Giacchero 1974, 210–11). JCF

Note

1. StoneContact.com, http://www.stonecontact.com/quarries-79/jebel-rouge-chemtou-marble-quarry, accessed September 12, 2017.

References

Gnoli 1988, 166–68; Khanoussi 1991; Fant 1993; Rakob 1993; Rakob 1994; Mackensen 2005

Granito bianco e nero

Cat. Nos. 8–11

Also known as: *m. Tibereum*
Source: Wâdi Barud, Eastern Desert of Egypt

Quartz diorite with large black hornblende crystals in a pure white groundmass. Quarried at Wâdi Barud, south of Mons Claudianus in the Eastern Desert of Egypt. This exotic stone was imported to Rome for a brief period and is not seen outside Rome.

Pliny (*Natural History* 36.55) attests the name *m. Tibereum* and his description closely matches a rare granodiorite found in Rome with the artisanal name *granito bianco e nero*. Ostraca found at the fortified camp in Wâdi Barud call it *Tiberiane* (Peacock and Maxfield 1997, 275–79), and an unpublished ostracon names the quarry area *metallon Tiberianón* (Cuvigny 2021, 1:13, O.Claud. inv. 6179), supporting the identification, which has been widely accepted (Hirt 2010, 259; Cuvigny 2021, 1:18 fig. 2, 20; Pensabene 2013, 252).

There are two quarry sites, the one known for some time, now renamed Wâdi Barud South, has larger lower workings, which are more convenient to an already long slipway down to the wâdi floor; and a smaller upper workings (Gnoli 1988, 151). Proposed nomenclature is *granito bianco e nero di Santa Prassede* (from two Roman columns reused in the Chapel of San Zenone in the Church of Santa Prassede in Rome), for the more common stone from the lower workings, and *granito bianco e nero del Cairo* for the upper workings (for Roman slabs reused in medieval mosques in Cairo, see Harrell, Brown, and Lazzarini 2002). Newly discovered (Harrell and Sidebotham 2018) workings on the north side of the ridge which projects eastward into Wadi Barud, now to be known as Wâdi Barud North, comprise a modest establishment quarrying rock of slightly different minerology. Three of the artifacts in the Kelsey collections appear to be from the lower workings.

The fort is considered generally contemporary with the late first-century fort of Mons Claudianus, and a sondage of a rubbish heap (*sebakh*) produced both Antonine pottery and ostraca, but hints of earlier occupation were noted (Peacock and Maxfield 1997, 277). *Tibereum* is a striking compositional element in the Domus Tiberiana (Peacock and Maxfield 1997, 275), in the Palatine nymphaeum of Nero's Domus Transitoria, and at Hadrian's Villa (Borghini 1989, 217). The first of these, along with its name, would be evidence that the quarries were already exporting in the reign of Tiberius, provided that it was original use; the entire complex of imperial palaces on the Palatine was occupied and renovated into late Antiquity.

For all its visual appeal, *Tibereum* had the drawbacks that large objects like columns could not be produced from its quarries, which were difficult of access even in terms of the Eastern Desert. But, as Gnoli had correctly surmised (1988, 150), the site was not summarily abandoned in favor of Mons Claudianus; pottery and ostraca show activity there at least into the mid-second century CE (Peacock and Maxfield 1997, 277–78). JCF

References

Mielsch 1985, 69; Gnoli 1988, 140–41; Peacock and Maxfield 1997, 277–78; Harrell and Lazzarini 2002; Harrell and Sidebotham 2018

Granito della colonna

Cat. Nos. 12–13

Source: Wâdi Umm Shegilat, Eastern Desert of Egypt

Pegmatitic diorite with large (up to 6 cm) jagged black hornblende crystals.

This stone is called *granito della colonna* from the column which Cardinal Colonna obtained in the Holy Land, said to be the column to which Christ was bound and flagellated. This column was brought to Rome and dedicated in the church of Santa Prassede (Gnoli 1988, 151). It is very rare and found most often in medieval reuse. But it has been securely identified as part of the cargo of the Porto Novo ship, dated by a coin of

Granito della colonna (cont.)

27 CE to the Tiberian or Claudian era (H. Bernard et al. 1998; Mazeran 1998). Allowing for the time to begin production, one can add Shegilat to the group of quarries prospected and developed in the early–mid Julio-Claudian period (earlier workings there cannot be ruled out since later Roman exploitation may have destroyed all traces (Harrell 2002, 241). The *granito della colonna* was in the form of pre-cut veneer panels along with ones of Lucullan, Phrygian, Numidian, Carystian, and Lunense marbles; the principal cargo was some 130 tons of high-quality Lunense blocks and columns. An embossed scabbard and buckle suggest the presence of a legionary officer and raise the possibility that the consignment was not going to a wealthy client in Gaul or Spain but to an official. Shegilat should not be dismissed as a minor quarry simply because it was small. JCF

References

Gnoli 1988, 150–52; Klemm and Klemm 1993; Brown and Harrell 1995; H. Bernard et al. 1998; Mazeran 1998; Maxfield 2001; Harrell, Brown, and Lazzarini 2002; Lazzarini 2002a, 230; Fant 2022

Granito del foro

Cat. Nos. 14–18, 703–704

Also known as: *m. Claudianum*
Source: Gebel Fatireh, Wâdi Umm Huyut, Eastern Desert of Egypt

Tonalite gneiss from Wâdi Umm Huyut and 130 individual quarry areas between Wâdi Fatireh and Wâdi Umm Diqal, collectively known as Mons Claudianus.

Despite its name, no evidence from the reign of Claudius has been found at Mons Claudianus. The fort's earliest phase is Domitianic, but the Hydreuma has yielded an ostracon of the time of Nero, raising the possibility that it was the original nucleus of the main quarry establishment (Peacock and Maxfield 1997, 148; Maxfield 2001, 148, citing unpublished O. Claud. inv. 8828). But the presence of small columns in Nero's Domus Transitoria palace (Peacock et al. 1994, 215; Peacock and Maxfield 1997, 304), which burned in the great fire of 64 CE, almost certainly restores the origins of Mons Claudianus to the reign of Claudius when the years required for prospecting and infrastructure development are factored in. The first quarrying activity in the area seems to have taken place at Wâdi Umm Huyut, slightly south of Mons Claudianus, where surface pottery is datable to the beginning of the Roman period and a building, which may have been administrative, presides over a row of workers' huts.[1]

Quarrying operations and manpower resources at Mons Claudianus are the most fully known in the Roman empire. A French-led team (1987–1993) surveyed and selectively excavated the standing buildings and studied the 130 identifiable quarry areas, but also recorded some 9,000 documents on ostraca (pot sherds), the largest cache from any Roman site. As a result, we know that at a time of high demand from the Trajanic and Hadrianic building program in Rome, the site was staffed with roughly 400 skilled Egyptian quarry workers (*paganoi*) and about 400 members of the imperial workforce in logistics and support roles (*familiares*) (Cuvigny 2021, ch. 11, O. Claud. inv. 1538). Sixty soldiers commanded by a centurion, an architect, artisans, and guards bring the total to 917. A typical crew consisted of a dozen *sklerourgoi* (workers in hard stone) and *akisklarioi* (pickmen) with a blacksmith and bellows man; other specialists arrived in phase with the work including wedge-splitting experts and transport workers (Bülow-Jacobsen 2009). Food had to be imported from the Nile Valley (supplemented by fish from the Red Sea and some greens grown at at forts with wells); wine and olive oil were brought in from producers across the Mediterranean (Tomber 1996; Van der Veen 1997). The skilled workers, who came from Aswan and Alexandria, were well paid and their families housed at Kaine (Qena) on the Nile at the head of the overland road to the quarry areas of Mons Claudianus and Porphyrites (Cuvigny 2021, chs. 9 and 14). There were no convict laborers at Claudianus; the Roman military presence served to maintain security from the Bedouin (called in the documents "barbarians").

Mons Claudianus granite was chiefly prized as a source of large monolithic columns. The 40-foot columns of the Basilica in the Forum of Trajan are a notable eye-catching feature of the present-day ruins. The 50-footers for the Temple of the Deified Trajan were most likely diverted from the Pantheon under construction, causing last-minute design changes (Davies, Hemsoll, and Jones 1987). Sixty-foot columns have been found at the quarries but apparently were impossible to transport. The output of Mons Claudianus was almost exclusively restricted to the imperial capital. JCF

Note

1. "Umm Huyut," on the personal website of Hans Bernard, www.barnard.nl/desert/huyut.html, accessed October 5, 2016 (cf. Sidebotham 1996, 186–89). The pottery was read by R. Tomber.

References

Corsi 1845, 213–14; Gnoli 1988, 148–52; Davies, Hemsoll, and Jones 1987; Klein 1988; Maxfield 2001; Peacock and Maxfield 1997; Bülow-Jacobsen 2009; Cuvigny 2021

Granito rosso and *granito nero di Siene* — Cat. Nos. 19–25

Also known as: *l. Syenites* or *m. Syenites*, *l. pyrrhopoikilos*, *l. Aethiopicus*, *l. Thebaicus*
Source: Locations between Aswan (ancient Syene) and Shellal District, Nile River Valley, Egypt

Coarse to fine-grained granite with pink to red potassium felspar crystals up to 3 cm, gray quartz and black biotite. Pliny (*Natural History* 36.63) says that *Syenites* was "formerly" called *pyrrhopoikilos*, an ancient Greek neologism meaning "red-spotted stone" alluding to its dappled fiery colors.

The massive deposits from which it was quarried are in the form of a dyke which surfaces at Aswan (ancient Syene) and forms the First Cataract of the Nile. Across a quarry field extending some 20 km between Aswan and Shellal District, its color and chemistry vary greatly. While the pink/red feldspar crystals of the traditional *Syenites* are coarse, reaching 3 cm, in the southern part of the area is a black granodiorite called *granito nero di Siene* whose grain is usually medium but sometimes fine and not easily visible to the naked eye (Brown and Harrell 1998; see also Strabo, *Geographica* 17.808, 818). Proximity to the river was ideal for practical reasons.

One of the oldest decorative stones and one of the first to be used for its symbolic as well as utilitarian and aesthetic properties, its use in Lower Egypt also betokened control of the river, that is, of Egypt itself, from its southern boundary northward, and so afforded the early kings visual support for their claim to be sovereigns of Upper and Lower Egypt. Used for the casing blocks of Menkaura's pyramid and the enormous blocks of his valley temple, canonical for obelisks, widespread in sculpture, it was the premier stone of ancient Egypt. After the Battle of Actium (31 BCE), it arrived in Rome on the wave of Egyptomania.

When original Egyptian obelisks ran out, this was not a fatal difficulty since the quarries were still very much in business, and Domitian was able to equip the Iseum Campense with a new obelisk, quarried in Egypt but inscribed in Rome with hieroglyphics celebrating his accession (Iversen 1968, 80–81; Ciampini 2004, 157–67).

By the late first century, if not earlier, Romans were ready to use Aswan granite in contexts not intended to evoke Egypt directly. Nerva's Temple of Peace had 50-foot monoliths in its pronaos (Fogagnolo 2002, 137). Basins were in the general form of the Dionysiac *lenos* (grape-crushing vat) and monolithic columns, like those of the interior of the Pantheon porch, in canonical Greco-Roman proportions. Only Mons Claudianus rivalled Aswan for its ability to produce 40- and 50-foot columns (e.g., Pompey's Pillar at Alexandria), but its distribution pattern is the polar opposite of that of Aswan, which in fact was sent all across the Mediterranean and is the most extensive in breadth, quantity, and chronological range, dwarfing even that of its local look-alikes (Williams Thorpe 2008). Similarly, Roman exploitation of the Aswan quarries in turn dwarfed that of earlier Egyptian regimes.[1] In the early third century CE, Romans were still working at extending the range of workable sources at Aswan, as we learn from a dedication near Philae boasting that "new quarries have been found and many large pilasters and columns extracted" (*novae lapidicinae adinventae, tractae que sunt parastaticae et columnae grandes et multae, CIL* III 75). JCF

Note

1. Röder (1965, 551–52) ventured broad estimates for the volumes of stone quarried from all the quarrying areas at Aswan: for the 4th–6th Dynasties, which used granite for their pyramids, ca. 45,000 cu. m finished products implying

Granito rosso and granito nero di Siene (cont.)

100,000 cu. m of raw quarried stone; for the rest of the pre-Roman period, the same; for the Roman period, ca. 1.5–2 million cu. m or roughly ten times all earlier production. Although his detailed assignment of dates based on the form of wedge holes has not been generally accepted (Nylander 1968), his overall sense of the quarry fields seems sound.

References

Röder 1965; Iversen 1968, 80–81; Mielsch 1985, 67, pl. 22, nos. 749–55; Gnoli 1988, 145–47; Borghini 1989, 225–26; Brown and Harrell 1998; Fogagnolo 2002, 137; Price 2007, 220–21; Klemm and Klemm 2008, 233–67; Williams-Thorpe 2008

Granito verde fiorito di bigio

Cat. No. 69

Source: Wâdi Umm Balad, Eastern Desert of Egypt

Quartz diorite from Wâdi Umm Balad, near Gebel Dokhân (Mons Porphyrites) in the Eastern Desert of Egypt.

This small quarry is within the immediate orbit of Mons Porphyrites and was probably an administrative dependency. There is a scatter of Roman-period pottery but none are closely datable.[1] The Balad quarries are connected to a fortlet just off the main Porphyrites–Coptos road. Its architecture is of a piece with that of the Antonine forts at Porphyrites and Claudianus. From it a mile-long quarry road, wide enough for two very large wagons to pass in opposite directions, leads to the quarries. The quarry consists of just a few extraction areas, but its products, although scarce and limited to the city of Rome, were highly prized. Gnoli (1988, 139, 160) has seen it in the elaborate *opus sectile* paving of the Domus Flavia triclinium and elsewhere in the Palatine palace complex. JCF

Note

1. Balad: possibly first through fourth centuries; Towat: century unknown (Brown and Harrell 1995, 224).

References

Gnoli 1988, 159–60; Borghini 1989, 229; Bülow-Jacobsen 2008

Granito verde della sedia di San Lorenzo, Granito verde della sedia di San Pietro

Cat. Nos. 26–27

Also known as: *ophites, ofite*

Sources: Wâdi Semna, Wâdi Umm Wikala (quarries proper), Eastern Desert of Egypt

Metagabbro. The quarries, in the Eastern Desert of Egypt, are usually said to be in Wâdi Semna but are actually in Wâdi Umm Wikala, which branches off northward from Semna (Sidebotham et al. 2001).

The rock, of complex chemistry, presents a speckled appearance with black, white, and green crystals ranging from light to dark green. The name *ophites*, "snakelike," refers to the dappled light and dark green appearance. Two varieties, *granito verde della sedia di San Lorenzo* and *granito verde della sedia di San Pietro*, are distinguished in the Italian craft tradition; *San Pietro* can have patches of much larger crystals. The names come from the best-known uses, a large disk set into the backrest of the papal throne in the church of San Lorenzo fuori le Mura in Rome (well illustrated in Borghini 1989, 232–33), and the veneered pedestal of the bronze statue of St. Peter in the Basilica di San Pietro in Rome, respectively (Gnoli 1988, 154ff.).

The well-known "curator" inscription of 11 CE found in Wâdi Umm Wikala attests, in tortured syntax (Tregenza 1951, with notes by A.H.M. Jones, followed by Gnoli), a Roman tribune of the Legion III Cyrenaica as prefect of Berenice and in charge of all mines and quarries in Egypt. Semna, known for gold mining, was also apparently Mons Ophiates (source of *ophites*). The 2001 survey of Sidebotham and team found ceramic evidence for activity from the early first century CE, and the dedication of a Paneion shows that the site was already staffed and busy by the date of the curator inscription.

> Pliny, *Natural History* 36.55: *non omnia autem in lapicidinis gignuntur, sed multa et sub terra sparsa,*

pretiosissimi quidem generis, sicut Lacedaemonium viride cunctisque hilarius, sicut et Augusteum ac deinde Tibereum, in Aegypto Augusti ac Tiberii primum principatu reperta. differentia eorum est ab ophite, cum sit illud serpentium maculis simile, unde et nomen accepit, quod haec maculas diverso modo colligunt, Augusteum undatim crispum in vertices, Tibereum sparsa, non convoluta, canitie.

Not all [of the most famous stones] are born in quarries, however; many, even some of the most prized in fact, are also dispersed under the soil; examples are the green Lacedaemonian, more cheerful in appearance than all others, the Augustan and then the Tiberian, first discovered in Egypt in the reigns of those emperors. They differ from *ophites* since it resembles the spots of snakes (whence its name), while they gather their markings in a different manner: the Augustan curls into waves but the Tiberian has gray markings that are dispersed, not rolled up. (trans. J.C. Fant)

Ophites is often considered to have been Pliny's *m. Augusteum* since it is so early and other obvious candidates are lacking, but Pliny here seems to make a distinction; his description of *Tibereum*, now securely identified as *granito bianco e nero* from Wâdi Barud (q.v.), is remarkably accurate while Pliny's curling waves of *Augusteum* do not fit *ophites* at all. The Neronian poet Lucan (*On the Civil War* 9.713) mentions the stone by name, *Thebanus ophites*, suggesting that it had become well known by then. Similarly, a bar façade at Pompeii has a centerpiece of *ophites*, which had come from a building wrecked in the earthquake of 62 CE (Fant, Russell, and Barker 2013, 19). JCF

References

Tregenza 1951; Gnoli 1971, 130–31; Mielsch 1985, pl. 24, nos. 804–05, 807; Gnoli 1988, 155–57; Borghini 1989, 232–24; Brown and Harrell 1995; Sidebotham et al. 2001; Fant, Russell, and Barker 2013, 193; Fant forthcoming

Greco scritto — Cat. Nos. 516–526

Sources: Cap de Garde and Mt. Filfila, Algeria; Hasançavuşlar, western Turkey

Greco scritto is a medium- to coarse-grained white marble with gray-blackish veining and dark gray-to-bluish spots distributed over a white background. Italian masons, who encountered a huge variety of coarse-grained gray marbles, distinguished *marmo greco scritto* from the wider *marmo greco* kaleidoscope by its distinctive lines and dots; and upon seeing markings like scribblings on stone with black ink, they likened it to ancient Greek writing. In some cases, *greco scritto* has a similar appearance to *bardiglio* from Carrara, or the banded white and blue marble from Marmara Island (ancient Proconnesus) when sawn in a particular direction. Given that even today *scalpellini* throughout Rome and Venice associate the stone more with its appearance than origin, perhaps it is unsurprising modern scholarship has identified several sources throughout the Mediterranean.

The quarries near the ancient town of Hippo Regius (modern Annaba) and Cap de Garde, Algeria, have traditionally been regarded as the most important sources of this type of marble (Gnoli 1971, 225; Pensabene 1976; Borghini 1989; Pensabene 2002). With North Africa and Egypt as the Roman playground for polychrome stones, the acquisition of this predominantly white marble was the exception. Its location on a coastal headland facilitated its shipment around the Mediterranean.

Analytical research on archaeological samples of *greco scritto* from local and regional North African towns, where the marble was used extensively in public buildings, indicates that the marble derived from places other than Cap de Garde (Antonelli et al. 2009, 2010). Recent fieldwork has confirmed this through the discovery of two new sources. Not far to the west of Cap de Garde, the slopes of Mt. Filfila produced a variety of *greco scritto* (Attanasio, Herrmann, et al. 2012). The streaky blue and white marble was also quarried in significant volume 22 km northeast of Ephesus, in several quarries around the village of Hasançavuşlar, Turkey (Yavuz, Bruno, and Attanasio 2011; Attanasio, Yavuz, et al. 2012). Similar to the distribution patterns of *greco scritto* in North Africa, the Ephesian source was also used locally for decoration. Although recent research may indicate use in civic buildings, those found in Rome and Italy usually appear in private contexts, such as in small thermae or domestic architecture, in the form of slabs and furniture. Chemical analyses have shown that discrimination between the Ephesian and Algerian sources is possible, and future research should bring

Greco scritto (cont.)

greater clarity on which source was the main exporter around the Mediterranean. LEL

References

Antonelli et al. 2009; Antonelli et al. 2010; Yavuz, Bruno, and Attanasio 2011; Attanasio, Herrmann, et al. 2012; Attanasio, Yavuz, et al. 2012

Lapis lazuli

Cat. No. 74

Source: Sar-i-Sang, Badakhshan, Afghanistan and Tajikistan

Metamorphosed blue lazurite and other feldspathoid minerals with specks of yellow pyrite and white calcite. Quarried in the Badakhshan region of modern Afghanistan and Tajikistan, remote today but on a trade route of the ancient world; chemically indistinguishable sources at Lake Baikal, Russia, are verified but unlikely to have been exported. Widely traded, it was prized by the Harrapan culture of the Indus Valley, China, and Egypt (e.g., the funeral mask of Tutankhamun), with which it was most closely associated in the ancient Mediterranean. JCF

References

Price 2007, 262–63; Zöldföldi and Kasztovszky 2009; Reeves 2015

Lumachella

Cat. Nos. 430–433, 434?, 435?, 436–442, 644, 660–661

Sources: Henchir el Kasbat (ancient Thuburbo Maius), Tunisia; Val Pantena near Verona, Italy; quarries across the Mediterranean

A wide variety of fossiliferous limestones go under the name of *lumachella*, not just those incorporating the namesake snail shells, and their sources are similarly common from Spain to Asia Minor. Particularly large shells were given the name *lumachellone* but are similar in other ways. Varieties of *lumachella* have attracted elaborate and often fanciful names; we have been conservative, noting only color, apart from well-established types like *lumachella orientale* or *d'Egitto* (cat. no. 440), (*c*)*astracane* (cat. nos. 660–661), both from Roman Africa, and the distinctive *lumachella di San Vitale* (cat. nos. 442, 644) from the Val Pantena area north of Verona. Some geological subtypes like *occhio di pavone* (q.v.), incorporating rudist bivalves, have their own artisanal identity. Gnoli (1988, 201–03) noted that the Romans generally favored gray and black *lumachelle*. JCF

References

Gnoli 1988, 201–03; Borghini 1989, 257–62; Price 2007, 151–57, 162–63; Lazzarini and Mariottini 2012; Lazzarini 2014

m. Lunense

Cat. Nos. 530–534, 536–541, 545–549, 553?, 557–562, 564?–565?

Also known as (white varieties): *Luna statuario, Luna ordinario, marmo di Carrara,* Carrara marble, Luna marble

Also known as (gray variety): *Luna bardiglio, bardiglio di Carrara*

Source: Carrara (ancient Luna), Italy

Fine-grained white and gray marble in massive deposits in the Apuan Alps with the quarries stretching south from Carrara (ancient Luna) along the three valleys of Torano, Miseglia, and Colonnata. The Romans seem to have referred to several varieties by the common geographical designation *marmor Lunense*. The gray variety, distinctive for its large and irregular masses of color, is called *bardiglio* in the traditional artisanal terminology (because grays nonetheless are hard to distinguish, we have been conservative about identifying artifacts as *bardiglio*). The gray coloration is caused by graphite originating in carbonaceous organic matter. Most white Luna marble has graphite streaks or blobs; the modern term is *ordinario*. The purest white, which Pliny (*Natural History* 36.14) said rivalled Parian marble for sculpture, is termed *statuario* and it is today very difficult to find on the market because it has been essentially exhausted since the Renaissance. There is also a black marble deposit in the valley of Colonnata.

The Romans began using marble imported from Greece in the Republican period, apparently unaware that Luna sheltered one of the largest concentrations of marble in the world. The irony in this is less than appears: the Etruscans used Luna marble, albeit in small quantities (Banti 1931; Mannoni 1987), and the Roman colony at Luna (founded 177 BCE) could not have been unaware of what the mountains behind it were made of. The obstacles presented by the vertical landscape were not overcome until Caesar sponsored the first efforts and then the huge demands of Augustan building campaigns justified the investment (Fant 1988).

The earliest exploitation seems to have begun in the area of Calagio and Gioia and expanded from there to sixteen quarry fields along the principal valleys of Torano, Miseglia, and Colonnata (Dolci 1988, 79). The first use of Luna marble known at Rome is in the renovation of the Regia in the Roman Forum in 36 BCE (Claridge 2010, 110), and imperial agents are attested in the temple of Apollo in Circo (Sosianus) in the 20s BCE (Pensabene 1998, 342) and the Forum of Augustus, dedicated 2 CE (Granino Cecere and Morizio 2007). A freedman of Augustus who made a dedication to Silvanus at Luna (*CIL* IX 6948), was probably expediting supplies for his master's building campaign. Yet it has also become very clear that most of the quarrying areas remained under the control of the colony and were exploited by prominent families of Luna, most conspicuously the Baebii, whose activities spread Luna marble across Gaul, Spain, and North Africa (Menella 1990; Pensabene 2013, 426–33).

Use of Luna marble began to fall off in the later first century and then markedly in the second century CE for reasons that are not fully understood. One factor may have been the progressive silting up of the Luna harbor, which seems to have opened the way for competition from the eastern quarries, particularly those of Proconnesus, which rivalled Luna for size and had the advantage of being in gentle terrain and much closer to a good harbor. That blocks of Proconnesian were being used to rebuild the temple of Venus at Pompeii before 79 CE, and Pentelic and Dokimeion marble also grow in use at Rome (Bruno, Conti, et al. 2002; Bruno, Cancelliere, et al. 2002), are another indication of Luna marble falling out of favor. JCF

References

Banti 1931; Mannoni 1987; Menella 1990; Dolci 1998; Pensabene 1998, 64–65; Bruno, Cancelliere, et al. 2002; Bruno, Conti, et al. 2002; Attanasio 2003, 165–70; Attanasio, Brilli, and Ogle 2006, 69–85; Granino Cecere and Morizio 2007; Price 2007; Pensabene 2013; Paribeni and Segenni 2015

Nero antico — Cat. Nos. 527–529

Sources: Ain el Ksir and Gebel Aziza/Gebel Oust, Tunisia; Mani Peninsula and island of Chios, Greece; Göktepe, near Aphrodisias, western Turkey

Nero antico is the name Italian marble workers give to a fine-grained limestone that is strongly colored black with little to no variation (although calcite veins and fossils are fairly common). The dividing line between *nero antico* and *bigio morato* (dark gray limestone, not represented in the Kelsey collections) is somewhat subjective, while *bigio antico* (q.v.) is a true marble with medium to coarse crystals in varying gray hues (Corsi 1845, 94, 106–07; Price 2007, 76). The source has traditionally been thought to be the quarries of Cape Taenarum in the Mani Peninsula of Greece, but the quarries of the black limestone there are very small and unlikely to have yielded significant exports. Hence Pliny (*Natural History* 36.135 and 158) and others seem to have been speaking instead of the *bigio antico* deposits. Nomenclature consistency in the literature, however, remains to be achieved, just as the rock in individual sources may oscillate between fine limestone and true crystallized marble.

The literary evidence has highlighted Cape Taenarum, but the stone is not a strong match and there is no sign of large-scale workings. There are also black limestones and marbles near Chemtou and at Gebel Aziza in Tunisia (Lazzarini, Agus, and Cara 2006), Lesbos, Chios (Lazzarini 2007, 151ff.), and alongside the *africano* quarries of Teos (modern Sığacık). A new research effort has identified significant sources at Göktepe near Aphrodisias in south-

Nero antico (cont.)

western Turkey and Belevi near Ephesus on the Aegean coast (Attanasio, Bruno, Prochaska, and Yavuz 2017).

Nero antico could be polished to a mirror finish resembling patinated bronze, making it valuable for sculpture. Well-known statues at Rome illustrate the complexities of provenance assignment. A trio of female herms from the House of Augustus now in the Antiquarium del Palatino (inv. nos. 1048, 1053, 1056) are consistent petrographically with *nero antico* from the Mani Peninsula (Bruno and Pallante 2002). But two centaurs (only inv. no. 658 was analyzed) in the Musei Capitolini bear the signatures of two sculptors from Aphrodisias, and a young centaur (inv. no. 656) has been assigned to the recently discovered Göktepe quarries near Aphrodisias (Attanasio, Bruno, and Yavuz 2009) and may emerge as an alternative source. However, black marbles may be more common than are currently reported. The discovery of black marble in the upper zones of the Belevi quarry at Ephesus, more famous for its likely use in the nearby construction of the Temple of Artemis, is a case in point. More work is needed to clarify the entire picture, and without a database with the full battery of tests performed on different sources, it is difficult to assign origins of *nero antico* and/or *bigio antico* with certainty. JCF & LEL

References

Corsi 1845, 94, 106–07; Pensabene and Lazzarini 1998; Bruno and Pallante 2002; Lazzarini, Agus, and Cara 2006; Price 2007, 76; Lazzarini 2007; Attanasio, Bruno, and Yavuz 2009; Brilli et al. 2010; Bruno et al. 2012; Attanasio, Bruno, Prochaska, and Yavuz 2017; Brilli, Giustini, and Kadıoğlu 2019

Occhio di pavone rosso and *pavonazzo* — Cat. Nos. 443–446

Also known as: *m. Triponticum*
Source: Kutluca and the Sakarya (ancient Sangarius) Valley, north central Turkey

"Eye of the peacock," a subtype of *lumachella* with rudist bivalve fossils often largely intact, alludes not to the peacock's actual eye but to the "eyes" in the male's tail feathers. Colors are variable, but the prized red-purple hues have been traced to valley of the Sakarya (ancient Sangarius) River and connected with the Roman *m. Triponticum*, mentioned in Diocletian's Price Edict but previously unidentified. JCF

References

Corsi 1828, 119; Gnoli 1971, 177–80; Giacchero 1974; Mielsch 1985, 161–64, nos. 149–58; Gnoli 1988, 206–10, 214; Borghini 1989, 257–61, nos. 103–07; Lazzarini 2002b

Palombino — Cat. Nos. 573?–580?, 583?, 584, 589?, 592?, 598, 600?

Also known as: *palombino bianco*
Source: Tolfa, Lazio, Italy

A compact, fine-grained calcitic or dolomitic limestone, buff or ivory in color, compared to the feathers of the ring-necked dove. This is not a rare lithotype. Corsi (1825, 24) hypothesized a source in Turkey following Pliny's description of *corallitícum,* a stone from Asia which he described as ivory-colored and not yielding pieces larger than two cubits (*Natural History* 36.62). Gnoli noted a similar stone in the Eastern Desert of Egypt (1988, 260). Given that *palombino* was heavily used in Italy (in pavements and particularly in small intricate shapes for intarsia, of which the Kelsey pieces are an excellent example), it would be unlikely that artisans would depend on imports from so far afield when good deposits were nearby at Tolfa in northern Lazio (MUSNAF, s.v. "Palombino"; accessed March 28, 2019). At Tolfa there is also a black *palombino*; it was used for the famous Lapis Niger, the so-called Tomb of Romulus, near the Comitium in the Roman Forum (Fornaseri et al. 1993). JCF

References

Corsi 1828, 89; Mielsch 1985, 52–53, nos. 398–401; Gnoli 1988; Bruno 2002c, 289b; Price 2007, 98

m. Parium

None in Collection[1]

Also known as: Parian marble
Source: Paros, Cyclades, Greece

Of the several varieties of marble quarried at several sites on the Cycladic island of Paros, the one called *lychnites*, quarried (at Marathi) in subterranean galleries by lamp light (Pliny, *Natural History* 36.14, quoting Varro; Korres 1995 and 2000), was the most sought after sculptural marble in antiquity: famous Greek artists preferred it, and sometimes would work exclusively in Parian marble. (*Lychnites* is now classified geologically as Paros-1.) The Augustus of Prima Porta was carved in *lychnites* (Pollini et al. 1998). Medium- to fine-grained, highly pure and deeply translucent, its consistency also allowed it to be worked in detail. Of the hundreds of extant marble blocks stockpiled at the marble stations at Portus and the Marmorata in Rome, Parian is the only one that arrived in shapeless lumps, suggesting that even the smallest and most irregular piece would be put to use (Baccini Leotardi 1989, 59–78).

In the Chorodaki Valley 1 km west of Marathi (today used as Paros town's garbage dump), are quarries of a coarser marble ranging in color from pure white to blue-gray, sometimes banded, now called Paros-2 (Germann et al. 1988). Paros-3, quarried at Agios Minas only a few hundred meters uphill from Marathi, is similar enough to Paros-2 that it may end up being assimilated to it as Paros-2(3). Paros-4, quarried in modest quantities at Karavos, several kilometers southeast of Marathi, is similar to *lychnites* but slightly coarser and less translucent. Formless blocks of Paros-2 were interspersed with ones of *lychnites* at the Portus *statio marmorum* and signed by the same quarry operators (Fant 2001, 182–84, table 6). The garden of the House of the Vettii at Pompeii contains some two dozen basins and fountains, of which seven are in Paros-2 and enjoy the most prominent positions. There are none in *lychnites*, suggesting that the more abundant Chorodaki marble could be deployed for less prestigious uses than sculpture but still probably benefit (commercially) from the Parian name. JCF

Note

1. Although there are no verified pieces in any of the Parian varieties in the Kelsey Museum collections, we include this Profile because Parian was so prestigious that it frequently comes up in identifications suggested to F. W. Kelsey by his informants.

References

Germann et al. 1988; Baccini Leotardi 1989, 59–78; Korres 1995; Korres 2000; Fant 2001; Fant et al. 2002; Attanasio, Brilli, and Ogle 2006, 121–35; Pensabene 2013, 276–81

Pavonazzetto

Cat. Nos. 161–183, 352?, 366?, 376?

Also known as: *m. Phrygium, m. Synnadicum, l. Dokimenos*, Phrygian marble, Dokimeion white marble
Sources: village of İscehisar (ancient city of Dokimeion or Docimium) and Upper Tembris Valley, Turkey

A fine-grained quite translucent white marble produced in part by contact metamorphism, quarried several kilometers east of the village of İscehisar, Turkey (ancient Dokimeion or Docimium). A small part of the quarry field contains marble with coloration in veins or breccia ranging from dark purple to a near gray, commonly referred to as "pavonazzetto" (Röder 1971), found also in a number of small quarries in the Upper Tembris Valley to the northwest (around the towns of Çakırsaz, Altıntaş, Obruk Tepe, and others).

The profusion of nomenclature is not as random as initially appears. The earliest mentions in Roman literature (second half of the first century BCE: Tibullus 3.3, 13, *quidve domus prodest Phrygiis innixa columnis*; Horace, *Carmina* 3.1.41, *Phrygius lapis*) call it simply Phrygian, which connects it with the area already well known to the Romans as the home of the Magna Mater. This remains the most frequent name in Roman sources. Strabo (12.8.14), writing in Greek in the late Augustan or early Tiberian period, is the earliest source to mention the quarries specifically. He passes on the name "*synnadikos lithos*, as the Romans call it; the locals call it *dokimites* and *dokimaion*." The name *Synnadicum* alludes to the administrative center of a network of imperial estates in central Phrygia at Synnada (Dalla Rosa 2016), while Dokimeion, a veterans' settlement of a general of Antigonas, provided the usual identifying adjective in Greek (or

Pavonazzetto **(cont.)**

Docimium in Latin). Strabo goes on to say, "at first the quarries produced only small pieces ... but now on account of the Romans' wealth large monolithic columns are taken out" (trans. J.C. Fant). It is clear that all these terms, as used by Roman writers, described the purple-veined form of the marble while *lithos Dokimenos*, used by Asians in Greek, included or even denoted predominantly the white marble. Thus Strabo's remarks provide a *terminus ante quem* for Roman administration of (at least a part of) the quarries (Waelkens 1982, 36 n. 23; Fant 1989, 9). For the sake of clarity, we use *pavonazzetto*, or Phrygian when sources use that, to designate the purple-veined variety exported to Rome.

It is not clear when or how Dokimeion came into the imperial orbit. Broughton (1959, 567) tentatively suggested that the quarries had been in the royal domains of the Attalids of Pergamum once Phrygia was annexed (late third century BCE). Although an attractive theory, it seems ruled out by the fact that Phrygian marble (white or veined) was not used in Attalid architecture, for example, the Great Altar and the Upper Agora Temple, where Proconessian was preferred (Cramer 2004, 213–35). Pensabene (1998b, 337) considers that they had become the property of Pompey during his campaigns in Asia Minor, having previously been owned by Lucullus, as had the *africano* marble at Teos named for him *m. Luculleum*. Again, while attractive, this is purely conjectural. The proposal continues with the suggestion that the Phrygian quarries later passed into the ownership of Agrippa (Bruzza 1870, no. 255; cf. Pensabene 1994, 281; 1998, 141ff.) and thence of Augustus as a bequest. The involvement of Agrippa, however, is probably a canard as it rests entirely on an imperfect reading of a monogram cut into a column at Rome, reported by the antiquarian Francesco Ficoroni and amended by Bruzza without seeing the artifact (Bruzza 1870, no. 255). This was then accepted by Hirschfeld (1905, 147 n. 1) and so entered the literature. But the monogram, even if correctly interpreted, should be understood only as marking ownership of that piece, not the quarries as a whole (Fant 1989, 8).

The chronology suggested in the nomenclature is supported by the record of imports to Rome. Strabo's "small pieces" could cover a trickle of objects like mosaic inserts (Guidobaldi and Salvatori 1988, 174) and table supports, which may have reached Rome in the Republican period although none are datable with certainty (Moss 1988). The white Phrygian marble was reaching the west in the middle of the first century BCE, as a portrait of Caesar carved in his lifetime shows (Blanc, Bromblet, and Leroux 2009). That this was only a trickle is further supported by the silence of Pliny the Elder, who vociferously decried scandalous displays of luxury in the late Republic involving Carystian, Lunense, Numidian, Lucullan, and Chian marbles (*Natural History* 36.46, 48, 49).

The situation changed suddenly and unmistakably in the early Augustan period, when Phrygian marble became a leading feature of major imperial projects. In the final phase of the temple of Apollo in Circo (Sosianus) in the early 20s BCE, *pavonazzetto* was one of the colored marbles used in the pavement of the cella (Viscogliosi 1996; De Nuccio 2002). Then it appeared in columns and statues in a victory monument in the Basilica Aemilia in its Augustan renovation initiated in 13 BCE, the temple of Concord as rebuilt by Tiberius in his consulship of 7 BCE, in the pavement and some interior columns of the temple of Mars Ultor (2 CE), and the paving of the Basilica Julia (Fant 1999). Unmistakable evidence for Roman emperors exploiting *pavonazzetto* at their disposal comes from an over life-size statue of a Dacian prisoner abandoned in the quarries; it had been bound for Trajan's Forum in Rome as part of the ensemble of Dacian prisoners (Waelkens 1985). The most striking example, but also thoroughly typical, is the pairing of Phrygian with Numidian to provide the columns of the niches in the Pantheon. Detailed quantitative studies of both luxury dwellings and massive but low-status reuse before the eruption of Vesuvius show that Phrygian was the scarcest of the famous "imperial" polychrome marbles (Barker and Fant 2018; Barker and Fant 2019a; Fant, Russell, and Barker 2013). This was not purely due to the place given the marble in imperial architecture but also to the constricted and costly supply caused by torturous transport from inland Phrygia, involving overland haulage (sometimes by requisitioned wagons; Mitchell 1976) to the marginally navigable headwaters of both the Sakarya (ancient Sangarius) and the Maeander rivers (Christol and Drew-Bear 2005, 199 n. 35).

The *pavonazzetto* quarries were the focus of intense imperial interest as shown by the large and growing harvest of administrative inscriptions in Latin. The precise systems of control and exploitation remain subject to discussion (Fant 1989; Christol and

Drew-Bear 1991 and 2005; Pensabene 2010 and 2013, 197–205; Russell 2013, 42–51; Fant 2016). Although the quarry inscriptions cease abruptly with the crisis of the third century (230s CE), continuing production on some scale is shown by archaeological evidence (Pensabene 2010; 2013, 197–205), corroborated by the testimony of Diocletian's Price Edict, where Phrygian, which we are confident means *pavonazzetto*, is given the second-highest permissible price at 200 denarii (Erim and Reynolds 1970).

In spite of the near obliteration of the ancient quarries by modern activity, fresh material continues to resurface. Newly identified ancient quarries have been found to the east of the main site at Bacakale, along with extensive areas of exploitation at Altıntaş, previously undocumented (Wielgosz-Rondolino et al. 2015). Another recent survey uncovered red painted inscriptions on several of the huge quarry faces at Bacakale that have been interpreted as extraction sequences (Bruno 2017). The largest record specific dates from top to bottom (February 13, March 4, April 29, July, 14 and 16, and August 12) and cover a 30-meter vertical span. A rare one-off among Roman quarries, the dates and depths provide exact measurements for how much marble was removed from a single location during a specified time period, and corroborate evidence elsewhere that volumes of marble extracted for export were strictly controlled.

It is noteworthy that the sources of white marble, far more abundant than the colored, were never under imperial control. They were widely used for funerary stelae and sarcophagi as well as other sculpture (Fant 1985; Waelkens 1982). The white marble was also preferred for architectural and sculptural work within Asia where the impetus was local, not imperial (Burrell 2012). The Dokimeion workshops must have been among the most active in Asia Minor. Artisans, identifying themselves with the ethnic Dokimeion, have been found on sculpture, funerary stelae, and sarcophagi throughout Asia Minor (a partial listing is most readily available in Hall and Waelkens 1982; Waelkens et al. 2002). A remarkable series of small sculptures of Jonah and the Good Shepherd and three pairs of statuettes of aristocratic couples, all from Asia Minor and in white Phrygian marble, show that local carving traditions were still flourishing in the third century CE (Cleveland Museum of Art, inv. 1965.237–241, 1965.242–247). Similarly, a honey-colored yellow and white breccia, commonly known in Turkey as *Afyon bal* ("Afyon honey"), and a blue and white breccia, called *kaplan postu* ("tigerskin"), located in geologic units below their more famous counterparts, seem to have achieved only a regional distribution (Long 2012). The use of these colored marbles as monolithic columns and revetment in the nearby cities of Sagalassos, Sardis, and Ephesus is not surprising given that these cities were located along trade routes already employed for the transport of its signature stone. JCF & LEL

References

Hirschfeld 1905; Broughton 1959; Erim and Reynolds 1970; Röder 1971; Mitchell 1976; Hall and Waelkens 1982; Waelkens 1982; Fant 1985; Waelkens 1985; Christol and Drew-Bear 1986; Waelkens 1986; Christol and Drew-Bear 1988; Moss 1988; Christol and Drew-Bear 1991; Viscogliosi 1993, 51–53; Pensabene 1998; De Nuccio 2002, 152; Waelkens et al. 2002; Cramer 2004; Christol and Drew-Bear 2005; Blanc, Bromblet, and Leroux 2012; Burrell 2012; Long 2012; Russell 2013; Barker, Fant, and Ward 2015; Dalla Rosa 2016; Fant 2016; Barker and Fant 2018; Barker and Fant 2019a

Pentelic

Cat. Nos. 494?, 498–504, 505?, 506?

Also known as: *Pentelico*, *m. Pentelicum*, Pentelic white marble
Source: Mt. Pentelikon, Attica, Greece

White, more rarely gray, fine-grained foliated calcitic marble, weathered iron minerals often gives it a slight yellowish tint, strata of schist sometimes create a "gorgonzola" look.

The marble par excellence of Athenian architecture and sculpture like the fifth-century Parthenon (22,000 metric tons), Pentelic was the material for Rome's first temples in a purely Greek style, erected in the second half of the second century BCE (Gros 1973, 1976; S.G. Bernard 2010). From the late Republic through the first century CE, Athenian export workshops using Pentelic supplied Rome and other Mediterranean coastal cities with volumes of "neo-Attic" sculpture and other objects like *oscilla* (Pensabene 2013, 269–78). It continued to be used for monumental architecture in Rome, such as the Arch

Pentelic (cont.)

of Titus and the Pantheon, as well as the Stadium of Herodes Atticus at Athens and his stadium at Delphi (Pausanias 1.19.6, 10.22.1; Romano, Pike, and Gazda 2012). According to Korres (1995), this massive exploitation exhausted the better deposits, although detailed provenance work on the mountain has not provided corroboration (this project has opened up the possibility of associating particular monuments and artifacts with specific quarrying areas on the mountain; Pike 1999; S.G. Bernard and Pike 2015). Quarrying and exports certainly did continue, as shown by the assemblage of columns in the "temple" of the *fabri navales* at Ostria, among others (Pensabene et al. 1999, 147–56). JCF

References

Gros 1973; Gros 1976; Korres 1995; Pensabene et al. 1999; Pike 1999; Attanasio, Brilli, and Ogle 2006, 91–103; S.G. Bernard 2010; Romano, Pike, and Gazda 2012; Pensabene 2013, 263–78; S.G. Bernard and Pike 2015

Porfido bigio

Cat. No. 70

Also known as: *granito a morviglione, granito bigio morviglione*, Estérelite; subvariety: *granito pediculare or pidocchioso*
Source: Dramont, Estérel, southern France

A gray-green porphyritic dacite from quarries of Boulouris, between Dramont and Fréjus in the Estérel massif, in southern France. The description *a morviglione* comes from Roman artisans of the nineteenth century who saw in the phenocrysts a resemblance to crawling insects, so named in Roman dialect (*pediculare/pidocchioso* is a subvariety, formerly thought to be from Egypt, with smaller phenocrysts suggesting lice). Its use was mostly local (notably in the amphitheater of Fréjus) apart from a few exports in the Severan period. A few columns have been identified at Ostia; Kelsey's column fragment, acquired at Ostia, may well have come from this source. JCF

References

Gnoli 1988, 140–41; Borghini 1989, 219; Antonelli 2002, 275–76; Price 2007, 210–11

Porfido rosso

Cat. Nos. 28–55

Also known as: *porfido rosso antico, m. porphyriticum, l. porphyrites, leptopsephos*, purple porphyry
Source: Gebel Dokhân (ancient Mons Porphyrites), Eastern Desert of Egypt

Andesite-dacite porphyry quarried at some forty sites near Gebel Dokhân (ancient Mons Porphyrites) in Egypt's Eastern Desert. One of the early foci of Roman activity in the Eastern Desert, the quarries of purple porphyry are extensive. While porphyritic rock is common, the purple tone of this one is apparently unique worldwide (Williams-Thorpe et al. 2001, 306), and it was unique too in that it had not been used, other than sporadically, before the Romans. The stone was prized because it represented in lithic form the colors of cloth dyed with Tyrian purple, well known since the Bronze Age. The purple color was associated with royalty, especially in Late Antiquity when Constantine's female family members chose to be buried in elaborate porphyry sarcophagi, but Nero, who restricted the use of porphyry to the imperial household because of its regal connotations, was the first emperor to be buried in a porphyry sarcophagus. It remained an important symbol of imperialism from the Byzantine emperors to Charlemagne—who is said to have been crowned king on a porphyry disk originally from St. Peter's Basilica—and even into the Middle Ages, when its reddish color became a symbol of the blood of Christ.

Black porphyry too has been found at a number of quarries in the Mons Porphyrites quarrying area. Its discovery at the Bradford quarry, where in 18 CE a Roman prospector named it specifically (Van Rengen 1995), puts it on an equal footing with purple porphyry and locates it at the beginning of the development of the area. In addition to the Bradford quarry, which was not worked long, black porphyry is known from Lepsius, the Foot Quarry, Lycabettus, and the Northwest quarries; furthermore, two columns in the Vatican Museums do not match any of these known sources (Williams-Thorpe et al. 2001). Another dark porphyry, not represented in our collections, is the Roman *lapis hieracitis* and artisans' *porfido verde egiziano* (Borghini 1989, 278; Harrell 2017, no. 12 a).

The prospecting that found purple and black porphyry was part of an extensive Roman effort to develop the lithic resources of Egypt's Eastern Desert. It seems to have grown out of ongoing quarrying of the red granite of Aswan (*granito rosso*), extended to reopening and expanding the quarries of *basanite* (*bekhen*-stone, *basalto verde* [*del antico*]) in Wâdi Hammâmât and thence to finding new lithotypes including *ophites*, *granito della sedia di San Lorenzo*, *granito della colonna*, *serpentina*, *granito bianco e nero* (*m. Tibereum*), and *granito del foro* (*m. Claudianum*) (Maxfield 2001). Of these, porphyry was the most successful and enduring, with quarrying extending into the Late Antique period (there were sporadic small-scale projects in the twentieth century).

Archaeological investigations conducted at Mons Porphyrites revealed one of the most remarkable industrial landscapes in the Roman world (Maxfield and Peacock 2001). Not only did the archaeologists find quarries untouched since antiquity, but extremely well-preserved workers' villages, complete with temples, baths and water systems, and stables for animals. Porphyry's first appearance in Rome was poorly received: *statuas ex eo Claudio Caesari procurator eius in urbem ex Aegypto advexit Vitrasius Pollio, non admodum probata novitate; nemo certe postea imitatus est*, "the procurator Vitrasius Pollio had statues from porphyry brought to the City [ca. 41 CE] for Claudius Caesar, a development which met with distinct disapproval; certainly no one imitated it later" (Pliny, *Natural History* 36.57; trans. J.C. Fant). Since the statues were presumably of Claudius himself, the grandiosity of the gesture would provoke a negative reaction on several levels. No such scruples prevented Nero from surrounding himself with porphyry and purple (Suetonius, *Nero* 25, 32, and 50; Cassius Dio 62.6.2; porphyry and the complementary colors of *rosso antico* and *pavonazzetto* were used in the fountain court of the Domus Transitoria), and its success was assured thenceforth (Delbrueck 1932; Lucci 1964)—but no statues are known from this period. In Diocletian's Edict on Maximum Prices, purple porphyry shares the highest price of 250 denarii with the other famous green porphyry from Lacedaemonia (Giacchero 1974, 210–11). JCF

References

Delbrueck 1932; Lucci 1964; Gnoli 1988, 122–33; Van Rengen 1995; Maxfield and Peacock 2001; Williams-Thorpe et al. 2001; Maxfield 2001; Maxfield and Peacock 2007

Porfido verde **Cat. Nos. 56–68, 702**

Also known as: *serpentino, porfido verde antico, porfido verde di Grecia, l. Krokeatis, l. Lacedaemonius, m. Lacedaemonium*, green porphyry
Source: Krokees (ancient Krokeai), Peloponnese, Greece

Andesite-diabase, the groundmass is colored green by epidote and chlorite while the phenocrysts are plagioclase feldspar and black pyroxenes. The general coloration is light green on dark green, but there can be considerable variation (Lazzarini 2007, 45). The source is thin outcrops near ancient Krokeai (now Krokees) in Laconia, Greece. It is best known under the medieval-modern artisanal name *serpentino*, alluding to its resemblance to a snake's skin, but today this is felt to be confusingly similar to the geological term serpentinite. Green porphyry is not a serpentinite but a metamorphic magmatite, hence the recent development of names like *porfido verde antico* or *di Grecia* to distinguish it also from the better-known Egyptian porphyry. For most of its long history in Greek it has been designated by a simple geographical name alluding to the region.

Like *rosso antico* (q.v.), green porphyry was known and used since the Bronze Age, particularly for small objects like vases and seals (Gnoli 1988, 141; Lazzarini 2007, 47). Colored stones were generally neglected in the revival of Greek cities, and the reappearance of green porphyry seems closely connected with Roman dominance of Greece, particularly the defeat of Sparta. Pavements at Ostia show its arrival in Italy by the middle of the first century BCE, among a number of polychrome stones now being imported. The elaborate *opus sectile* floors from the Domus Augustana, now on display at the Palatine Antiquarium, are typical of the high end. Although the deposits are in thin layers, maximum use was sometimes made of them, as shown

Porfido verde (cont.)

by the thick paving blocks of the Domus Augustana in Rome (just south of the "Aula Isiaca") and the thresholds stockpiled in the house of N. Popidius Priscus at Pompeii (VIII.2.20; Bragantini et al. 1986, 75). Pliny declared it "more cheerful in appearance than all others" (*Natural History* 36.55, trans. J.C. Fant). Small pieces, frequently combined with purple porphyry, could lend prestige to floors that did not display large expanses of marble; this was notable particularly in aspiring villas in central Italy (Jongste et al. 1992). Like Egyptian porphyry, its durability favored constant reuse, most spectacularly in the medieval pavements of the Cosmati workshops but also right into the Baroque period. JCF

References

Ellis, Higgins, and Simpson 1968; Bragantini et al. 1986; Gnoli 1988, 414–42; Borghini 1989, 279–81; Jongste et al. 1992; Lazzarini 2007, 45–69; Price 2007, 206–07

Portasanta

Cat. Nos. 204–268, 278?

Also known as: *m. Chium*
Source: Latomi, island of Chios, Greece

Fine-grained limestone, mostly brecciated with clasts of greatly varying size. Color is predominantly rose but highly variable, from gray with or without red veining to a dark liver hue. The best and most accessible examples of the color range in Rome today are the north and south fountains in Piazza Navonna, which is a Renaissance creation from ancient blocks found at Portus (Gnoli 1988, 173; Lazzarini 2007, 121), and the fountain in Piazza Colonna. The name derives from the marble's use in the door frame of the Porta Santa ("holy door") in St. Peter's Basilica in the Vatican, which is opened ceremonially only in Jubilee years (Ficoroni 1744, 188). The quarries, on the island of Chios just west of Chios town at a place called Latomi ("quarry"), were worked on a modest scale in the Classical and Hellenistic periods. In the Roman period exploitation was so massive that the mountaintop was cratered as if resembling a volcano. Modern exploitation was supplanted by a military zone (for its excellent view of the Turkish coast) until the 1980s.

There is no clear date for the beginning of Roman imports, which Gnoli is ready to date generally to the late Republic, in parallel with those of the other major "imperial" polychrome marbles (Numidian, Lucullan, Carystian, and Phrygian; Gnoli 1988). When Cicero visited the island (probably while he was studying on Rhodes in the early 70s BCE), the Chians proudly showed off their walls of Chian marble. He replied snidely, "I would be much more impressed if you had made them of travertine" (as related by Pliny, *Natural History* 36.46; trans. J.C. Fant). The point of the anecdote is that Romans were impressed by exotic origins and distance of origin, not so much by the intrinsic quality of the stone. It does not prove that Cicero was not already familiar with Chian marble, but such is likely as imports were already known at Rome (Lazzarini 2007, 122, agreeing with Gnoli's general surmise).

As with other marbles, imports and use grew exponentially in the Augustan period. In fact, the house (if such it was) of Augustus on the Palatine displayed columns of Chian marble. Thenceforth Chian was the most extensively used in Rome and widely distributed throughout both the western and eastern Mediterranean. Its representation in the Kelsey collections is similarly disproportionate: 68 of Kelsey's 442 pieces collected are of Chian. A variety of factors were at work, such as the extensive reuse of marble in the Baths of Diocletian, one of Kelsey's major sources. Overall, this is probably a valid reflection of the ancient reality. An abundance of inscriptions associated with imperial agents points to at least partial control from the beginning of the empire (Chios was captured by Lucullus at Sulla's orders in 86 BCE; Lucullus' interest in marble is already on record; see *Africano*). Its absence in the Diocletianic Price Edict suggests that exploitation had lapsed, although other evidence points to activity in the early Byzantine period. JCF

References

Ficoroni 1744; Gnoli 1988, 172–73; Borghini 1989, 285–87; Lazzarini 2007, 119

Portoro **Cat. Nos. 666–668, 669?**

Also known as: *mischio giallo e nero,* Portovenere marble, black and gold
Source: Porto Venere and Palmaria, La Spezia, Liguria, Italy

Portoro is one of the best known Italian black limestones because of its striking gold-yellow veins. Although much used in medieval Genoa, and from the seventeenth century widely in Italy, it enjoyed substantial use in the Roman period, chiefly in the nearby city of Luna. There it is found in the amphitheater and in columns now in the La Spezia museum. JCF

References

Price 2007, 86; Fratini et al. 2015

m. Proconnesium **Cat. No. 509?**

Also known as: *m. Proconnense, m. Cyzicenum, marmo di Proconneso*, Proconnesian white marble
Source: Marmara Island (ancient Prokonnesos); Sea of Marmara (ancient Propontis), Turkey

Proconnesian white marble ranges in color from dull white to light gray and in figure from uniform to strongly banded in even horizontal bands. The marble is easy to identify from the whiff of organic hydrogen sulfide the crystals emit when crushed. The quarries extend east and west of the present town of Saraylar on the northern coast of the island of Marmara (ancient Prokonnesos). The gentle slopes and good natural harbor, combined with Marmara's easy access to the west coast of Turkey and the Aegean in general, and particularly to Constantinople/Istanbul in later times, explain why Proconnesian marble was exported early. The source may have been known as early as the Archaic period, when it was apparently used for building the first Artemision in Ephesus (Vitruvius 10.2.15); its use in the Mausoleum of Mausolus in Halicarnassus in the fourth century BCE is attested by Vitruvius (2.8.20) and the Elder Pliny (*Natural History* 36.6). The Great Altar of Pergamon (early second century BCE) is another outstanding example. A shipwreck (early first century BCE) excavated in the early 2000s was carrying freshly quarried architectural elements of Proconnesian marble for the temple of Apollo at Claros, on the coast west of Ephesus (Carlson and Aylward 2010). Today the quarries of Marmara Island remain the largest operation in Turkey, with nearly a quarter of the island literally quarried away (even visible on Google Earth) as a result of its intensive extraction over the millennia. Numerous archaeological and archaeometric surveys have been conducted in the quarries (Monna and Pensabene 1977; Asgari 1978; Attanasio, Brilli, and Ogle 2006) up to the present day (Wielgosz-Rondolino, forthcoming), but the ancient topography is virtually unrecognizable.

In the Roman Imperial period, Proconnesian marble began to reach the western Mediterranean despite the development of the vast resources of Luna marble in northern Italy. The earliest appearances are at the Temple of Venus and the House of the Vettii at Pompeii (Bruno, Cancelliere, et al. 2002; Fant et al. 2002; Walker 1988). Increasingly in the second century, it became prevalent over Luna marble on the east coast of Italy and coastal cities all around the Mediterranean, presumably due to the economic advantage of shipping. It was largely responsible for what Ward-Perkins called the "marble style," a homogeneous monumentalization of city centers that signaled alignment with the capital (Dodge and Ward-Perkins 1992, ch. 5). Although quarry inscriptions are plentiful, there is no evidence that the Saraylar quarries were under imperial control or protection (Asgari and Drew-Bear 2002).

In the later Empire, Proconnesian marble came to be perceived almost as a colored or figured stone. This is evident as early as the mid-second century CE in the Hadrianeum in Rome. Here the column drums were quarried not horizontally, following the bedding plane, and so the gray-blue banding, but vertically, so that when erected the bands remained horizontal (Claridge 2010, 223–26). Beyond architecture, Proconnesian was predominantly used for sarcophagi, and to a lesser extent sculpture and architectural elements, such as the Corinthian capitals and column shafts now abandoned in the quarries (Asgari 1988, 1990, 1992). Although the Proconnesian workshops exported raw material, they specialized in manufacturing half-finished objects; the mass and serial production of items is visible in the open-air museums set up by Asgari throughout the quarries.

m. Proconnesium (cont.)

One workshop, situated near the port, roughed out blanks in popular shapes thus reducing both shipping weight and work left for workshops on the receiving end (Asgari 1990). JCF & LEL

References

Monna and Pensabene 1977; Asgari 1977, 1978, 1988, 1990, 1992; Walker 1988; Dodge and Ward-Perkins 1992, ch. 5; Asgari and Matthews 1995; Asgari and Drew-Bear 2002; Bruno, Cancelliere, et al. 2002; Cramer, Germann, and Heilmeyer 2002; Fant et al. 2002; Carlson and Aylward 2010; Claridge 2010, 223–26; Pensabene 2013, 317–48; Wielgosz-Rondolino forthcoming

Rosso antico — Cat. Nos. 447–474, 674

Also known as: *m. Taenarium*
Source: Mani Peninsula, Greece

Fine-grained, non-fossiliferous, red to purplish-red or mottled light and dark red marble (the color is due to hematite and, more rarely, manganese) with occasional white calcite inclusions or veins and some fine black veins. It is interbedded with white and gray marble so that dimensions are limited. It is odd that the ancient sources, whey they speak of color, mention only black, although this stone from Cape Taenarum was not significant while the red congenor was famous. This may be explained by the predominance of geographical designations, which should be understood to subsume the red stone (Gnoli 1988, 189; Lazzarini 2007, 72). There are many outcrops in the Mani Peninsula of southern Greece (Lazzarini 2007, 77, fig. 24).

Rosso antico, alongside *serpentino* (q.v.), has the longest history of use in the ancient world outside of Egypt, starting with the Treasury of Atreus at Mycenae (Ellis, Higgins, and Simpson 1968) and stretching into the Byzantine period. It was one of the earliest imports to Italy. It was prominent in the décor of the House of the Faun at Pompeii coeval with the First Style painting and was part of the insert floor of the house under the Ludus Magnus (Gnoli 1988, 189 n. 4). It was often combined in floors with Italian slate, green limestone, and other local stones, but unlike them it found a resurgence in the early Imperial period even as marble displaced most local colored lithotypes. *Rosso antico* was widely employed for moldings and small strips (*listelli*) in *opus sectile*, wall incrustation, and delicate intarsia work (see ch. 2). From the Augustan period to the Hadrianic it was also used for pilaster capitals, often inlaid with other marbles. Certainly its color favored *rosso antico* since it was easily associated with the deep colors of Tyrian purple and its lithic embodiment in *porfido rosso* (q.v.). In the Domus Transitoria nymphaeum under the Domitianic triclinium of the Palatine palace, a molding of *rosso antico* tops a dado of Phrygian *pavonazzetto* and a dado of purple porphyry (Nero was obsessed with purple). It was also prized for sculptures associated with Bacchus and his followers since the stone evokes the color of wine. JCF

References

Corsi 1845, 93–94; Ellis, Higgins, and Simpson 1968; Mielsch 1985, 59, nos. 596–99; Gnoli 1988, 187–91; Borghini 1989, 288; Lazzarini 2007, 119–36; Price 2007, 122

Rosso Verona — Cat. Nos. 675–676

Also known as: *rosso Veronese, broccato rossa di Verona*
Source: Lessini Mountains near Verona, Italy

Pink to red marly nodular limestone with dense fossils and stylolites. It was and is quarried in the Monti Lessini of Val Pantena and Valpolicella in the hinterland of Verona. In the Roman period it was used prevalently in Verona itself (most notably in the amphitheater) and in the Tenth Augustan Region, but Bruno (2002c) cites a number of examples at Ostia, which, however, may have come from different outcroppings of the same *rosso ammonitico* formation in the Apennines. JCF

References

Mielsch 1985, 52, nos. 395–97; De Vecchi and Lazzarini 1994; Bruno 2002c, 277; Price 2007, 115

Sperone

Cat. No. 595

Also known as: *l. Gabinus*, Gabii tufa, sometimes *peperino*
Source: Gabii, Italy

A volcanic conglomerate from the Quaternary period eruptions of the Latial volcanoes southeast of Rome and quarried at the Osteria dell'Osa at Gabii, 20 km southeast of Rome. These quarries are on the other side of the same crater that created the beds of *l. Albanus* (*peperino*) at Marino. This explains the strong similarities that lead to the Gabii stone sometimes being called *peperino*. Both are even-textured, but *sperone* is coarser and has larger inclusions of dark lava and limestone. Both are resistant to the elements and to fire, but Alban *peperino*'s fine texture allowed it to be carved and to hold coats of stucco. Recent research has shown the three most common peperino stones (*l. Gabinus*, *l. Albanus*, and *tufo del Palatino*) can fruitfully be distinguished from one another by trace element analysis. *Lapis Gabinus*, however, had an economic advantage over *l. Albanus* in that it could be barged to Rome with only 6 km of land transport, while from Marino 24 km of cartage was necessary (Blake 1947, 34–39). *Sperone* was introduced into Roman architecture in the second half of the second century BCE (e.g., the Aqua Marcia, 144 BCE) and was used particularly for the outer faces of buildings. Both stones had a brief revival in use after Nero's fire of 64 CE. Notable uses are in the Pons Aemilius, the Tabularium, the Theater of Pompey, and the fire wall of the Forum of Augustus. JCF

References

Blake 1947, 38–39; Jackson and Marra 2006; Claridge 2010, 39; Farr, Marra, and Terrenato 2015

m. Thasium

Cat. Nos. 510–512, 566?

Also known as: Thasian white marble, *marmo tasio*
Source: island of Thasos, Greece

There are four principal varieties of marble on the island of Thasos in the northern Aegean. The acropolis of ancient Thasos town (modern Limenas) was used mostly for local building (but there is an unfinished colossus in the local museum). Quarries at Cape Phanari, close to Limenas, were also opened early but remained modest in size. Both marbles are white to grayish, sometimes with veining. The Saliara quarries at Cape Vathy, to the southeast, also opened very early but were active on a large scale through the Roman period; a quarry block has been identified at the *statio marmorum* at Portus (Baccini Leotardi 1989, no. 123).

The marble is very white with medium-sized crystals of great brilliance and dolomitic, that is, calcium-magnesium carbonate. It was used chiefly for sculpture, including sarcophagi (extensively documented by J.J. Herrmann and colleagues: Herrmann and Newman 1995; Herrmann, van den Hoek, and Newman 2002). Aliki in the southeast was a very large quarrying area at sea level; cranes in each bay could load ships directly. The marble is calcitic, with large, brilliant white crystals. Pliny (*Natural History* 36.44), remarking that Thasos marble was favored by ancient sculptors after that of the Cyclades, was almost certainly speaking of Cape Vathy stone, whether he was aware of it or not. The same applies to Seneca's remark (*Epistulae* 86.6) on Thasian marble in baths as one index of contemporary decadent luxury. Thasian's low price of 40 denarii in the Diocletianic Price Edict probably reflects its advantageous location at seaside and abundant supplies rather than low esteem (Giacchero 1974, 210–11). JCF

References

Giacchero 1974; Sodini, Lambraki, and Koželj 1980, 5–63; Gnoli 1988, 262–63; Herrmann and Newman 1995; Herrmann, Barbin, and Mentzos 1999; Herrmann, van den *m. Thasium* (cont.)

Hoek, and Newman 2002; Attanasio, Brilli, and Ogle 2006, 163–43; Pensabene 2013, 281–90

Travertino

Cat. Nos. 597, 599, 647?

Also known as: *travertino di Tivoli, l. Tiburtinus*, travertine
Source: Tivoli, Italy

Travertine is a fresh-water sedimentary limestone formed from cooling waters saturated with calcium carbonate from hot springs. Since it is a depositional rock formed by precipitation from hot water, fossils are not present. Travertine (called by geologists tufa but not to be confused with volcanic tufo or tuff) is not rare. On the plain below ancient Tibur the sulphur springs of Bagni di Tivoli are heated by the same ancient vulcanism that produced *l. Gabinus* and *l. Albanus* (see *Sperone*, above). This travertine is of unusual purity and density; since travertine is formed on the surface rather than by subterranean pressure, it tends to be quite porous, but the beds at Tivoli are very deep because the deposition has been continuous for long periods.

Travertine began to be quarried in the third century BCE but only appears in substantial quantities late in the century (e.g., Pons Milvius, 109 BCE). It must have been known to the Romans much earlier, but the impetus of monumental architecture executed in imported—but scarce—Greek marble (see *Pentelic*, above) led to infrastructure development since travertine was seen as a plausible substitute for, or at least companion of, marble. It could be transported to Rome on barges from the Anio (now Aniene) to the Tiber (Blake 1947, 44). Notable buildings in Rome built in travertine are the Theater of Marcellus and the Colosseum. Lanciani (1897, 35–38) paints a vivid word picture of the Tivoli quarries' large-scale exploitation for building the new capital around the turn of the twentieth century. JCF

References

Lanciani 1897, 35–38; Blake 1947, 44–48; Davies 2017, 158–67

Verde antico

Cat. Nos. 380–386, 387?, 388–394

Also known as: *m. Thessalicum, l. Atracius*
Source: Kasambali near Larissa, Thessaly, Greece

Metamorphic ophycarbonate breccia of green serpentinite and white marble clasts of widely varying size and shape in a serpentinite-calcite matrix. The shades of green present can also vary greatly although in general two tonal groups, a light green and a very dark green from the color of the matrix, are distinguished.

The names are all late, sixth century CE and onward. Paul the Silentiary in his description of Hagia Sophia mentions the village Atrax (near Larissa in Thessaly, Greece) as the source, and *Thessalian* originates with the lexicographer Pollux (Gnoli 1988, 162 n. 2). The Italian artisanal name *verde antico* today invites confusion with the widely used pure serpentinite from Vermont called *verde antique* (see cat. nos. 635–636).

While green stones had a long history in the ancient world, *m. Thessalicum* was unusual in being available in blocks large enough for column production. Its introduction to Rome is firmly attributed to Hadrian, who used it for his villa at Tivoli. Thereafter its popularity grew and reached a climax with the forty-eight columns of Hagia Sophia in Constantinople under Justinian (537 CE). JCF

References

Corsi 1845, 160; Mielsch 1985, 63–64, nos. 683–91; Gnoli 1988, 162–65; Lazzarini 2007, 223–44; Price 2007, 186–87

Chapter 5

CATALOGUE OF MARBLES

J. Clayton Fant, Leah E. Long, and Lynley J. McAlpine

The decorative marbles in the Kelsey Museum collection are organized in this catalogue by lithotypes gathered into ten levels of scarcity and prestige in the Roman world. This scale was developed to categorize the 8,000+ marble fragments decorating fifty-seven streetside bars studied at Pompeii and Herculaneum (Fant, Russell, and Barker 2013; Fant 2009), and it has proven useful also at luxury villas (Barker, Fant, and Ward 2015). The hope is that this "Prestige Index" will help to make this large catalogue accessible to anyone wishing to learn more about a particular type of stone or to identify a piece of their own. In addition to providing an organizing principle here, it supports a consideration of the interests of the collectors who formed the Kelsey collection. In descending order, the lithotypes are:

Prestige Level 10	Granites, porphyries, other hard stones
Prestige Level 9	Polychrome marbles most closely associated with imperial construction: *africano, giallo antico, pavonazzetto*
Prestige Level 8	*Cipollino* (all varieties except *rosso*), *portasanta*, jasper
Prestige Level 7	Alabasters and onyx: *cotognino, fiorito, a pecorella,* etc. Breccias: *Aleppo, corallina, medicea, Serravezza, Settebasi*; *verde antico*
Prestige Level 6	*Broccatello, broccatellone, cipollino rosso, fior di pesco, lumachella* including *occhio di Pavone, rosso antico*, other breccias and limestones
Prestige Level 5	"Island" (Greek and Anatolian) white marbles (i.e., large crystals), *greco scritto, nero antico*
Prestige Level 4	Luna *bardiglio* (gray), *bigio antico*, and other gray marbles
Prestige Level 3	Luna white, especially *ordinario*, travertine, Caserta limestone, *palombino*, sandstones
Prestige Level 2	Basalt for paving, limestones, metamorphic clay, sandstones, stones chiefly associated with late Republican decoration
Prestige Level 1	Casual collections, surface finds, unworked samples
Modern and/or non-Mediterranean Samples	
Unidentifiable Lithotypes	

The order of lithotypes within the levels is alphabetical by the most common nomenclature, usually the traditional Italian artisanal name. Within lithotypes, artifacts are grouped by collector in order of number of pieces, and within collector groups, by chronology where possible. The aim is to give some feel for the activities and tastes of the collectors. Any scheme leaves much to be desired in organizing such a large and heterogeneous collection, and in the penultimate category, Modern and/or non-Mediterranean Samples, it seemed more useful to organize by lithotype name where one is established, and then by stone type.

Catalogue Entries

Catalogue entries present these items of information in the following order and format:

Catalogue Number
The identifying number in this catalogue is followed by the Kelsey Museum accession number (KM).

Figure and Plate Numbers
The intent is to illustrate every artifact. We photographed top (front) and bottom (back) of every piece possible; bottoms (backs) are illustrated only when they convey useful information or present an interesting original label.

Description of the Object as Artifact
Pieces that are not simply basic paving or veneer are discussed in chapter 2. The object description also fleshes out the artifact condition, describes evidence of tool marks, and reports evidence of reuse or treatment such as the application of wax to the polished surfaces.

Lithotype Identification
Lithotype listings in the catalogue are ordered and alphabetized by the most widely used modern name, which is generally the Italian artisanal name since the great majority of the artifacts were collected in Italy and were often appraised by local workers and experts whom Kelsey consulted habitually. Where there is no Italian name, we give the most common name in circulation, in whatever language. More extensive discussions about individual lithotype names in provided in Chapter 4, Lithotype Profiles, and a full listing can be found in Appendix A, Gazetteer of Quarries.

A brief indication of geographical origin is given for each lithotype; for more detail, consult the relevant lithotype profile. Comments and elaboration following the identification, parenthetical or not, are ours.

Our identifications rely on our own field experience, fortified by the many well-published collections now in print (Dolci and Nista 1992; Mielsch 1985; Price 2007; several each by Pensabene and Lazzarini) and online collections (the collection of the Museo di Storia Naturale dell'Accademia dei Fisiocritici, or MUSNAF, overseen by Lorenzo Lazzarini; and the Corsi Collection of the Oxford Museum of Natural History, or OUMNH, curated by Monica Price. See page xi for web addresses). For white marbles, we had funding for a limited set of samples where there was an archaeological context. Otherwise, where we make a firm identification, we append "(visual)" to it. Where a maximum grain size (MGS) is indicated but a full analysis has not been carried out, the MGS measurements were done in the collections by Fant and Long with a hand lens (our thanks to Professor Harrell for instruction; he bears no responsibility for accurate deployment). Inferences based on our MGS measurements are based on the database in Attanasio, Brilli, and Ogle 2006. We have also drawn upon the expertise of Drs. Simon Barker, Matthias Bruno, Simona Perna, and Anna Gutiérrez Garcia-Moreno. We particularly thank Dr. Lorenzo Lazzarini (Laboratory for Analysis of Ancient Materials at Iuav University of Venice), who answered numerous pleas for help with identifications outside our own sphere of knowledge, especially the lithotypes of northern Italy; and Dr. James A. Harrell (Department of Environmental Science, University of Toledo, Ohio), who kindly not only answered emailed questions but made several visits to the collections in person.

Description in Accession
Museum accession materials convey original suggested lithotype identifications. These mostly derive from labels on the pieces themselves or the bags that hold them, the originals of which are almost all lost, and are generally transcribed into the accession books. Although we consider many of these erroneous, we wished to preserve their context and flavor, but without sowing confusion between old identifications and terminology and ones made by us in the light of the last fifty years of research. Thus we have included the old identifications in the catalogue entry line "Description in Accession."

Van Deman's labels show a deep involvement with the language of the Italian artisanal tradition. Candler's pieces sometimes have the original paper labels still attached, as would be suitable for display in a small private collection as conversation pieces. In the case of Kelsey's collections, a note in the accession book at KM 1996 (cat. no. 559) provides an explicit description:

> In 1930 the material contained in several lots of loose sheets and cards was copied here to save the information. Had it not been for his long illness in 1902 Professor Kelsey would have brought all this together himself. The various titles seem to have been added by various stone experts as the writing

varies. The one used first here was the one which stood first on the slip.

These notes are often inaccurate, and we have not relied on them. Identifications are frequently credited to named individuals whom Kelsey apparently consulted regularly and in some cases purchased from. One of Kelsey's informants, a certain "Orlandi," probably the artist/collector Aristide Orlandi (see ch. 1), was reliably erroneous in his attributions. "Boni," likely the pioneering archaeologist Giacomo Boni who was working in the Roman Forum during Kelsey's term as professor at the American School of Classical Studies at Rome (1900–1901) and whom Kelsey engaged to lecture to his students, is not much more on target. Esther Van Deman's notes are more sparse, limited to labels still affixed to objects (see fig. 1.4); however, they carried more weight with us since she was an expert in her own right (although we suspect that she took pleasure in elaborating descriptions in Italian that sometimes go well beyond the traditional artisanal nomenclature). She delighted also in alternate spellings, such as *affricano*, which may be rooted in *Romanesco*, the dialect of the *Romani di Roma*, which she had doubtless learned in her trampings in the Roman countryside. Unfortunately, none of her pieces identify provenance.

Collection

We have departed from the customary catalogue entry format to give as much context about the collectors' activities as possible. Thus we quote extensively from labels on artifacts, notes on plastic museum bags which are transcriptions from earlier labels or Kelsey's own slips (quotation marks report the exact words).

Entries are as specific as possible about the place and circumstances of collection; we have not hesitated to include our research and conjecture. This vein of information is richest for Kelsey's activity, and, unfortunately, almost barren for Van Deman's since it seems likely that she would keep good notes about her forays in the Roman *campagna* and elsewhere. We have transmitted such indications from the accession books even though there is some suspicion that notes like "Rome" or "Italy" may be mere guesses from probability.

Collection dates are frequently noted in Kelsey's pieces (presumably originating from his own slips, as described in Orma Fitch Butler 1930). Accession dates are also given where known. Esther Van Deman's collection was accessioned all together in 1938 when it arrived in Ann Arbor from Baltimore, pursuant to her will.

Accession book entries for Kelsey's pieces, deriving from his original slips (see above, pp. 5 and 15), include a good number (86) which provide not only a date in the "When Collected" column (sometimes, probably in error, "When Received," i.e., in Ann Arbor), but also one immediately following the place named in the "Locality" column. This second date must also have originated from Kelsey's contemporaneous slips, as is clear from pieces that he picked up in the street outside the Methodist palazzo at 38 Via Firenze (cat. nos. 188 and 451) or in the works for the Quirinal tunnel (cat. no. 554). In most cases, this date is repeated in the "When Collected" column. The inference could be drawn that Kelsey's habit was to write up his slips every evening, so as to record the pertinent information while it was fresh.

In eight clear cases and another likely one, however, the dates do not agree. Apart from one simple mistake (cat. no. 204, "Oct. 1901" under "When Received," when Kelsey was already back in Ann Arbor), they seem to be carefully considered; in no case is the first date later than the second. A group of seven pieces has January 1901 in the "Locality" column and July 1901 in the second occurrence (cat. nos. 18, 30, 56, 82, 217, 273, 631). All are from the Baths of Diocletian; it is noteworthy that while F. Romano supplied the bulk of pieces from the Baths, he does not figure in any of these records, implying perhaps that Kelsey gathered these himself, as is also likely the case for the two pieces found in the works at 38 Via Firenze, with January 4 (cat. no. 188) and January 21 (cat. no. 451) in the "Locality" column and a second date of July 1901. It may be that Kelsey was distracted from writing up slips for these pieces and perhaps came upon them the following summer as he was preparing to pack up for his return and reconstructed the slips from memory.

As there is no further information, these suggestions can be no more than a hypothesis. Hence in the catalogue entries we have not indicated double dates where they are repetitive, but in the cases discussed here, we include both dates, as "January and July 1901" without further comment.

Bibliography

References are given only for challenging identifications or information that is not routine. None of the artifacts has been previously published, with the exception of twenty-two objects included in the Kelsey Museum's 2016 Oplontis exhibition (Gazda and Clarke

2016). General references for lithotypes can be found in the Lithotype Profile essays on each.

Notes
When further explanation about a piece is needed, it is given in a separate Note section.

Abbreviations and Conventions

Entries in the Kelsey Museum's handwritten accession books are excerpted within quotation marks, as mentioned. Francis Kelsey's pieces often have several conflicting identifications that are attributed to specific individuals (Orlandi, Boni, Romano, etc.). Many of these, as noted above, are erroneous and even propose origins that are geologically wholly inconsistent.

Degrees of Certainty about Date and Place
For many of Francis Kelsey's pieces we have, or can reconstruct, the date and place of acquisition, and this is what is recorded in the catalogue. Other dates of accession are given but do not shed much light on the collection process: Van Deman's collection came to the museum in one shipment as enjoined by her will (see ch. 2), and the process of building it over her many years in Rome is invisible. Burton's eclectic collecting is also obscured, with the exception of cat. no. 587, which attests to her visit (or that of a proxy) to the island of Delos.

Degrees of uncertainty about date and place of collection are indicated in this fashion: Rome / (Rome) / (Rome?).

A term given without parentheses (e.g., "Rome" reports a provenance directly based on a label or accession book notation.

A term given in parentheses (e.g., "(Rome)") means that we have inferred the place or date of collection, but with a fair degree of confidence. A good example is cat. no. 175, where no place of collection is indicated but the note "Bought of F. Romano" is given in the accession book, drawn from Kelsey's slip. Since Romano was a worker and/or dealer who was active in the construction around Piazza Esedra in Rome, it is probable, but not certain, that the piece came from there.

A term followed by a question mark and encased in parentheses (e.g., "(Rome?)") indicates a conjecture based on similar entries but which is less than certain. For instance, J.D. Candler picked up a number of pieces of porphyry at Rome, as indicated by labels such as the one on cat. no. 46; this is the basis for suggesting Rome as the provenance of cat. no. 45.

Catalogue Entries

Prestige Level 10

Granites, porphyries, and other hard stones

Basanite

Source: Wâdi Hammâmât, Eastern Desert of Egypt

1. **KM 2214** **fig. 2.36a; pl. 1**
Description: Sculpture fragment, knee or thigh, L. 18.5, W. 10.8, D. 6.5 cm, two preserved edges.
Lithotype: Basanite, basalto verde, bekhen-stone.
Collection: F. W. Kelsey, purchased from F. Romano, (Rome), (1900–1901). Accessioned (1901).

2. **KM 6860** **fig. 2.36b; pl. 1**
Description: Sculpture fragment, fingers or hand holding object, L. 8.3, W. 4.0, D. 4.4 cm, one preserved edge.
Lithotype: Basanite, basalto verde, bekhen-stone.
Description in Accession: "basalt."
Collection: F. W. Kelsey, "Europe," no collection date. Accessioned 1938 (posthumous).

3. **KM 2215** **fig. 2.37; pl. 1**
Description: Cylinder with smooth surface (drill core?), L. 16.1, D. 9.1 cm, broken across the diameter at both ends.
Lithotype: Basanite, basalto verde, bekhen-stone, but intermediate with the *breccia verde d'Egitto* (*hexatontolithos*), the quarries of which are coterminous in Wâdi Hammâmât.
Collection: F. W. Kelsey, purchased from F. Romano, "Al Porto Antico," Ostia, February 1901. Accessioned 1901.
Note: Core drilling was an ancient practice, as early as the Minoan period, and rock cylinders similar to this one are found in Egypt (J. A. Harrell, personal communication, August 12, 2018). For another possible drill core, see cat. no. 552.

Gabbro eufotide

Source: Wâdi Maghrabîya, Eastern Desert of Egypt

4. **KM 2209** **pl. 1**
Description: Paving or veneer, L. 10.5, W. 7.8, D. 4.6 cm, three preserved edges.
Lithotype: Gabbro eufotide, variant form *granito verde plasmato.*
Collection: F. W. Kelsey, purchased from F. Romano, "Al Porto Antico," Ostia, February 1901. Accessioned 1901.

5. **KM 2217** **pl. 1**
Description: Paving or veneer, L. 16.6, W. 12.0, D. 5.4 cm, top and bottom surfaces preserved.
Lithotype: Gabbro eufotide.
Collection: F. W. Kelsey, "Purchased of Vecchio," (Italy), (1900–1901). Accessioned 1901.

6. **KM 6859** **pl. 1**
Description: Paving, veneer, or *opus sectile*, L. 8.1, W. 8.1, D. 0.91–1.0 cm, four preserved edges.
Lithotype: Gabbro eufotide.
Collection: F. W. Kelsey, "Europe," no collection date. Accessioned 1938 (posthumous).

7. **KM 2205** **pl. 1**
Description: Paving or veneer, L. 10.0, W. 9.2, D. 2.3 cm, two preserved edges, wax on polished surface.
Lithotype: Gabbro eufotide.
Collection: F. W. Kelsey, no provenance, (1900–1901). Accessioned (1901).

Granito bianco e nero

Source: Wâdi Barud, Eastern Desert of Egypt

8. **KM 2202** **pl. 1**
Description: Paving or veneer, L. 10.1, W. 8.2, D. 1.4 cm, no preserved edges, wax on polished surface.
Lithotype: Granito bianco e nero, m. Tibereum.
Collection: F. W. Kelsey, "Purchased of Vecchio," (Italy), (1900–1901). Accessioned (1901).

9. **KM 6858** **fig. 2.6a; pl. 1**
Description: Opus sectile triangle, L. 9.4, W. 6.8, D. 1.8 cm, one preserved edge, wax on polished surface.
Lithotype: Granito bianco e nero, m. Tibereum.
Collection: F. W. Kelsey, (Rome?), no collection date. Accessioned 1938 (posthumous).
Note: This piece probably came from a floor of an imperial building, perhaps on the Palatine.

10. **KM 6794** **pl. 2**
Description: Paving or veneer, L. 17.2, W. 14.5, D. 1.6 cm, one preserved edge.
Lithotype: Granito bianco e nero (Wâdi Barud North), *m. Tibereum.*
Collection: E. B. Van Deman, (Italy), no collection date. Accessioned 1938.

11. **KM 93568** **pl. 2**
Description: Paving or veneer, L. 5.18, W. 4.15, D. 1.35 cm, no preserved edges.
Lithotype: Granito bianco e nero, m. Tibereum.
Collection: J. D. Candler, (Rome? This stone was not used outside Rome and central Italy), 1860–1890. Accessioned 1964.

Granito della colonna

Source: Wâdi Umm Shegilat, Eastern Desert of Egypt

12. KM 2355 **fig. 2.34a; pl. 2**

Description: Fragment of a vessel rim, L. 14.0, W. 6.2, D. 4.36 cm, no preserved edges. Extrapolated vessel diameter ca. 36 cm.

Lithotype: *Granito della colonna.*

Collection: Kelsey, Rome, 1901. Accessioned (1901).

Note: Possibly Predynastic, more likely Roman since such vessels became popular décor for baths.

13. KM 2072 **pl. 2**

Description: Paving or veneer, L. 11.1, W. 5.6, D. 5.8 cm, no preserved edges, wax on polished surface.

Lithotype: *Granito della colonna.*

Collection: F.W. Kelsey, "purchased of old man," Rome, March 13, 1901 (written inaccurately in accession book as "March 13, 1900"; Kelsey was not in Italy in March 1900). Accessioned 1902.

Granito del foro

Source: Gebel Fatireh, Wâdi Umm Huyut, Eastern Desert of Egypt

14. KM 2198 **pl. 2**

Description: Paving or veneer, L. 21.0, W. 8.2, D. 8.2 cm, two preserved edges.

Lithotype: *Granito del foro, m. Claudianum.*

Collection: F.W. Kelsey, purchased of F. Romano, (Rome), (1900–1901). Accessioned 1901.

15. KM 2343 **pl. 2**

Description: Paving or veneer, L. 15.4, W. 12.5, D. 3.7 cm, no preserved edges, appears to be waxed.

Lithotype: *Granito del foro, m. Claudianum.*

Collection: F.W. Kelsey, (Rome), (1900–1901). Accessioned 1901.

16. KM 2354 **pl. 2**

Description: Paving or veneer, L. 18.3, W. 8.0, D. 3.9 cm, no preserved edges, surface waxed.

Lithotype: *Granito del foro, m. Claudianum.*

Collection: F.W. Kelsey, Rome, 1901. Accessioned (1901).

17. KM 2332 **pl. 2**

Description: Paving or veneer, L. 25.0, W. 20.0, D. 4.54 cm, no preserved edges, surface waxed.

Lithotype: *Granito del foro, m. Claudianum.*

Collection: F.W. Kelsey, "Tor Paterno, Castel Fusano," Italy, (1900–1901). Accessioned 1901.

18. KM 2546 **pl. 2**

Description: Modern sample plaque, L. 9.5, W. 5.0, D. 4.3 cm, three preserved edges, polished on two long sides and one end.

Lithotype: Tonalite gneiss typical of *Granito del foro, m. Claudianum.*

Collection: E. Jones, (Rome), 1885. Accessioned 1912.

Note: The plaque must be a reworked ancient piece since this granite was not quarried after antiquity, and in Jones's era its source was not known.

Granito rosso and granito nero di Siene

Source: locations between Aswan (ancient Syene) and Shellal District, Nile River Valley

19. KM 2204 **pl. 2**

Description: Paving or veneer, L. 15.2, W. 13.0, D. 3.63 cm, one preserved edge, top and bottom surfaces finished smooth.

Lithotype: *Granito rosso, l. Syenites.*

Collection: F.W. Kelsey, Baths of Diocletian, Rome, January and July 1901. Accessioned (1901).

20. KM 2199 **fig. 2.33a; pls. 2–3**

Description: Fragment of a large *labrum* (basin), L. 10.5, W. 21.0, D. 8.0 cm, one preserved edge, original surface waxed.

Lithotype: *Granito rosso, l. Syenites.*

Collection: F.W. Kelsey, Baths of Diocletian, Rome, (1900–1901). Accessioned 1901.

21. KM 2196 **fig. 2.33b; pl. 3**

Description: Fragment of a vessel, possibly the bottom, L. 18.4, W. 14.4, D. 7.1 cm, one preserved edge, wax on polished surface.

Lithotype: *Granito rosso, l. Syenites.*

Collection: F.W. Kelsey, "purchased of old man," (Rome), March 13, 1901. Accessioned 1901.

22. KM 2200 **pl. 3**

Description: Paving or veneer, L. 15.9, W. 15.0, D. 3.4 cm, one preserved edge, wax on polished surface, traces of saw marks.

Lithotype: *Granito nero di Siene, l. Syenites* (see Profile for *granito rosso*).

Collection: F.W. Kelsey, purchased of F. Romano, Ostia, March 2, 1901. Accessioned 1901.

23. KM 2197 **pl. 3**

Description: Paving or veneer, L. 23.4, W. 12.0, D. 8.06 cm, two preserved edges.

Lithotype: *Granito nero di Siene, l. Syenites* (see under *granito rosso*).

Collection: F.W. Kelsey, (Italy), (1900–1901). Accessioned 1901.

24. KM 93520 **pl. 3**

Description: Unknown, L. 11.4, W. 6.3, D. 3.6 cm, no preserved edges.

Lithotype: *Granito rosso, l. Syenites.*

Collection: J.D. Candler, "From the Tombs near the great Sphinx, Cairo, Egypt," (September 1867). Accessioned 1964.

Note: Candler was in Ephesus on September 6, 1867 (see cat. no. 291), and probably continued thence to Egypt on his Grand Tour.

25. KM 29519 **pl. 3**
Description: Plaque, L. 23.5, W. 23.3, D. 1.86 cm, three preserved edges. The parallel edges and the third, gently curving side are set off by a 3 mm groove.
Lithotype: Granito rosso, l. Syenites.
Collection: E. B. Van Deman, Ostia, Italy, no collection date. Accessioned 1938.

Granito verde della sedia di San Pietro and Granito verde della sedia di San Lorenzo

Sources: Wâdi Semna, Wâdi Umm Wikala (quarries proper), Eastern Desert of Egypt

26. KM 2335 **pl. 3**
Description: Paving or veneer, L. 19.7, W. 14.4, D. 6.76 cm, one preserved edge.
Lithotype: Granito verde della sedia di San Pietro, ophites.
Collection: F. W. Kelsey, no provenance, (1900–1901). Accessioned 1901.

27. KM 2203 **pl. 3**
Description: Paving or veneer, L. 12.4, W. 11.0, D. 2.8 cm, no preserved edges, wax on polished surface.
Lithotype: Granito verde della sedia di San Lorenzo, ophites.
Collection: F. W. Kelsey, "Tor Paterno," Italy, (1900–1901). Accessioned (1901).

Porfido rosso

Source: Gebel Dokhân (ancient Mons Porphyrites), Eastern Desert of Egypt

28. KM 2211 **pl. 3**
Description: Paving or veneer, L. 11.6, W. 4.6, D. 1.8 cm, four preserved edges.
Lithotype: Porfido rosso, m. porphyriticum.
Collection: F. W. Kelsey, purchased from F. Romano, (Rome), (1900–1901). Accessioned (1901).

29. KM 2208 **pl. 3**
Description: Paving or veneer, L. 13.3, W. 12.9, D. 3.9 cm, no preserved edges, wax on polished surface.
Lithotype: Porfido rosso, m. porphyriticum.
Collection: F. W. Kelsey, no provenance, (1900–1901). Accessioned (1901).

30. KM 2210 **pl. 4**
Description: Paving or veneer, L. 10.9, W. 8.2, D. 2.7 cm, one preserved edge.
Lithotype: Porfido rosso, m. porphyriticum.
Description in Accession: "modern" (Boni).
Collection: F. W. Kelsey, Baths of Diocletian, Rome, January and July 1901. Accessioned (1901).
Note: Boni's "modern" designation is unlikely since the quarries at Gebel Dokhân were abandoned after the Roman period, apart from sporadic and small twentieth-century efforts (Maxfield and Peacock 2001, 5, 8).

31. KM 2352 **pl. 4**
Description: Paving or veneer, L. 14.5, W. 10.0, D. 3.2 cm, no preserved edges, wax on polished surface.
Lithotype: Porfido rosso, m. porphyriticum.
Collection: F. W. Kelsey, no provenance, (1900–1901). Accessioned (1901).

32. KM 6853 **pl. 4**
Description: Paving or veneer, L. 9.6, W. 5.0, D. 1.0 cm, no preserved edges.
Lithotype: Porfido rosso, m. porphyriticum.
Collection: F. W. Kelsey, "Europe," no collection date. Accessioned 1938 (posthumous).

33. KM 6854 **pl. 4**
Description: Paving or veneer, L. 9.5, W. 3.9, D. 0.9 cm, no preserved edges.
Lithotype: Porfido rosso, m. porphyriticum.
Collection: F. W. Kelsey, "Europe," no collection date. Accessioned 1938 (posthumous).

34. KM 6855 **pl. 4**
Description: Paving or veneer, L. 3.4, W. 2.9, D. 2.8 cm, one preserved edge.
Lithotype: Porfido rosso, m. porphyriticum. The sample is partly black porphyry, which comes from at least four quarry sites at Gebel Dokhân (Mons Porphyrites), including the earliest one (see lithotype profile for *porfido rosso*)
Collection: F. W. Kelsey, "Europe," no collection date. Accessioned 1938 (posthumous).

35. KM 6856 **pl. 4**
Description: Paving or veneer, L. 6.7, W. 5.2, D. 1.0 cm, no preserved edges.
Lithotype: Porfido rosso, m. porphyriticum.
Collection: F. W. Kelsey, "Europe," no collection date. Accessioned 1938 (posthumous).

36. KM 29719 **pl. 4**
Description: Paving or veneer, L. 9.1, W. 3.6, D. 2.0 cm, no preserved edges.
Lithotype: Porfido rosso, m. porphyriticum.
Collection: De Criscio, Pozzuoli or vicinity, "Part of 2nd purchase of De Criscio collection 1922–23." Accessioned 1923.

37. KM 29725 **pl. 4**
Description: Paving or veneer, L. 4.0, W. 18.0, D. 1.2 cm, four preserved edges.
Lithotype: Porfido rosso, m. porphyriticum.
Collection: G. De Criscio, Pozzuoli or vicinity, no collection date. Accessioned 1923.

38. KM 29726 **pl. 4**
Description: Paving or veneer, L. 7.6, W. 7.3, D. 2.8 cm, no preserved edges.
Lithotype: Porfido rosso, m. porphyriticum.
Collection: G. De Criscio, Pozzuoli or vicinity, no collection date. Accessioned 1923.

Porfido rosso (cont.)

39. KM 29727 **pl. 4**
Description: Paving or veneer, L. 9.9, W. 3.8, D. 2.4 cm, no preserved edges.
Lithotype: Porfido rosso, m. porphyriticum.
Collection: G. De Criscio, Pozzuoli or vicinity, no collection date. Accessioned 1923.

40. KM 29731 **pl. 4**
Description: Paving or veneer, L. 3.6, W. 3.4, D. 4.0 cm, one preserved edge.
Lithotype: Porfido rosso, m. porphyriticum.
Collection: G. De Criscio, Pozzuoli or vicinity, 1922. Accessioned 1923.

41. KM 29732 **pl. 4**
Description: Paving or veneer, L. 4.5, W. 2.3, D. 1.42 cm, no preserved edges.
Lithotype: Porfido rosso, m. porphyriticum.
Collection: G. De Criscio, Pozzuoli or vicinity, 1922. Accessioned 1923.

42. KM 29733 **pl. 5**
Description: Paving or veneer, L. 2.7, W. 2.6, D. 0.5 cm, no preserved edges.
Lithotype: Porfido rosso, m. porphyriticum.
Collection: G. De Criscio, Pozzuoli or vicinity, 1922. Accessioned 1923.

43. KM 29734 **pl. 5**
Description: Paving or veneer, L. 9.1, W. 3.6, D. 2.0 cm, no preserved edges.
Lithotype: Porfido rosso, m. porphyriticum.
Collection: G. De Criscio, Pozzuoli or vicinity, 1922. Accessioned 1923.

44. KM 29605 **pl. 5**
Description: Paving or veneer, L. 2.4, W. 2.0, D. 1.7 cm, no preserved edges.
Lithotype: Porfido rosso, m. porphyriticum.
Collection: Candler, (Rome?), (1860s?). Accessioned 1964.

45. KM 29612 **pl. 5**
Description: Paving or veneer, L. 9.6, W. 2.3, D. 0.8 cm, no preserved edges.
Lithotype: Porfido rosso, m. porphyriticum.
Collection: J. D. Candler, (Rome?), (1860s?). Accessioned 1964.

46. KM 29623 **pl. 5**
Description: Paving or veneer, L. 4.4, W. 4.2, D. 1.0 cm, no preserved edges.
Lithotype: Porfido rosso, m. porphyriticum.
Collection: J. D. Candler, Rome, (1860s?). Accessioned 1964.

47. KM 93566 **pl. 5**
Description: Paving or veneer or possibly *opus sectile*, L. 5.19, W. 4.18, D. 0.72 cm, no preserved edges.
Lithotype: Porfido rosso, m. porphyriticum, with some pink phenocrysts.
Collection: J. D. Candler, (Rome?), (1860s?). Accessioned 1964.

48. KM 93567 **fig. 2.21a; pl. 5**
Description: Fillet strip, L. 6.9, W. 3.4, D. 1.2 cm, two preserved edges; three segmental curves, width of fully preserved central curve is 3.43 cm.
Lithotype: Porfido rosso, m. porphyriticum, with some pink phenocrysts.
Collection: J. D. Candler, (Rome?), (1860s?). Accessioned 1964.

49. KM 2543 **pl. 5**
Description: Paving or veneer, L. 9.3, W. 8.6, D. 2.8 cm, no preserved edges.
Lithotype: Porfido rosso, m. porphyriticum.
Collection: E. Jones, "Roman Forum," Rome, (1885). Accessioned (1912).

50. KM 93901 **pl. 5**
Description: Paving or veneer, L. 5.5, W. 6.8, D. 2.5 cm.
Lithotype: Porfido rosso, m. porphyriticum.
Collection: F. W. Kelsey (posthumous), "Material brought back from the U-M expedition to Antioch in Pisidia," directed by D. M. Robinson, 1924. Accession date unknown.

51. KM 1969.5.1 **pl. 5**
Description: Paving or veneer, L. 7.4, W. 7.4, D. 2.5 cm, one preserved edge.
Lithotype: Porfido rosso, m. porphyriticum.
Collection: J. G. Winter, Rome, "near Arch of Titus, 1922." Accessioned 1969.

52. KM 91752 **fig. 2.21b; pl. 5**
Description: Fillet, L. 11.5, W. 2.4, D. 1.6 cm, both ends broken, concave profile.
Lithotype: Porfido rosso, m. porphyriticum.
Collection: E. S. McCartney, "Argiva," Sicily, no collection date. Accessioned 1937.
Note: We were not able to locate Argiva.

53. KM 29695 **fig. 2.34b; pl. 6**
Description: Vessel fragment, L. 4.0, W. 2.3, D. 1.0 cm, no preserved edges, curved profile.
Lithotype: Porfido rosso, m. porphyriticum, with pinkish clasts.
Collection: P. O. S. Burton, no provenance, no collection date. Accessioned 1961.

54. KM 29875 **pl. 6**
Description: Paving or veneer, L. 9.9, W. 8.0, D. 1.4 cm, one preserved edge.
Lithotype: Porfido rosso, m. porphyriticum.
Collection: Collector, provenance, collection date unknown. Accession date unknown.

55. KM 29876 **pl. 6**
Description: Paving or veneer, L. 4.5, W. 4.3, D. 1.5 cm, no preserved edges.
Lithotype: Porfido rosso, m. porphyriticum.

Collection: Collector, provenance, collection date unknown. Accession date unknown.

Porfido verde

Source: Krokees (ancient Krokeai), Peloponnese, Greece

56. **KM 2065** **pl. 6**
Description: Paving or veneer, L. 13.5, W. 10.3, D. 4.1 cm, one preserved edge.
Lithotype: Porfido verde, l. Lacedaemonius.
Collection: F. W. Kelsey, Baths of Diocletian, Rome, January and July 1901. Accessioned 1902.

57. **KM 2066** **pl. 6**
Description: Paving or veneer, L. 14.5, W. 9.9, D. 3.6 cm, no preserved edges.
Lithotype: Porfido verde, l. Lacedaemonius.
Collection: F. W. Kelsey, Rome, 1901. Accessioned 1902.

58. **KM 2067** **pl. 6**
Description: Paving or veneer, L. 11.8, W. 10.9, D. 5.1 cm, no preserved edges.
Lithotype: Porfido verde, l. Lacedaemonius.
Collection: F. W. Kelsey, Rome, 1901. Accessioned 1902.

59. **KM 2068** **pl. 6**
Description: Paving or veneer, L. 19, W. 15, D. 9 cm, no preserved edges.
Lithotype: Porfido verde, l. Lacedaemonius.
Collection: F. W. Kelsey, Rome, 1901. Accessioned 1902.

60. **KM 2070** **pl. 6**
Description: Paving or veneer, L. 21.3, W. 8.9, D. 5.9 cm, no preserved edges.
Lithotype: Porfido verde, l. Lacedaemonius (but possibly a gabbro with green felspar and muscovite, not Egyptian).
Collection: F. W. Kelsey, Rome, 1901. Accessioned 1902.

61. **KM 2071** **pl. 6**
Description: Paving or veneer, L. 6.8, W. 3.9, D. 1.6 cm, no preserved edges.
Lithotype: Porfido verde, l. Lacedaemonius.
Collection: F. W. Kelsey, Astura, Italy, 1901. Accessioned 1902.

62. **KM 2069** **pl. 6**
Description: Paving or veneer, L. 13.0, W. 10.8, 2.2 cm, one preserved edge.
Lithotype: Porfido verde, l. Lacedaemonius, with unusually dark phenocrysts.
Description in Accession: "Breccia di Egitto minuto" (Orlandi).
Collection: F. W. Kelsey, Rome, 1901. Accessioned 1902.

63. **KM 6846** **pl. 6**
Description: Paving or veneer, L. 7.0, W. 8.4, D. 1.6 cm, no preserved edges.
Lithotype: Porfido verde, l. Lacedaemonius.
Collection: F. W. Kelsey, "Europe," no collection date. Accessioned 1938 (posthumous).

64. **KM 93848** **pl. 7**
Description: Paving or veneer, L. 8.2, W. 3.4, D. 0.9 cm, no preserved edges.
Lithotype: Porfido verde, l. Lacedaemonius.
Collection: G. De Criscio, Pozzuoli or vicinity, no collection date. Accessioned 1923.

65. **KM 776** **pl. 7**
Description: Paving or veneer, L. 5.3, W. 4.7, D. 0.4 cm, no preserved edges.
Lithotype: Porfido verde, l. Lacedaemonius.
Collection: C. L. Meader, "Purchased for the University by C. L. Meader in the spring of 1898," Rome. Accessioned 1901.

66. **KM 1963.5.12** **pl. 7**
Description: Opus sectile fragment, L. 4.12, W. 1.96, D. 0.86 cm, two preserved edges.
Lithotype: Porfido verde, l. Lacedaemonius.
Collection: A. E. R. Boak; gift of Mrs. Boak, "probably from Fayoum, Egypt," no collection date. Accessioned 1963.

67. **KM 29614** **pl. 7**
Description: Modern sample from ancient artifact or quarry debris? L. 6.4, W. 5.5, D. 0.9 cm, no preserved edges, bottom surface marked with heavy striations ca. 5 in 1 cm width, surface a uniform terracotta red.
Lithotype: Porfido verde, l. Lacedaemonius, color shades from green to brown.
Description in Accession: "Regular porphyry, shading to bluish-green."
Collection: J. D. Candler, no provenance, (1860s?). Accessioned 1964.
Note: Green porphyry was widely used, and around its source small pieces could easily be collected on the ground. A sample plaque could easily be fashioned for sale to Grand Tourists.

68. **KM 29615** **pl. 7**
Description: Paving or veneer. L. 7.5, W. 3.5, D. 1.7 cm, no preserved edges.
Lithotype: Porfido verde, l. Lacedaemonius.
Description in Accession: "Excellent rich green coloring; fragments and ground."
Collection: J. D. Candler, no provenance, (1860s?). Accessioned 1964.

Granito verde fiorito di bigio

Source: Wâdi Umm Balad, Eastern Desert of Egypt

69. **KM 2207** **fig. 2.6b; pl. 7**
Description: Opus sectile triangle, L. 15.5, W. 14.1, D. 2.8 cm, three preserved edges, wax on polished surface.
Lithotype: Granito verde fiorito di bigio.
Collection: F. W. Kelsey, "Purchased of Vecchio," (Italy), (1900–1901). Accessioned (1901).

Other igneous and high-value stones from other locations

70. KM 2206 **fig. 2.29a; pl. 7**
Description: Column fragment, L. 11.4, W. 9.0, D. 5.3 cm, ca. one quarter of a circumference.
Lithotype: *Porfido bigio*. From Dramont, Estérel, southern France.
Collection: F.W. Kelsey, purchased from F. Romano, Ostia, March 2, 1901. Accessioned 1901.

71. KM 2316 **pl. 7**
Description: Paving or veneer, L. 19.4, W. 11.0, D. 5.12 cm, no preserved edges.
Lithotype: Granite (epidote), not from Egypt.
Collection: F.W. Kelsey, purchased from F. Romano, (Rome), (1900–1901). Accessioned 1901.

72. KM 2330 **pl. 7**
Description: Unworked cobble, L. 16.8, W. 11.0, D. 8.3 cm, no preserved edges? Cut and polished one side.
Lithotype: Granite, not further identifiable.
Collection: F.W. Kelsey, purchased from F. Romano, (Rome), (1900–1901). Accessioned 1901.

73. KM 93786 **pl. 7**
Description: Paving or veneer? L. 5.6, W. 4.2, D. 2.1 cm, one preserved edge.
Lithotype: Granite, not further identifiable.
Collection: E. Jones, (Rome), (1885). Accessioned (1912).

74. KM 29720 **pls. 7–8**
Description: Unknown. L. 6.0, W. 4.7, D. 2.1 cm, polished and waxed one side.
Lithotype: Lapis lazuli. From Sar-i-Sang, Badakhshan, Afghanistan and Tajikistan.
Collection: E. Jones, (Rome), (1885). Accessioned (1912).

Prestige Level 9

Polychrome marbles most closely associated with imperial building: *africano, giallo antico, pavonazzetto*

Africano

Source: Sığacık (ancient Teos), western Turkey

75. KM 2083 **pl. 8**
Description: Paving or veneer, L. 14.4, W. 10.4, D. 4.8 cm, no preserved edges, partial saw mark on back.
Lithotype: *Africano, m. Luculleum.*
Collection: F.W. Kelsey, Baths of Diocletian, Rome, January 1901. Accessioned 1902.

76. KM 2080 **pl. 8**
Description: Paving or veneer, L. 17.1, W. 14.6, D. 4.1 cm, no preserved edges.
Lithotype: *Africano, m. Luculleum.*
Collection: F.W. Kelsey, Baths of Diocletian, Rome, February 8, 1901. Accessioned 1902.

77. KM 2087 **pl. 8**
Description: Paving or veneer, L. 21.0, W. 15.0, D. 2.6 cm, no preserved edges.
Lithotype: *Africano, m. Luculleum.*
Collection: F.W. Kelsey, Baths of Diocletian, Rome, February 8, 1901. Accessioned 1902.

78. KM 2092 **pl. 8**
Description: Paving or veneer, L. 19.4, W. 15.0, D. 8.6 cm, one preserved edge, wax on polished surface.
Lithotype: *Africano, m. Luculleum.*
Collection: F.W. Kelsey, Baths of Diocletian, Rome, February 8, 1901. Accessioned 1902.

79. KM 2084 **pl. 8**
Description: Paving or veneer, L. 10, W. 7, D. 1 cm, one preserved edge. Displays what the Romans prized most in *africano*: rich red markings with gray and black veins in the background.
Lithotype: *Africano, m. Luculleum.*
Collection: F.W. Kelsey, purchased from F. Romano, Baths of Diocletian, Rome, August 10, 1901. Accessioned 1902.

80. KM 2085 **pl. 8**
Description: Paving or veneer, L. 10.5, W. 9.5, D. 1.3 cm, two preserved edges.
Lithotype: *Africano, m. Luculleum.*
Collection: F.W. Kelsey, "Palestrina, Hadrian's Villa & adjoining sites" (evidently Kelsey's itinerary for the day), March 2, 1901. Accessioned 1902.

81. KM 2086 **pl. 8**
Description: Paving or veneer, L. 20, W. 14, D. 6 cm, no preserved edges.
Lithotype: *Africano, m. Luculleum.*
Collection: F.W. Kelsey, purchased from F. Romano, Baths of Diocletian, Rome, March 9 and August 10, 1901. Accessioned 1902.

82. KM 2088 **pl. 8**
Description: Paving or veneer, L. 14.3, W. 15.5, D. 3.5 cm, no preserved edges.
Lithotype: *Africano, m. Luculleum.*
Collection: F.W. Kelsey, Baths of Diocletian, Rome, January and July 1901. Accessioned 1902.

83. **KM 2089** **pl. 8**
Description: Paving or veneer, L. 25.0, D. 16.1, D. 2.77 cm, no preserved edges.
Lithotype: *Africano, m. Luculleum.*
Collection: F.W. Kelsey, purchased from F. Romano, "Al Porto Antico," Ostia, (1901). Accessioned 1902.
Note: Accession note: "Bought of F. Romano marmorista in new bldg. on part of site of Baths of Diocletian."

84. **KM 2091** **pl. 8**
Description: Paving or veneer, L. 20.7, W. 11.5, D. 4.9 cm, one preserved edge.
Lithotype: *Africano, m. Luculleum.*
Collection: F.W. Kelsey, (Italy), (1901). Accessioned 1902.

85. **KM 2094** **pl. 8**
Description: Paving or veneer, L. 15.0, W. 17.8, D. 3.6 cm, no preserved edges.
Lithotype: *Africano, m. Luculleum.*
Collection: F.W. Kelsey, purchased from F. Romano, Rome, (1901). Accessioned 1902.

86. **KM 2164** **pl. 9**
Description: Paving or veneer, D. 15.9, W. 14.2, D. 8.5, no preserved edges, wax on top much browned.
Lithotype: *Africano, m. Luculleum.*
Collection: F.W. Kelsey, "Purchased of vecchio," (Italy), (1900–1901). Accessioned (1901).

87. **KM 2049** **pl. 9**
Description: Paving or veneer, L. 15.0, W. 13.7, D. 2.9 cm, no preserved edges, pick marks on bottom.
Lithotype: *Africano, m. Luculleum.*
Collection: F.W. Kelsey, purchased from F. Romano, (Rome), October 1900. Accessioned 1902.

88. **KM 2258** **pl. 9**
Description: Paving or veneer, L. 15.3, W. 10.4, D. 2.8 cm, no preserved edges, wax on polished surface.
Lithotype: *Africano, m. Luculleum.*
Collection: F.W. Kelsey, "From Romano," Baths of Diocletian, Rome, August 10, 1901. Accessioned (1901).

89. **KM 2347** **pl. 9**
Description: Paving or veneer, L. 24.3, W. 14.6, D. 8.0 cm, no preserved edges, mortar on bottom surface.
Lithotype: *Africano, m. Luculleum.*
Collection: F.W. Kelsey, "Bought of old man," Rome, (1900–1901). Accessioned 1901.

90. **KM 2350** **pl. 9**
Description: Paving or veneer, L. 15.1, W. 6.0, D. 2.2 cm, three preserved edges.
Lithotype: *Africano, m. Luculleum.*
Collection: F.W. Kelsey, "From F. Romano," Baths of Diocletian, Rome, August 10, 1901. Accessioned (1901).

91. **KM 2382** **pl. 9**
Description: Paving or veneer, L. 11.0, W. 11.0, D. 2.3 cm, one preserved edge, marks from point chisel covered by thin iron encrustation.
Lithotype: *Africano, m. Luculleum.*
Collection: F.W. Kelsey, "Tor Paterno, Castel Fusano," Italy, 1901. Accessioned (1901).

92. **KM 2384** **pl. 9**
Description: Paving or veneer, L. 21.0, W. 12.2, D. 4.0 cm, no preserved edges, point chisel on back.
Lithotype: *Africano, m. Luculleum.*
Collection: F.W. Kelsey, Baths of Diocletian, Rome, December 1900. Accessioned (1901).

93. **KM 2385** **pl. 9**
Description: Paving or veneer, L. 13, W. 13, D. 4 cm, no preserved edges, point chisel marks on back.
Lithotype: *Africano, m. Luculleum.*
Collection: F.W. Kelsey, Baths of Diocletian, Rome, January 1901. Accessioned (1901).

94. **KM 2394** **pl. 9**
Description: Paving or veneer, L. 16.0, W. 15.0, D. 4.5 cm, two preserved edges but heavily abraded.
Lithotype: *Africano, m. Luculleum.*
Collection: F.W. Kelsey, purchased from F. Romano, "Al Porto Antico," Ostia, March 2, 1901. Accessioned (1901).
Bibliography: Gazda and Clarke 2016, 229, no. 141.

95. **KM 2397** **pl. 9**
Description: Paving or veneer, L. 23.7, W. 14.8, D. 2.8 cm, one preserved edge.
Lithotype: *Africano, m. Luculleum.*
Collection: F.W. Kelsey, purchased from F. Romano, "Al Porto Antico," Ostia, March 2, 1901. Accessioned (1901).

96. **KM 2398** **pl. 10**
Description: Paving or veneer, L. 14.0, W. 14.0, D. 3.8 cm, no preserved edges.
Lithotype: *Africano, m. Luculleum.*
Collection: F.W. Kelsey, purchased from F. Romano, (Rome), (1901). Accessioned (1901).

97. **KM 2399** **fig. 2.29b; pl. 10**
Description: Column fragment, unfluted. L. 22.4, W. 10.4, D. 8.0 cm; reconstructed diameter is ca. 117 cm; one preserved edge, modern cut and polish.
Lithotype: *Africano, m. Luculleum.*
Collection: F.W. Kelsey, purchased from F. Romano, (Rome), (1901). Accessioned (1901).

98. **KM 2400** **pl. 10**
Description: Original object not identifiable, L. 18.4, W. 6.6, D. 8.4 cm, one preserved edge, wax on polished surface.
Lithotype: *Africano, m. Luculleum.*
Description in Accession: "Marble, unidentified."
Collection: F.W. Kelsey, (Italy), (1901). Accessioned (1901).

Africano (cont.)

99. KM 2415 **pl. 10**
Description: Paving or veneer, L.34.4, W. 20.1, D. 7.6 cm, no preserved edges.
Lithotype: Africano, m. Luculleum.
Collection: F.W. Kelsey, Rome, 1901. Accessioned (1901).

100. KM 2434 **pl. 10**
Description: Paving or veneer, L. 13.2, W. 7.4, D. 2.9 cm, one preserved edge.
Lithotype: Africano, m. Luculleum.
Collection: F.W. Kelsey, Baths of Diocletian, Rome, 1901. Accessioned (1901).

101. KM 2043 **pl. 10**
Description: Paving or veneer, L. 19.2, W. 22.5, D. 4.0 cm, no preserved edges.
Lithotype: Africano, m. Luculleum. The characteristic red of *africano* is present only in small and scarce patches, so this stone could also be classified as the *bigio* of Teos.
Collection: F.W. Kelsey, "Purchased of Vecchio," "Tor Paterno," Italy, 1901. Accessioned 1902.

102. KM 2078 **pl. 10**
Description: Paving or veneer, L. 21.1, W. 7.0, D. 6.8 cm, no preserved edges.
Lithotype: Africano, m. Luculleum.
Collection: F.W. Kelsey, purchased from F. Romano, Rome, 1901. Accessioned 1902.

103. KM 2079 **pl. 10**
Description: Paving or veneer, L. 10.7, W. 9.9, D. 3.0 cm, one preserved edge. Traces of mortar on edge.
Lithotype: Africano, m. Luculleum.
Collection: F.W. Kelsey, purchased from F. Romano, Rome, August 10, 1901. Accessioned 1902.

104. KM 2081 **pl. 10**
Description: Paving or veneer, L. 11.0, W. 6.5, D. 1.0 cm, one preserved edge.
Lithotype: Africano, m. Luculleum.
Collection: F.W. Kelsey, purchased from F. Romano, Rome, 1901. Accessioned 1902.
Bibliography: Gazda and Clarke 2016, 225, no. 134.

105. KM 2082 **fig. 2.38a; pl. 10**
Description: Paving or veneer, L. 15.5, W. 9.5, D. 4.1 cm, no preserved edges, pick marks on bottom.
Lithotype: Africano, m. Luculleum.
Collection: F.W. Kelsey, "Isola Farnese near Fiumicino," Italy, 1901. Accessioned 1902.
Note: "Isola Farnese" may be a mistake for Isola Sacra.

106. KM 25567 **pl. 10**
Description: Paving or veneer, L. 4.9, W. 3.6, D. 1.9 cm, one preserved edge.
Lithotype: Africano, m. Luculleum.
Collection: G. De Criscio, "purchase 1901," Pozzuoli or vicinity. No accession date.
Note: The accession book notation "purchase 1901" is likely an error. There were four purchases of De Criscio materials, in 1898/99, 1905, 1909, and 1922.

107. KM 25591 **pl. 11**
Description: Paving or veneer, L. 7.2, W. 5.0, D. 1.8 cm, one preserved edge.
Lithotype: Africano, m. Luculleum.
Collection: F.W. Kelsey, "Hadrian's Villa and adjoining sites, Palestrina," March 2, 1901. Accessioned (1901 or 1902).
Bibliography: Gazda and Clarke 2016, 228, no. 136.

108. KM 6833 **pl. 11**
Description: Paving or veneer, L. 7.9, W. 5.2, D. 1.0 cm, no preserved edges.
Lithotype: Africano, m. Luculleum.
Collection: F.W. Kelsey, "Europe," no collection date. Accessioned 1938 (posthumous).
Bibliography: Gazda and Clarke 2016, 227, no. 132.

109. KM 6834 **fig. 2.20a; pl. 11**
Description: Dado or orthostat, L. 6.8, W. 5.6, D. 1.7 cm, one preserved edge.
Lithotype: Africano, m. Luculleum.
Collection: F.W. Kelsey, "Europe," no collection date. Accessioned 1938 (posthumous).

110. KM 93858 **fig. 2.6c; pl. 11**
Description: Opus sectile triangle, L. 15.4, W. 13.0, D. 1.14–1.48 cm, one preserved edge.
Lithotype: Africano, m. Luculleum.
Collection: G. De Criscio, Pozzuoli or vicinity, no collection date. Accessioned 1923.

111. KM 93913 **pl. 11**
Description: Paving or veneer, L. 7.7, W. 5.7, D. 2.7 cm, three preserved edges.
Lithotype: Africano, m. Luculleum.
Collection: G. De Criscio, Pozzuoli or vicinity, no collection date. Accessioned 1923.

112. KM 93790 **fig. 2.6d; pl. 11**
Description: Opus sectile triangle, L. 2.53, W. 2.32, D. 0.38 cm, equilateral triangle with one corner broken.
Lithotype: Africano, m. Luculleum.
Collection: F.W. Kelsey, Pompeii, April 1893. No accession date.
Note: Not same module as cat. no. 113.

113. KM 93791 **fig. 2.6e; pl. 11**
Description: Opus sectile triangle, L. 2.99, W. 2.51, D. 0.48 cm, two preserved edges.
Lithotype: Africano, m. Luculleum.
Collection: F.W. Kelsey, Pompeii, April 1893. No accession date.
Note: Not same module as cat. no. 112.

114. KM 6788 **pl. 11**
Description: Paving or veneer, L. 14.2, D. 8.6, D. 0.9 cm, no preserved edges.
Lithotype: Africano, m. Luculleum.
Collection: E. B. Van Deman, (Italy), no collection date. Accessioned 1938.
Bibliography: Gazda and Clarke 2016, 229, no. 140.

115. KM 6791 **pl. 11**
Description: Paving or veneer, L. 13.7, W. 11.5, D. 0.8 cm, one preserved edge, wax on polished surface.
Lithotype: Africano, m. Luculleum.
Collection: E. B. Van Deman, (Italy), no collection date. Accessioned 1938.

116. KM 2544 **pl. 11**
Description: Opus sectile element, L. 12.7, W. 3.9, D. 2.2 cm, all edges preserved, all surfaces polished except bottom.
Lithotype: Africano, m. Luculleum.
Collection: E. Jones, (Rome?), (1885). Accessioned (1912).

117. KM 1969.5.6 **pl. 11**
Description: Opus sectile element, L. 2.58, W. 2.28, D. 0.54 cm, no preserved edges.
Lithotype: Africano, m. Luculleum.
Collection: J. G. Winter, no provenance, no collection date. Accessioned 1969.

Giallo antico

Source: Chemtou (ancient Simitthus), Tunisia

118. KM 2175 **fig. 2.8a; pl. 11**
Description: Opus sectile lozenge. L. 11.3, W. 19.5, D. 3.9 cm, four preserved edges, broken edge given modern polish.
Lithotype: Giallo antico, m. Numidicum.
Collection: F. W. Kelsey, "Purchased of Vecchio," (Italy), (1900–1901). Accessioned (1901).

119. KM 2177 **fig. 2.30e; pl. 12**
Description: Column fragment, fluted, with finely cut molding, L. 36.5, W. 13.0, D. 8.0 cm, one preserved edge, broken surface given modern polish to serve as sample. Interval between flutes, 5.5 cm; depth of flutes, 3.0 cm; between filets, 9.0 cm; possibly part of a 40-foot (ca. 11.8 m) column. Reconstructed diameter is 52 cm.
Lithotype: Giallo antico, m. Numidicum.
Collection: F. W. Kelsey, purchased from F. Romano, Baths of Diocletian, Rome, March 9, 1901. Accessioned 1901.

120. KM 2179 **fig. 2.8b; pl. 12**
Description: Opus sectile lozenge, L. 21.3 cm overall, W. 14.4, D. 3.4 cm, each edge ca. 15 cm allowing for damage, three preserved edges, mortar on bottom.
Lithotype: Giallo antico, m. Numidicum.
Description in Accession: "Macchia di giallo brecciato" (Orlandi).
Collection: F. W. Kelsey, Baths of Diocletian, Rome, January 1901. Accessioned 1901.

121. KM 2181 **fig. 2.30a; pl. 12**
Description: Column fragment, fluted, L. 8.5, W. 11.3, D. 6.3 cm, one flute and one interval, no preserved edges.
Lithotype: Giallo antico, m. Numidicum.
Collection: F. W. Kelsey, "found in front of 38 Via Firenze," Rome, January 21, 1901. Accessioned 1901.

122. KM 2256 **fig. 2.30b; pl. 12**
Description: Column fragment reprocessed as veneer or paving, L. 23.0, W. 11.3, D. 4.2 cm, no preserved edges, mortar on back; reconstructed diameter is 83 cm (see ch. 2).
Lithotype: Giallo antico, m. Numidicum.
Collection: F. W. Kelsey, "street in front of 38 Via Firenze," Rome, January 21, 1901 ("An excavation for laying electric light brought these to light"). Accessioned (1901).

123. KM 2182 **fig. 2.30f; pl. 12**
Description: Column fragment, L. 22.2, W. 10.8, D. 6.8 cm, one flute with two moldings, one preserved edge; mortar on surface of flute suggests reuse as aggregate.
Lithotype: Giallo antico, m. Numidicum.
Collection: F. W. Kelsey, "found in front of 38 Via Firenze," Rome, January 23, 1901. Accessioned 1901.

124. KM 2183 **figs. 2.31a, 2.37a; pls. 12–13**
Description: Fragment of a *columna rudentata*, L. 15.5, W. 17.0, D. 6.8 cm, parts of two flutes with roping; reconstructed diameter is 44 cm (see ch. 2). One preserved edge, polished modern cut surface (originally a broken interior surface).
Lithotype: Giallo antico, m. Numidicum.
Collection: F. W. Kelsey, "found in front of 38 Via Firenze," Rome, January 23, 1901. Accessioned 1901.

125. KM 2185 **pl. 13**
Description: Paving or veneer, L. 13.0, W. 12.8, D. 2.5 cm, one preserved edge, saw mark preserved on edge.
Lithotype: Giallo antico, m. Numidicum?
Collection: F. W. Kelsey, "found in front of 38 Via Firenze," Rome, January 23, 1901. Accessioned 1901.
Bibliography: Gazda and Clarke 2016, 226, no. 126.

126. KM 2186 **fig. 2.31b; pl. 13**
Description: Fragment of one flute of a *columna rudentata*, L. 22.5, W. 10.6, D. 7.4 cm; one preserved edge, cement in grooves of flute. Reconstructed diameter is 65.5 cm (see ch. 2).
Lithotype: Giallo antico, m. Numidicum.
Collection: F. W. Kelsey, "found in front of 38 Via Firenze," Rome, January 23, 1901. Accessioned 1901.

127. KM 2184 **pl. 13**
Description: Paving or veneer, L. 12.5, W. 10.9, D. 2.6 cm but varies, no preserved edges.
Lithotype: Giallo antico, m. Numidicum.
Collection: F. W. Kelsey, "ad Porto" (presumably Portus but possibly Ostia), no collection date. Accessioned 1901.

Giallo antico (cont.)

128. KM 2188 **pl. 13**
Description: Paving or veneer, L. 16.5, W. 16.0, D. 6.0 cm, no preserved edges, preserved mortar encrustations, one side sawn and given modern polish.
Lithotype: Giallo antico, m. Numidicum.
Collection: F.W. Kelsey, Baths of Diocletian, Rome, December 1, 1900. Accessioned 1901.
Bibliography: Gazda and Clarke 2016, 225, no. 121.

129. KM 2189 **fig. 2.39b; pl. 13**
Description: Paving or veneer, L. 12.5, W. 14.0, D. 3.8 cm, two preserved edges, point chisel marks on back.
Lithotype: Giallo antico, m. Numidicum.
Collection: F.W. Kelsey, Baths of Diocletian, Rome, February 1, 1901. Accessioned 1901.

130. KM 2190 **pl. 13**
Description: Paving or veneer, L. 12.7, W. 16.5, D. 2.5 cm, one preserved edge with saw marks.
Lithotype: Giallo antico, m. Numidicum? Visually, it could be also *giallo di Siena*, but *giallo antico* is strongly supported by the piece's provenance from the Baths of Diocletian.
Description in Accession: "Giallo canario, Giallo di siena moderna."
Collection: F.W. Kelsey, Baths of Diocletian, Rome, February 8, 1901. Accessioned 1901.

131. KM 2192 **pl. 13**
Description: Paving or veneer, L. 18.6, W. 10.5, D. 2.6 cm, no preserved edges.
Lithotype: Giallo antico, m. Numidicum.
Collection: F.W. Kelsey, (Rome), no collection date. Accessioned 1901.
Bibliography: Gazda and Clarke 2016, 225, no. 123.

132. KM 2193 **fig. 2.20b; pl. 13**
Description: Dado or orthostat with incised molding, L. 6.9, W. 10.5, D. 2.9 cm, two preserved edges.
Lithotype: Giallo antico, m. Numidicum.
Collection: F.W. Kelsey, no provenance, no collection date. Accessioned 1901.

133. KM 2253 **pl. 14**
Description: Paving or veneer, L. 9.0, W. 7.5, D. 1.0 cm, two preserved edges.
Lithotype: Giallo antico, m. Numidicum.
Collection: F.W. Kelsey, purchased from F. Romano, (Rome), (1900–1901). Accessioned 1901.

134. KM 2353 **pl. 14**
Description: Paving or veneer (paving more likely than veneer because of thickness), L. 20.0, W. 20.0, D. 8.9 cm, no preserved edges.
Lithotype: Giallo antico, m. Numidicum.
Description in Accession: "Marmor macchiato."
Collection: F.W. Kelsey, no provenance, (1900–1901). Accessioned (1901).
Bibliography: Gazda and Clarke 2016, 227, no. 128.

135. KM T2008.62 **pl. 14**
Description: Paving or veneer, L. 8.3, W. 5.7, D. 1.3 cm, no preserved edges.
Lithotype: Giallo antico, m. Numidicum.
Collection: (F.W. Kelsey, purchased from Père R. P. Delattre, Carthage, Tunisia, March 2, 1893; originally from Cap de Garde, Algeria? See note below.) Accessioned 2008.
Note: Cat. nos. 135, 185, 295, 352, 520, and 530 were apparently discovered in 2008 when the museum's collections were moved from their old storage location in Newberry Hall to the basement of the newly constructed Upjohn Wing. The pieces were given temporary accession numbers and all pieces are marked with masking tape on the bottom surface. T2008.58 (cat. no. 295) reads "Cap du Garde 2," while the others carry numbers from 3 to 6 without further indications. These probably come from Kelsey's 1893 trip to Carthage, where Butler (1930, 1) noted that Père R. P. Delattre "helped him toward a realization of his plans by selling him 109 duplicate specimens of lamps, vases, and building materials from his own Museum."
Bibliography: Gazda and Clarke 2016, 228, no. 133.

136. KM 25573 **pl. 14**
Description: Paving or veneer, L. 10.8, W. 8.7, D. 1.5 cm, no preserved edges.
Lithotype: Giallo antico, m. Numidicum.
Collection: F.W. Kelsey, purchased from Père Delattre, Carthage, Tunisia, March 2, 1893. Accessioned 1893.
Bibliography: Gazda and Clarke 2016, 229, no. 139.

137. KM 2314 **pl. 14**
Description: Paving or veneer, L. 13.3, W. 7.8, D. 13.1 cm, no preserved edges.
Lithotype: Giallo antico, m. Numidicum.
Description in Accession: "Breccia tigrata."
Collection: F.W. Kelsey, Rome, (1900–1901). Accessioned 1901.

138. KM 2333 **fig. 2.39a; pl. 14**
Description: Paving or veneer, L. 16.5, W. 29.5, D. 5.1 cm, no preserved edges, point chisel marks on back.
Lithotype: Giallo antico, m. Numidicum.
Description in Accession: "Giallo brecciato."
Collection: Kelsey, no provenance, (1900–1901). Accessioned 1901.

139. KM 6864 **pl. 14**
Description: Paving or veneer, L. 10.2, W. 7.3, D. 1.5 cm, two preserved edges.
Lithotype: Giallo antico, m. Numidicum.
Description in Accession: "Marble unidentified"; "Rosastro, or heated giallo antico."
Collection: F.W. Kelsey, "Europe," no collection date. Accessioned 1938 (posthumous).

140. KM 6829 **pl. 14**
Description: Paving or veneer, L. 10.2, W. 7.3, D. 1.5 cm, two preserved edges.
Lithotype: *Giallo antico, m. Numidicum.*
Collection: F. W. Kelsey, "Europe," no collection date. Accessioned 1938 (posthumous).

141. KM 7109 **pl. 14**
Description: Paving or veneer strip, L. 12.0, W. 5.8, D. 2.7 cm, all edges preserved.
Lithotype: *Giallo antico, m. Numidicum.*
Collection: F. W. Kelsey, Ostia, no collection date. Accessioned 1924.
Note: According to the accession book at KM 7109, this is one of "three fgts marbles . . . found in desk of F. W. K." The other fragments are cat. nos. 252 and 586; all are from Ostia. "1924" is given in the "When Collected" column of the accession book, but this is when the fragments were found in Kelsey's desk, not when he collected them. The three artifacts are of different dimensions and marbles but share notably beveled and finished edges. As they are quite short for original strips intended for borders or dividers in a composition, it seems likely that they had been specially shaped for insertion as repair elements. That they were found as the only(?) marble pieces in Kelsey's desk makes it tempting to speculate that he had them ready to make a point to his students about repair techniques.
Bibliography: Gazda and Clarke 2016, 228, no. 137.

142. KM 93543 **fig. 2.12a; pl. 14**
Description: *Opus sectile* figural element, L. 8.37, W. 4.6, D. 0.82 cm, three (of four?) preserved edges of a leaf(?) with incised edge lines and vegetal stalk.
Lithotype: *Giallo antico, m. Numidicum.*
Collection: J. D. Candler, (Rome), 1860–1890. Accessioned 1964.

143. KM 93544 **fig. 2.12b; pl. 14**
Description: *Opus sectile* figural element, L. 9.82, W. 4.62, D. 4.2–8.3 cm, two (of three?) preserved edges, scalloped edge, incised lines of vegetal design.
Lithotype: *Giallo antico, m. Numidicum.*
Collection: J. D. Candler, no provenance, 1860–1890. Accessioned 1964.

144. KM 93553 **fig. 2.12c; pl. 14**
Description: *Opus sectile* figural element, L. 6.65, W. 3.91, D. 0.97 cm, two preserved edges, incised edge lines and vegetal stalk.
Lithotype: *Giallo antico, m. Numidicum.*
Collection: J. D. Candler, no provenance, 1860–1890. Accessioned 1964.

145. KM 93554 **fig. 2.12g; pl. 15**
Description: *Opus sectile* figural element, L. 2.78, W. 1.6, D. 0.42 cm, four preserved edges, wedge-shaped part of a circle of extrapolated diameter 5 cm.
Lithotype: *Giallo antico, m. Numidicum.*
Collection: J. D. Candler, no provenance, 1860–1890. Accessioned 1964.

146. KM 93555 **fig. 2.12d; pl. 15**
Description: *Opus sectile* figural element, L. 4.8, W. 5.58, D. 0.56 cm, three (of four) preserved edges.
Lithotype: *Giallo antico, m. Numidicum.*
Collection: J. D. Candler, no provenance, 1860–1890. Accessioned 1964.

147. KM 93556 **fig. 2.12e; pl. 15**
Description: *Opus sectile* figural element, L. 6.20, W. 4.05, D. 0.51 cm, two preserved edges, incised border and curving vegetal stalk.
Lithotype: *Giallo antico, m. Numidicum.*
Collection: J. D. Candler, no provenance, 1860–1890. Accessioned 1964.

148. KM 93552 **fig. 2.12f; pl. 15**
Description: *Opus sectile* rectangular element, L. 5.56, W. 0.96, D. 0.89 cm, three preserved edges.
Lithotype: *Giallo antico, m. Numidicum.*
Collection: J. D. Candler, no provenance, 1860–1890. Accessioned 1964.

149. KM 93557 **pl. 15**
Description: Paving or veneer fragment, L. 4.58, W. 1.86, D. 0.97 cm.
Lithotype: *Giallo antico, m. Numidicum.*
Collection: J. D. Candler, no provenance, 1860–1890. Accessioned 1964.

150. KM 93558 **fig. 2.12h; pl. 15**
Description: *Opus sectile* rectangular element, L. 5.56, W. 0.96, D. 0.89 cm, three preserved edges.
Lithotype: *Giallo antico, m. Numidicum.*
Collection: J. D. Candler, no provenance, 1860–1890. Accessioned 1964.

151. KM 2552 **pl. 15**
Description: Paving or veneer, L. 9.0, W. 8.0, D. 1.5 cm, one preserved edge (modern shaping).
Lithotype: *Giallo antico, m. Numidicum.*
Collection: E. Jones, Aventine Hill, Rome, (1885). Accessioned (1912).
Bibliography: Gazda and Clarke 2016, 228, no. 135.

152. KM 2556 **fig. 2.6f; pl. 15**
Description: *Opus sectile* triangle, L. 10.4, W. 9.0, D. 1.2 cm, three preserved edges.
Lithotype: *Giallo antico, m. Numidicum.*
Collection: E. Jones, "Tablinum of House of Vestals," Forum, Rome, (1885). Accessioned (1912).

153. KM 2558 **fig. 2.9a; pl. 15**
Description: *Opus sectile* element, L. 6.9, W. 6.9, D. 0.9 cm, complete quarter circle, three preserved edges.
Lithotype: *Giallo antico, m. Numidicum.*
Collection: E. Jones, (Rome), (1885). Accessioned (1912).

Giallo antico (cont.)

154. KM 29729 **fig. 2.28c; pl. 15**
Description: Cornice molding, L. 8.7, W. 3.1, D. 2.8 cm, two preserved edges.
Lithotype: Giallo antico, m. Numidicum.
Collection: G. De Criscio, Pozzuoli or vicinity, 1922. Accessioned 1923.

155. KM 6729 **pl. 15**
Description: Architectural element, L. 18.5, W. 8.0, D. 3.8 cm, one preserved edge. Probably the first slice from a block as the exterior surface (bottom of the extant piece) is very roughly picked. Surfaces given modern polish. The rosy tint probably results from exposure to fire.
Lithotype: Giallo antico, m. Numidicum.
Collection: E. B. Van Deman, "A. b. Concreta of Domitian" (it is not clear what structure she intended by this), (Italy), no collection date. Accessioned 1938.

156. KM 6730 **fig. 2.30g; pl. 16**
Description: Column fragment with deep flutes and fine moldings, D. 11.7, W. 10.7, D. 2.4 cm; reconstructed diameter is 43 cm (see ch. 2).
Lithotype: Giallo antico, m. Numidicum.
Collection: E. B. Van Deman, (Italy), no collection date. Accessioned 1938.

157. KM 6772 **fig. 2.9c; pl. 16**
Description: Opus sectile circle, L. 23.0, W. 12.6, D. 1.2 cm, two pieces glued together, one preserved edge, wax on polished surface.
Lithotype: Giallo antico that has been in a hot fire, shifting the color toward red.
Description in Accession: "Giallo brecciato brucrato(?) [bruciato?]"; Ward-Perkins: "Giallo antico."
Collection: E. B. Van Deman, (Italy), no collection date. Accessioned 1938.
Note: For Ward-Perkins at the University of Michigan, see note at cat. no. 409.

158. KM 6813 **pl. 16**
Description: Paving or veneer Internal fragment or piece of quarry debris, L. 12.0, W. 6.9, D. 2.6 cm, no preserved edges or original surfaces.
Lithotype: Giallo antico, m. Numidicum.
Collection: E. B. Van Deman, (Italy), no collection date. Accessioned 1938.

159. KM 775 **pl. 16**
Description: Paving or veneer, L. 4.6, W. 4.3, D. 0.4 cm, one preserved edge.
Lithotype: Giallo antico, m. Numidicum.
Collection: C. L. Meader, Rome, spring of 1898. Accessioned 1898.
Note: Accession note: "Purchased for the University by C. L. Meader in the spring of 1898," when he also bought lamps from dealers in Rome.

160. KM 1969.5.3a–c **pl. 16**
Description: Paving or veneer, three of four pieces inventoried together:
a: L. 6.90, W. 4.00, D. 1.28 cm;
b: L. 5.05, W. 4.10, D. 1.29 cm;
c: L. 4.76, W. 4.04, D. 1.29 cm.
Lithotype: Giallo antico, m. Numidicum with unusual pink cast.
Collection: J. G. Winter, no provenance, no collection date. Accessioned 1969.
Note: For the fourth piece, labeled d, see cat. no. 592.

Pavonazzetto

Sources: village of İscehisar (ancient city of Dokimeion or Docimium) and Upper Tembris Valley, Turkey

161. KM 2060 **pl. 16**
Description: Paving or veneer, L. 12.8, 12.5, D. 4.5 cm, no preserved edges, pick dressed on bottom surface.
Lithotype: Pavonazzetto, m. Phrygium.
Collection: F. W. Kelsey, "Isola Farnese near Fiumicino," Italy, 1901. Accessioned 1902.
Note: "Isola Farnese" may be a mistake for Isola Sacra.
Bibliography: Gazda and Clarke 2016, 228, no. 138.

162. KM 2095 **pl. 16**
Description: Paving or veneer, L. 21.0, W. 13.5, D. 5.5 cm, two preserved edges, small point chisel on back, small point chisel with claw chisel on edges.
Lithotype: Pavonazzetto, m. Phrygium.
Collection: F. W. Kelsey, Rome, 1901. Accessioned 1902.

163. KM 2104 **pl. 16**
Description: Paving or veneer, L. 18.0, W. 10.9, 3.83 cm, one preserved edge, ancient edge has end of saw-cut ridge, lower surface broken and not original.
Lithotype: Pavonazzetto, m. Phrygium.
Collection: F. W. Kelsey, "purchased of old man," Rome, March 13, 1901. Accessioned 1901.

164. KM 2229 **pl. 16**
Description: Paving or veneer, L. 17.7, W. 13.7, D. 1.8 cm, two pieces glued together, mortar on bottom surface, no preserved edges.
Lithotype: Pavonazzetto, m. Phrygium.
Collection: F. W. Kelsey, purchased from F. Romano (who said he found it at the Forum), Rome, (1900–1901). Accessioned (1901).
Note: Accession note: "Forum, F. Romano said he found it there."

165. KM 2231 **pl. 17**
Description: Paving or veneer, L. 24.0, W. 16.0, D. 3.8 cm, one preserved edge.
Lithotype: Pavonazzetto, m. Phrygium.
Collection: F. W. Kelsey, "purchased of vecchio," (Italy), (1900–1901). Accessioned (1901).

166. KM 2232 **pl. 17**
Description: Paving or veneer, L. 15.3, W. 12.1, D. 4.1 cm, no preserved edges, wax on polished surface.
Lithotype: *Pavonazzetto, m. Phrygium.*
Collection: F.W. Kelsey, (Italy), (1900–1901). Accessioned (1901).

167. KM 2233 **pl. 17**
Description: Paving or veneer, L. 33.5, W. 9.0, D. 2.6 cm, no preserved edges, mortar on bottom surface.
Lithotype: *Pavonazzetto, m. Phrygium.*
Collection: F.W. Kelsey, purchased from F. Romano, (Rome), February 1901. Accessioned 1901.

168. KM 2235 **pl. 17**
Description: Paving or veneer, L. 17.0, W. 15.0, D. 1.7 cm, one preserved edge, wax on polished surface.
Lithotype: *Pavonazzetto, m. Phrygium.*
Collection: F.W. Kelsey, (Italy), (1900–1901). Accessioned (1901).

169. KM 2236 **pl. 17**
Description: Paving or veneer, L. 32.0, W. 11.5, D. 3.1 cm, one preserved edge, saw marks, point chisel marks on edge.
Lithotype: *Pavonazzetto, m. Phrygium.*
Collection: F.W. Kelsey, (Italy), (1900–1901). Accessioned (1901).

170. KM 2366 **pl. 17**
Description: Paving or veneer, L. 15.0, W. 10.0, D. 4.1 cm, no preserved edges, pick marks on bottom surface.
Lithotype: *Pavonazzetto, m. Phrygium.*
Collection: F.W. Kelsey, Rome, 1901. Accessioned (1901).

171. KM 2367 **pl. 17**
Description: Paving or veneer, L. 16.6, W. 11.9, D. 2.6 cm, no preserved edges, pick marks on bottom surface, wax on polished surface.
Lithotype: *Pavonazzetto, m. Phrygium.*
Collection: F.W. Kelsey, Rome, 1901. Accessioned (1901).

172. KM 2359 **pl. 17**
Description: Paving or veneer, L. 16.0, W. 6.5, D. 5.3 cm, three preserved edges, pick marks on bottom surface, piece perhaps recut in modern period. Labeled "TEAH," significance unknown.
Lithotype: *Pavonazzetto, m. Phrygium.*
Collection: F.W. Kelsey, Rome, 1893. Accessioned 1901.

173. KM 2406 **pl. 17**
Description: Paving or veneer, L. 9.5, W. 4.5, D. 2.0 cm, no preserved edges.
Lithotype: *Pavonazzetto, m. Phrygium.*
Collection: F.W. Kelsey, (Rome?), (1900–1901). Accessioned (1901).
Bibliography: Gazda and Clarke 2016, 227, no. 129.

174. KM 2326 **pl. 17**
Description: Paving or veneer, L. 16.1, W. 11.4, D. 4.8 cm, three preserved edges.
Lithotype: *Pavonazzetto, m. Phrygium,* or possibly *breccia di Settebasi.*
Collection: F.W. Kelsey, "Tor Paterno, Castel Fusano," Italy, 1900–1901. Accessioned 1901.

175. KM 2151 **pl. 17**
Description: Paving or veneer, L. 13.0, W. 11.5, D. 2.0 cm, one preserved edge.
Lithotype: *Pavonazzetto, m. Phrygium.*
Collection: F.W. Kelsey, purchased from F. Romano, (Rome), 1900–1901. Accessioned 1901.

176. KM 25801 **fig. 2.18a; pl. 18**
Description: Pilaster fragment? L. 13.3, W. 3.2, D. 2.1 cm, no preserved edges.
Lithotype: *Pavonazzetto, m. Phrygium.*
Collection: F.W. Kelsey, Astura, Italy, March 2, 1901. Accessioned (1901).
Note: This is one of ten architectural fragments originally accessioned together as KM 2444. At some point the lot was broken up and the pieces assigned new KM numbers: KM 25594 (cat. no. 585), KM 25595 (not in catalogue), KM 25596 (cat. no. 509), KM 25597 (not in catalogue), KM 25598 (cat. no. 540), KM 25599 (not in catalogue), KM 25802 (not in catalogue), and KM 93792 (not in catalogue).

177. KM 1845 **pl. 18**
Description: Paving or veneer, L. 26.7, W. 23.0, D. 2.4 cm, one preserved edge.
Lithotype: *Pavonazzetto, m. Phrygium.*
Collection: F.W. Kelsey, "Summer palace of Herod at Jericho," January 1920. Accessioned 1921.
Note: Density of honey color suggests origin from Bacakale II, the new quarry east of Bacakale opened in the second century CE (Fant 1989, 46–47; see also cat. no. 182).

178. KM 6830 **fig. 2.18b; pl. 18**
Description: Pilaster fragment, L. 10.6, W. 6.0, D. 1.4 cm, two preserved edges.
Lithotype: *Pavonazzetto, m. Numidicum.*
Collection: F.W. Kelsey, "Europe," no collection date. Accessioned 1938 (posthumous).
Bibliography: Gazda and Clarke 2016, 227, no. 131.

179. KM 93884 **pl. 18**
Description: Paving or veneer, L. 12.0, W. 10.0, D. 1.5 cm, no preserved edges.
Lithotype: *Pavonazzetto, m. Phrygium.*
Collection: F.W. Kelsey (posthumous), "Material brought back from the U-M expedition to Antioch in Pisidia," directed by D. M. Robinson, 1924. Accession date unknown.

180. KM 29636 **pl. 18**
Description: Fragmentary *opus sectile* element, L. 5.0, W. 4.8, D. 0.9 cm, three preserved edges, two edges are slightly

Pavonazzetto (cont.)

beveled to aid insertion. Two of the sides meet at 90 degrees, the other angle is 45 degrees.
Lithotype: Pavonazzetto, m. Phrygium.
Description in Accession: "Breccia bianca e amethysto."
Collection: J. D. Candler, no provenance, 1860–1870. Accessioned 1964.

181. KM 29652 **pl. 18**
Description: Paving, L. 5.9, W. 5.7, D. 1.9 cm, no preserved edges.
Lithotype: Pavonazzetto, m. Phrygium?
Collection: J. D. Candler, "Portion of the floor of St. Peter's, Rome," "1867?" (see also cat. no. 571). Accessioned 1964.

182. KM 29635 **pl. 18**
Description: Paving or veneer, L. 7.2, W. 4.1, D. 1.3 cm, no preserved edges.
Lithotype: Pavonazzetto, m. Phrygium?
Description in Accession: "Pink w/ violet markings."
Collection: J. D. Candler, "From the Odeion of Ephesus," Turkey, (September 1867; cf. cat. no. 291). Accessioned 1964.
Note: This marble's coloration points to Bacakale II, opened to quarrying in the second century CE (Fant 1989, 46–47), and so the piece is likely to come from a building of that date (see also cat. no. 177).

183. KM 6783 **fig. 2.20c; pl. 18**
Description: Dado or orthostat with molding, L. 12.5, W. 16.0, D. 0.5 cm, one preserved edge.
Lithotype: Probably *pavonazzetto toscano* from the Apuan Alps (Istituto commercio estero 1982, 28, fig. 24).
Description in Accession: "Macchia bianca di Seravezza"
Collection: E. B. Van Deman, (Italy), no collection date. Accessioned 1938.
Note: The accession note "Macchia bianca di Seravezza" probably goes back to Van Deman, and so the piece should be considered Italian.

Prestige Level 8

Cipollino mandolato, cipollino verde and *grigio, portasanta, diaspro* (jasper)

Cipollino mandolato

Source: Haute Pyrenees and Campan, France

184. KM 2131 **pl. 18**
Description: Paving or veneer, L. 29.9, W. 12.0, D. 2.3 cm, one preserved edge, traces of mortar on one edge.
Lithotype: Cipollino mandolato, cipollino dei Pirenei.
Description in Accession: "mandolato."
Collection: F. W. Kelsey, purchased from F. Romano, (Rome), (1901). Accessioned 1901.

Cipollino verde, cipollino grigio

Source: Karystos and Styra, Euboea, Greece

185. KM T2008.59 **pl. 18**
Description: Paving or veneer, L. 6.3, W. 3.6, D. 0.9 cm, one preserved edge.
Lithotype: Cipollino verde, m. Carystium.
Collection: (F. W. Kelsey, purchased from Père R. P. Delattre, Carthage, Tunisia, March 2, 1893). Accessioned 2008.
Note: This piece may originally have come from Cap de Garde, Algeria. See note at cat. no. 135.

186. KM 2127 **pl. 18**
Description: Paving or veneer, L. 27.6, W. 14.5, D. 2.4 cm, no preserved edges.
Lithotype: Cipollino verde, m. Carystium.
Collection: F. W. Kelsey, purchased from F. Romano, (Rome), (1901). Accessioned 1901.

187. KM 2128 **pl. 19**
Description: Paving or veneer, L. 19.7, W. 13.6, D. 2.2 cm, no preserved edges, wax on polished surface.
Lithotype: Cipollino verde, m. Carystium.
Collection: F. W. Kelsey, "Tor Paterno, Castel Fusano," Italy, (1901). Accessioned 1901.

188. KM 2129 **pl. 19**
Description: Paving or veneer, L. 22.7, W. 11.7, D. 5.5 cm, one preserved edge, mortar on sides and back.
Lithotype: Cipollino verde, m. Carystium.
Collection: F. W. Kelsey, "found [...] in the street in front of 38 Via Firenze," Rome, January 4 and 21, 1901. Accessioned 1901.

189. KM 2130 **pl. 19**
Description: Paving or veneer, L. 19.1, W. 18.9, D. 3.4 cm, no preserved edges.
Lithotype: Cipollino verde, m. Carystium.
Collection: F. W. Kelsey, (Rome?), (1901). Accessioned 1901.
Note: A collection date in 1901 is assumed because the series number, whether from a marble preparer in Rome or a Kelsey Museum curator, is consistent with others of that year.

190. KM 2133 **pl. 19**
Description: Paving or veneer, L. 25.7, W. 8.8, D. 6.5 cm, one preserved edge, pick marks on edge, wax on polished surface.
Lithotype: Cipollino verde, m. Carystium.

Collection: F.W. Kelsey, Baths of Diocletian, Rome, February 1, 1901. Accessioned 1901.

191. KM 2336 **pl. 19**
Description: Paving or veneer, L. 22.3, W. 11.2, D. 3.4 cm, no preserved edges, polished surface.
Lithotype: Cipollino verde, m. Carystium.
Collection: F.W. Kelsey, no provenance, (1900–1901). Accessioned 1901.

192. KM 2358 **pl. 19**
Description: Paving or veneer, L. 18.4, W. 12.8, D. 4.1 cm, no preserved edges.
Lithotype: Cipollino verde, m. Carystium.
Collection: F.W. Kelsey, Rome, 1901. Accessioned (1901).

193. KM 2319 **pl. 19**
Description: Quarry debris, L. 13.4, W. 15.0, D. 3.8 cm, no worked surfaces.
Lithotype: Cipollino verde, m. Carystium.
Collection: F.W. Kelsey from Arthur Hill, "new quarry, north side," location not specified (Euboea?), no collection date. Accessioned 1901.
Note: Arthur Hill apparently owned quarries on Euboea in addition to the one attested on Mt. Pentelikon; see cat. no. 500).

194. KM 2442 **pl. 19**
Description: Paving or veneer, L. 8.0, W. 5.9, D. 3.0 cm, no preserved edges.
Lithotype: Cipollino verde, m. Carystium.
Collection: F.W. Kelsey, "Greece," (1901). No accession date.

195. KM 25592 **fig. 2.18c; pl. 19**
Description: Pilaster fragment, L. 11.5, W. 4.6, D. 3.8 cm, one preserved edge.
Lithotype: Cipollino verde, m. Carystium.
Collection: F.W. Kelsey, "Hadrian's Villa and adjoining sites, Palestrina," March 2, 1901. Accessioned (1901 or 1902).

196. KM 2446 **pl. 20**
Description: Paving or veneer, L. 6.7, W. 7.4, D. 2.63 cm, no preserved edges, modern polish.
Lithotype: Cipollino grigio? Gray is as common on Euboea as the green variety but was mostly avoided by Roman quarrymen; many other white marbles with gray streaks, like Hymettus, also possible. See Lithotypes s.v. *cipollino verde*.
Collection: F.W. Kelsey, Athens, "Mar. 2 1901." Accessioned (1901).
Note: The collection date as recorded in the accession book is suspect since Kelsey's party did not land at Patras to begin their Greek tour until March 19 (Pedley 2012, 67).

197. KM 2077 **pl. 20**
Description: Paving or veneer, L. 12.2, W. 16.8, D. 2.0 cm, no preserved edges.
Lithotype: Cipollino grigio? Other white marbles with gray streaks, like Hymettus, also possible. See Lithotypes s.v. *Cipollino verde*.
Description in Accession: "africano verde brecciato."
Collection: F.W. Kelsey, Rome, March 2, 1901. Accessioned 1902.

198. KM 25589 **pl. 20**
Description: Paving or veneer, L. 8.0, W. 6.1, D. 2.4 cm, no preserved edges.
Lithotype: Cipollino verde, m. Carystium.
Collection: F.W. Kelsey, "Hadrian's Villa and adjoining sites, Palestrina," March 2, 1901. Accessioned (1901 or 1902).

199. KM 6737 **pl. 20**
Description: Paving or veneer, L. 25.3, W. 13.6, D. 1.5 cm, one preserved edge, wax on polished surface.
Lithotype: Cipollino verde, m. Carystium.
Collection: E. B. Van Deman, (Italy), no collection date. Accessioned 1938.

200. KM 6744 **pl. 20**
Description: Paving or veneer, L. 18.3, W. 15.4, D. 1.9 cm, one preserved edge.
Lithotype: Cipollino verde, m. Carystium marino.
Description in Accession: "bigio listato."
Collection: E. B. Van Deman, (Italy), no collection date. Accessioned 1938.
Note: Borghini and Gnoli (1989, 206) attribute *cipollino marino* to "Italia," but see cat. no. 196.

201. KM 93565 **pl. 20**
Description: Thin paving or veneer or *opus sectile*, L. 4.97, W. 2.66, D. 0.78 cm, no preserved edges.
Lithotype: Cipollino verde, m. Carystium.
Collection: J. D. Candler, no provenance, 1860–1890. Accessioned 1964.

202. KM 93847 **pl. 20**
Description: Paving or veneer, fragment of an original square, L. 11.17, W. 11.50, D. 7.20 cm, four preserved edges.
Lithotype: Cipollino verde, m. Carystium.
Collection: G. De Criscio, Pozzuoli or vicinity, no collection date. Accessioned 1923.

203. KM 93787 **pl. 20**
Description: Paving or veneer, L. 6.4, W. 3.6, D. 1.9 cm, no preserved edges.
Lithotype: Cipollino verde, m. Carystium.
Collection: E. Jones, (Rome), (1885). Accessioned (1912).

Portasanta

Source: Latomi, island of Chios, Greece

204. KM 2058 **pl. 20**
Description: Paving or veneer, L. 28.4, W. 12.9, D. 3.2 cm, no preserved edges, mortar on bottom.
Lithotype: Portasanta, m. Chium.

Portasanta (cont.)

Collection: F.W. Kelsey, Baths of Diocletian, Rome, "Oct. 1901" (a mistake for October 1900, since Kelsey was already back in Ann Arbor in October 1901). Accessioned 1902.

205. KM 2048 **pl. 20**
Description: Paving or veneer, L. 18.0, W. 12.5, D. 2.0 cm, no preserved edges.
Lithotype: *Portasanta, m. Chium.*
Collection: F.W. Kelsey, Baths of Diocletian, Rome, October 1900. Accessioned 1902.

206. KM 2050 **pl. 20**
Description: Paving or veneer, L. 21.4, W. 13.5, D. 2.0 cm, no preserved edges.
Lithotype: *Portasanta, m. Chium.*
Collection: F.W. Kelsey, Baths of Diocletian, Rome, January 1901. Accessioned 1902.
Bibliography: Gazda and Clarke 2016, 227, no. 130.

207. KM 2029 **pl. 20**
Description: Paving or veneer, L. 20.5, W. 10.7, D. 2.7 cm, no preserved edges.
Lithotype: *Portasanta, m. Chium.*
Description in Accession: "Diaspro di Sicilia"; "Porta santa pavonazza."
Collection: F.W. Kelsey, Baths of Diocletian, Rome, January 1901. Accessioned 1902.

208. KM 2053 **pl. 21**
Description: Paving or veneer, L. 14.4, W. 14.0, D. 3.2 cm, one preserved edge, pick marks on bottom, saw marks on edge.
Lithotype: *Portasanta, m. Chium.*
Collection: F.W. Kelsey, Baths of Diocletian, Rome, January 1901. Accessioned 1902.

209. KM 2054 **fig. 2.39c; pl. 21**
Description: Paving or veneer, L. 15.5, W. 13.6, D. 4.8 cm, one preserved edge, pick marks, point chisel marks, and traces of mortar on bottom.
Lithotype: *Portasanta, m. Chium.*
Collection: F.W. Kelsey, Baths of Diocletian, Rome, February 1901. Accessioned 1902.

210. KM 2055 **pl. 21**
Description: Paving or veneer, L. 16.7, W. 15.6, D. 4.6 cm, no preserved edges, pick marks on bottom.
Lithotype: *Portasanta, m. Chium.*
Collection: F.W. Kelsey, Baths of Diocletian, Rome, February 1901. Accessioned 1902.

211. KM 2061 **pl. 21**
Description: Paving or veneer, L. 13.5, W. 12.5, D. 3.0 cm, no preserved edges.
Lithotype: *Portasanta, m. Chium.*
Collection: F.W. Kelsey, Baths of Diocletian, Rome, February 1, 1901. Accessioned 1902.

212. KM 2174 **pl. 21**
Description: Paving or veneer, L. 14.4, W. 10.0, D. 1.9 cm, no preserved edges, wax on polished surface.
Lithotype: *Portasanta, m. Chium.*
Collection: F.W. Kelsey, purchased from of F. Romano, "Al Porto Antico, Ostia," March 2, 1901. Accessioned (1901).

213. KM 2028 **fig. 2.32a; pl. 21**
Description: Fragment of a *columna rudentata*, L. 19.0 (preserved height of column), W. 17.23, D. 3.17 cm, one preserved edge; reconstructed diameter is 56.5 cm (see ch. 2).
Lithotype: *Portasanta, m. Chium.*
Collection: F.W. Kelsey, purchased from F. Romano, Baths of Diocletian, Rome, March 9, 1901. Accessioned 1902.

214. KM 2051 **pl. 21**
Description: Paving or veneer, L. 18.0, W. 15.0, D. 2.6 cm, one preserved edge slightly convex.
Lithotype: *Portasanta, m. Chium?*
Description in Accession: "Porta santa variata (marmorina)."
Collection: F.W. Kelsey, purchased from F. Romano, Baths of Diocletian, Rome, March 9, 1901. Accessioned 1902.

215. KM 2135 **pl. 21**
Description: Paving or veneer, L. 21.7, W. 12.4, D. 1.9 cm, two pieces glued together in perfect join, no preserved edges, saw marks on edge.
Lithotype: *Portasanta, m. Chium.*
Collection: F.W. Kelsey, purchased from F. Romano (who said he found it at the Forum), Rome, March 9, 1901. Accessioned 1901.

216. KM 2178 **fig. 2.29c; pl. 21**
Description: Column fragment, L. 17.5, W. 8.6, D. 4.0 cm, reconstructed diameter is 62 cm (see ch. 2); one preserved edge, modern cutting and polish.
Lithotype: *Portasanta, m. Chium.*
Collection: F.W. Kelsey, "Tor Paterno, Castel Fusano," Italy, no collection date. Accessioned 1901.

217. KM 2057 **fig. 2.24c; pl. 22**
Description: Composite molding, L. 23.4, W. 10.3, D. 15.2 cm, four preserved edges.
Lithotype: *Portasanta, m. Chium.*
Collection: F.W. Kelsey, Baths of Diocletian, Rome, January and July 1901. Accessioned 1902.

218. KM 2030 **pl. 22**
Description: Paving or veneer, L. 15.1, W. 10.5, D. 2.1 cm, no preserved edges.
Lithotype: *Portasanta, m. Chium.*
Collection: F.W. Kelsey, purchased from F. Romano, (Rome), January 1901. Accessioned 1902.

219. KM 2037 **pl. 22**
Description: Paving or veneer, L. 9.1, W. 5.1, D. 1.5 cm, two preserved edges.
Lithotype: *Portasanta, m. Chium.*

Collection: F.W. Kelsey, "Ruins of villa," "Place marked Rudei [Ruderi] on staff map 145-III (Cassoli) S.E. of Roviano & E of Anticoli Corrado near Mola Nuova," province of L'Aquila, Italy, February 15, 1901. Accessioned 1902.
Note: Same provenance and collection date as cat. nos. 452 and 600.

220. KM 2042 **pl. 22**
Description: Paving or veneer, L. 18.0, W. 13.0, D. 2.8 cm, no preserved edges.
Lithotype: *Portasanta, m. Chium.*
Collection: F.W. Kelsey, "Isola Farnese near Fiumicino," Italy, 1901. Accessioned 1902.
Note: "Isola Farnese" may be a mistake for Isola Sacra.

221. KM 2062 **pl. 22**
Description: Paving or veneer, L. 19.9, W. 9.6, D. 2.8 cm, no definite preserved edges.
Lithotype: *Portasanta, m. Chium.*
Collection: F.W. Kelsey, Rome, 1901. Accessioned 1902.

222. KM 2147 **pl. 22**
Description: Paving or veneer, L. 11.1, W. 7.8, D. 1.8 cm, one preserved edge, wax on surface.
Lithotype: *Portasanta, m. Chium.*
Collection: F.W. Kelsey, "Purchased of Vecchio," (Italy), 1901. Accessioned 1901.

223. KM 2047 **fig. 2.23a; pl. 22**
Description: Molding of a simple type, L. 17.5, W. 6.0, D. 2.5 cm, four preserved edges. Profile identical to that of cat. no. 254 except for slightly greater width at bottom.
Lithotype: *Portasanta, m. Chium.*
Collection: F.W. Kelsey, purchased from F. Romano, Baths of Diocletian, Rome, January 1901. Accessioned 1902.

224. KM 2364 **pl. 22**
Description: Paving or veneer, L. 19.4, W. 9.4, D. 2.1 cm, two preserved edges, two parallel incisions near one edge.
Lithotype: *Portasanta, m. Chium.*
Description in Accession: "Breccia"
Collection: F.W. Kelsey, Rome, 1901. Accessioned (1901).

225. KM 2187 **fig. 2.30c; pl. 22**
Description: Fragment of fluted column, L. 17.5, W. 10.0, D. 8.4 cm, one column flute preserved.
Lithotype: *Portasanta, m. Chium.*
Collection: F.W. Kelsey, "Found in front of 38 Via Firenze," Rome, January 21, 1901. Accessioned 1901.

226. KM 2254 **fig. 2.29d; pl. 22**
Description: Column fragment, L. 13.9, W. 13.0, D. 7.9 cm, reconstructed diameter is 41.5 cm (see ch. 2); one preserved edge, wax on polished surface.
Lithotype: *Portasanta, m. Chium.*
Collection: F.W. Kelsey, "Purchased from Vecchio," (Italy), (1900–1901). Accessioned (1901).

227. KM 2264 **pl. 23**
Description: Paving or veneer? L. 13.6, W. 12.6, D. 11.4 cm, no preserved edges, wax on polished surface.
Lithotype: *Portasanta, m. Chium.*
Description in Accession: "Serravezza."
Collection: F.W. Kelsey, "Tor Paterno," Italy, (1900–1901). Accessioned (1901).

228. KM 2313 **pl. 23**
Description: Paving or veneer, L. 13.4, W. 10.7, D. 2.7 cm, no preserved edges, pick marks on bottom.
Lithotype: *Portasanta, m. Chium.*
Collection: F.W. Kelsey, no provenance, (1900–1901). Accessioned 1901.

229. KM 2338 **pl. 23**
Description: Paving or veneer, L. 11.0, W. 10.2, D. 2.2 cm, one preserved edge, wax on polished surface.
Lithotype: *Portasanta, m. Chium.*
Collection: F.W. Kelsey, "Isola Farnese near Fiumicino(?)," Italy, (1900–1901). Accessioned 1901.
Note: "Isola Farnese" may be a mistake for Isola Sacra.

230. KM 2360 **pl. 23**
Description: Paving or veneer, L. 11.5, W. 6.0, D. 2.9 cm, no preserved edges.
Lithotype: *Portasanta, m. Chium* or possibly *breccia corallina.*
Collection: F.W. Kelsey, Rome, 1901. Accessioned (1901).

231. KM 2362 **pl. 23**
Description: Paving or veneer, L. 20.1, W. 17.1, D. 4.2 cm, two modern edges.
Lithotype: *Portasanta, m. Chium.*
Collection: F.W. Kelsey, Rome, 1901. Accessioned (1901).

232. KM 2369 **pl. 23**
Description: Paving or veneer, L. 12.2, W. 10.6, D. 1.9 cm, no preserved edges, pick marks on back, wax on polished surface.
Lithotype: *Portasanta, m. Chium.*
Collection: F.W. Kelsey, Rome, 1901. Accessioned (1901).

233. KM 2372 **pl. 23**
Description: Paving or veneer, L. 24.5, W. 16.9, D. 3.8 cm, no preserved edges, pick marks on bottom.
Lithotype: *Portasanta, m. Chium.*
Collection: F.W. Kelsey, Rome, 1901. Accessioned (1901).

234. KM 2375 **pl. 23**
Description: Paving or veneer, L. 15.0, W. 10.4, D. 13.0 cm, no preserved edges, mortar on sides.
Lithotype: *Portasanta, m. Chium.*
Collection: F.W. Kelsey, Rome, 1901. Accessioned (1901).

235. KM 2388 **pl. 23**
Description: Paving or veneer, L. 16.6, W. 16.1, D. 2.1 cm, variable, no preserved edges.
Lithotype: *Portasanta, m. Chium.*

Portasanta (cont.)

Collection: Kelsey, purchased from F. Romano, (Rome), (1900–1901). Accessioned (1901).

236. KM 2404 **pl. 23**
Description: Paving or veneer, L. 22, W. 15, D. 7 cm, no preserved edges, pick marks on bottom, wax on modern polished surface.
Lithotype: Portasanta, m. Chium.
Collection: F.W. Kelsey, Rome, 1901. Accessioned (1901).

237. KM 2032 **pl. 23**
Description: Paving or veneer, L. 14.6, W. 12.5, D. 4.4 cm, one preserved edge.
Lithotype: Portasanta, m. Chium.
Collection: F.W. Kelsey, purchased from F. Romano, (Rome), January 1901. Accessioned 1902.

238. KM 2033 **pl. 23**
Description: Paving or veneer, L. 16.0, W. 9.5, D. 2.2 cm, no preserved edges.
Lithotype: Portasanta, m. Chium.
Collection: F.W. Kelsey, purchased from F. Romano, Centocelle ("Cento Celli"), Rome, January 1901. Accessioned 1902.

239. KM 2027 **pl. 24**
Description: Paving or veneer, L. 16.8, W. 8.6, D. 1.8 cm, no preserved edges.
Lithotype: Portasanta, m. Chium.
Collection: F.W. Kelsey, Centocelle, ("Cento Celli"), Rome, 1901. Accessioned 1902.

240. KM 2026 **pl. 24**
Description: Paving or veneer, L. 13.9, W. 9.5, D. 4.0 cm, no preserved edges.
Lithotype: Portasanta, m. Chium.
Collection: F.W. Kelsey, "Purchased of Vecchio," (Italy), 1901. Accessioned 1902.

241. KM 2031 **pl. 24**
Description: Paving or veneer, L. 12.4, W. 10.0, D. 2.1 cm, no preserved edges.
Lithotype: Portasanta, m. Chium.
Collection: F.W. Kelsey, "Isola Farnese near Fiumicino," Italy, 1901. Accessioned 1902.
Note: "Isola Farnese" may be a mistake for Isola Sacra.

242. KM 2035 **pl. 24**
Description: Paving or veneer, L. 17.0, W 14.5, D 2.5 cm, no preserved edges.
Lithotype: Portasanta, m. Chium.
Collection: F.W. Kelsey, purchased from F. Romano, (Rome), 1901. Accessioned 1902.

243. KM 2036 **pl. 24**
Description: Paving or veneer, L. 19.4, W. 13.9, D. 4.5 cm, no preserved edges.
Lithotype: Portasanta, m. Chium.
Collection: F.W. Kelsey, (Italy?), 1901. Accessioned 1902.

244. KM 2038 **fig. 2.38b; pl. 24**
Description: Paving or veneer, L. 13.3, W. 9.1, D. 1.4 cm, no preserved edges, pick marks on bottom, wax on polished surface.
Lithotype: Portasanta, m. Chium.
Collection: F.W. Kelsey, "Isola Farnese near Fiumicino," Italy, 1901. Accessioned 1902.
Note: "Isola Farnese" may be a mistake for Isola Sacra.

245. KM 2039 **pl. 24**
Description: Paving or veneer, L. 7.4, W. 6.4, D. 1.4 cm, no preserved edges.
Lithotype: Portasanta, m. Chium.
Collection: F.W. Kelsey, (Italy?), 1901. Accessioned 1902.

246. KM 2040 **pl. 24**
Description: Paving or veneer, L. 13.6, W. 7.7, D. 3.5 cm, no preserved edges, pick marks on back.
Lithotype: Portasanta, m. Chium.
Collection: F.W. Kelsey, (Italy?), 1901. Accessioned 1902.

247. KM 2044 **pl. 24**
Description: Paving or veneer, L. 18, W. 13, D. 4 cm, one preserved edge.
Lithotype: Portasanta, m. Chium.
Collection: F.W. Kelsey, purchased from F. Romano, (Rome), January 1901. Accessioned 1902.

248. KM 2045 **pl. 24**
Description: Paving or veneer, L. 12.5, W. 9.2, D. 1.6 cm, no preserved edges, quarry pick marks on back.
Lithotype: Portasanta, m. Chium.
Collection: F.W. Kelsey, "Tor Paterno, Castel Fusano," Italy, 1901. Accessioned 1902.

249. KM 2052 **pl. 24**
Description: Paving or veneer, L. 15.2, 17.5, 3.8 cm, two preserved edges.
Lithotype: Portasanta, m. Chium.
Collection: F.W. Kelsey, "Fiumicino near Ostia," Italy, 1901. Accessioned 1902.

250. KM 2056 **fig. 2.29e; pl. 25**
Description: Column fragment, L. 19.0, W. 12.0, D. 5.17 cm, original diameter estimated to be 36.5 cm (see ch. 2); one preserved edge.
Lithotype: Portasanta, m. Chium.
Collection: F.W. Kelsey, purchased from F. Romano, (Rome), 1901. Accessioned 1902.

251. KM 2059 **pl. 25**
Description: Paving or veneer, L. 25.3, W. 24.1, D. 6.6 cm, no preserved edges, pick marks on bottom.
Lithotype: Portasanta, m. Chium.
Collection: F.W. Kelsey, "Purchased of Vecchio," Ostia, 1901. Accessioned 1902.

252. KM 23398 **fig. 2.21c; pl. 25**

Description: Fillet fragment, L. 18.5, W. 5.3, D. 3.1 cm, four preserved edges, both long edges beveled slightly at lower half, both short edges finished carefully on top but more roughly lower in the taper.

Lithotype: Portasanta, m. Chium.

Collection: F. W. Kelsey, Ostia, no collection date. Accessioned 1924.

Note: For possible use as a teaching tool, see note at cat. no. 141.

253. KM 25566 **pl. 25**

Description: Paving or veneer, L. 10.7, W. 4.9, D. 1.1 cm, no preserved edges.

Lithotype: Portasanta, m. Chium.

Collection: F. W. Kelsey, (Italy), "FWK gift," 1901. Accessioned 1902.

254. KM 2034 **figs. 2.23b, 2.41a; pl. 25**

Description: Molding of a simple type, L. 19.0, W. 5.4, D. 3.1 cm, four preserved edges, saw marks on back. Profile nearly identical to that of cat. no. 223.

Lithotype: Portasanta, m. Chium.

Description in Accession: "Diasforo radiculato tenero antico" (Orlandi).

Collection: F. W. Kelsey, purchased from F. Romano, (Rome), January 1901. Accessioned 1902.

255. KM 2180 **pl. 25**

Description: Paving or veneer, L. 16.1, W. 12.2, D. 5.1 cm, two preserved edges, wax on polished surface.

Lithotype: Portasanta, m. Chium, or possibly *giallo antico.*

Collection: F. W. Kelsey, "Purchased of Vecchio," (Italy), (1900–1901). Accessioned 1901.

256. KM 2383 **pl. 25**

Description: Paving or veneer, L. 14.5, W. 10.8, D. 2.0 cm, no preserved edges.

Lithotype: Portasanta, m. Chium.

Collection: F. W. Kelsey, Baths of Diocletian, Rome, October 1900. Accessioned (1901).

257. KM 2341 **pl. 25**

Description: Paving or veneer, L. 18.2, 11.9, D. 5.3 cm, no preserved edges.

Lithotype: Portasanta, m. Chium.

Description in Accession: "Fior di Persico" (Boni).

Collection: F. W. Kelsey, purchased from F. Romano, Baths of Diocletian, Rome, August 10, 1901. Accessioned 1901.

258. KM 93885 **pl. 26**

Description: Paving or veneer, L. 8, W. 8, D. 3 cm, no preserved edges.

Lithotype: Portasanta, m. Chium.

Collection: F. W. Kelsey (posthumous), "Material brought back from the U-M expedition to Antioch in Pisidia," directed by D. M. Robinson, 1924. Accession date unknown.

259. KM 2243 **pl. 26**

Description: Paving or veneer, L. 19.3, W. 7.9, D. 1.6 cm, two pieces glued together, one preserved edge, saw marks on edge.

Lithotype: Portasanta, m. Chium.

Description in Accession: "modern" (Orlandi).

Collection: F. W. Kelsey, purchased from F. Romano, (Rome), (1900–1901). No accession date.

260. KM 29607 **pl. 26**

Description: Paving or veneer, L. 7.8, W. 5.6, D. 1.7 cm, no preserved edges, wax on polished surface.

Lithotype: Portasanta, m. Chium.

Description in Accession: "Lumachella rosso e giallo [*sic*]"; "Giallo del Fiorito."

Collection: J. D. Candler, no provenance, (1860s?). Accessioned 1964.

261. KM 29646 **pl. 26**

Description: Paving or veneer, L. 22.4, W. 12.9, D. 0.9–1.8 cm, no preserved edges, thickness varies with saw blade angle.

Lithotype: Portasanta, m. Chium.

Description in Accession: "Breccia rosastro."

Collection: J. D. Candler, no provenance, (1860s?). Accessioned 1964.

262. KM 29647 **pl. 26**

Description: Paving or veneer, L. 17.8, W. 12.6, D. 2.6 cm, no preserved edges.

Lithotype: Portasanta, m. Chium.

Description in Accession: "Lumachella breccia rosso"; "Breccia di Mont—" (the rest of the description is destroyed).

Collection: J. D. Candler, no provenance, (1860s?). Accessioned 1964.

263. KM 29625 **pl. 26**

Description: Opus sectile element, L. 7.6, W. 5.3, D. 0.73 cm, no preserved edges.

Lithotype: Pink marble or crystalline limestone, possibly *portasanta.*

Description in Accession: "Semi-translucent pink stone . . . (Gypsum? Alabaster?)."

Collection: J. D. Candler, Ephesus, Turkey, (1867?). Accessioned 1964.

264. KM 93856 **pl. 26**

Description: Paving or veneer, L. 17.5, W. 5.9, D. 1.76 cm, no preserved edges.

Lithotype: Portasanta, m. Chium.

Collection: G. De Criscio, Puzzuoli or vicinity, no collection date. Accessioned 1923.

265. KM 93857 **fig. 2.38c; pl. 26**

Description: Paving or veneer, L. 14.1, W. 9.0, D. 1.73 cm, two preserved edges, bottom pick dressed.

Lithotype: Portasanta, m. Chium.

Collection: G. De Criscio, Puzzuoli or vicinity, no collection date. Accessioned 1923.

Portasanta (cont.)

266. KM 29696 **pl. 26**

Description: Paving or veneer? Irregular, weathered and possibly unworked. L. 9.3, W. 7.9, D. 2.2 cm, no preserved edges.

Lithotype: Portasanta, m. Chium.

Collection: P. O. E. Burton, no provenance, no collection date. Accessioned 1961.

267. KM 29677 **pl. 26**

Description: Paving or veneer, L. 11.3, W. 6.8, D. 2.5 cm, one preserved edge.

Lithotype: Portasanta, m. Chium.

Collection: P. O. E. Burton, no provenance, no collection date. Accessioned 1961.

268. KM 93785 **pls. 26–27**

Description: Modern sample? But, given the period of Jones's collecting, probably fashioned from an ancient artifact. L. 6.7, W. 5.7, D. 3.0 cm.

Lithotype: Portasanta, m. Chium.

Collection: E. Jones, (Rome), (1885). Accessioned (1912).

***Diaspro* (jasper)**

Sources: San Pietro di Rostino, Corsica; western Sicily

269. KM 2245 **pl. 27**

Description: Paving or veneer, L. 12.0, W. 6.0, D. 3.5 cm, no preserved edges.

Lithotype: Diaspro, jasper (pebbly sandstone but small grains are jasper).

Description in Accession: "porfido andesitico-basaltico"; "Breccia di Egitto sanguigno" (Orlandi); "Verde fiorito rosso Greco."

Collection: F. W. Kelsey, purchased from F. Romano, Baths of Diocletian, Rome, August 10, 1901. Accessioned (1901).

270. KM 2063 **pl. 27**

Description: Paving or veneer, L. 17.2, W. 12.4, D. 3.7 cm, one preserved edge, pick marks on back.

Lithotype: Diaspro, jasper di Sicilia.

Description in Accession: "porta santa."

Collection: F. W. Kelsey, Ostia, Italy, 1901. Accessioned 1902.

271. KM 2224 **pl. 27**

Description: Paving or veneer, L. 17.2, W. 12.4, D. 3.7 cm, one preserved edge.

Lithotype: Diaspro di Corsica, Corsican jasper.

Description in Accession: "Verde mandolata."

Collection: F. W. Kelsey, "Al Porto Antico," Ostia, March 2, 1901. Accessioned 1901.

Note: Based on the place and date of collection, this piece possibly purchased from F. Romano.

272. KM 2240 **pl. 27**

Description: Paving or veneer, L. 18.1, W. 13.6, D. 2.0 cm, no preserved edges. Note in accession book, "Cut with copper wheel by Romano who polished it"; this work probably led to the designation "modern" in the accession book.

Lithotype: Diaspro, jasper?

Description in Accession: "Broccatellone"; "Diasporo [*sic*] di Sicilia, modern."

Collection: F. W. Kelsey, purchased from F. Romano, Baths of Diocletian, Rome, March 9, 1901. Accessioned (1901).

273. KM 2090 **pl. 27**

Description: Paving or veneer, L. 20.1, W. 8.2, D. 7.8 cm, one preserved edge.

Lithotype: Diaspro, jasper, probably from area of Palermo, Sicily.

Description in Accession: "Diaspro radicellato."

Collection: F. W. Kelsey, purchased from F. Romano, Baths of Diocletian, Rome, January and July 1901. Accessioned 1902.

274. KM 2244 **pl. 27**

Description: Paving or veneer, L. 11.5, W. 8.9, D. 1.6 cm, no preserved edges.

Lithotype: Diaspro di Sicilia, Sicilian jasper.

Collection: F. W. Kelsey, "Purchased from Vecchio," (Italy), (1900–1901). Accessioned (1901).

275. KM 2138 **pl. 27**

Description: Paving or veneer, L. 8.9, W. 8.9, D. 2.7 cm, one preserved edge with saw marks.

Lithotype: Diaspro di Sicilia, Sicilian jasper.

Description in Accession: "Diaspro di Sicilia."

Collection: F. W. Kelsey, Baths of Diocletian, Rome, January and July 1901. Accessioned 1901.

Note: The two dates given in the accession book are "Jan 1900" and "July 1901"; 1900 must be a mistake for 1901, since Kelsey was not in Italy in January of 1900.

276. KM 2165 **pl. 27**

Description: Paving or veneer, D. 17.5, W. 8.9, D. 2.8 cm, no preserved edges.

Lithotype: Diaspro di Sicilia, Sicilian jasper.

Description in Accession: "Diaspro radiculato giallo di Serravezza" (Orlandi); "Agata (Diaspro radiculato, giallo antico)"; "The words 'Di valor poco settore [*sic*]' on slip."

Collection: F. W. Kelsey, purchased from F. Romano, (Rome), (1900–1901). Accessioned (1901).

277. KM 6865 **pl. 27**

Description: Paving or veneer, L. 6.6, W. 3.5, D. 1.6 cm, two preserved edges, wax on polished surface.

Lithotype: Diaspro di Sicilia, Sicilian jasper.

Collection: F. W. Kelsey, no provenance, no collection date. Accessioned 1938 (posthumous).

278. KM 6845 **pl. 27**

Description: Paving or veneer, L. 9.3, W. 4.5, D. 1.1 cm, no preserved edges.

Lithotype: Diaspro, jasper? Or possibly red agate or *portasanta*.

Description in Accession: "Agata rossa."

Collection: F. W. Kelsey, no provenance, no collection date. Accessioned 1938 (posthumous).

279. KM 6750 **pl. 28**
Description: Paving or veneer, L. 11.9, W. 5.4, D. 1.0 cm, no preserved edges.
Lithotype: *Diaspro di Sicilia*, Sicilian jasper.
Description in Accession: "Diaspro radicellato Sorato."
Collection: E. B. Van Deman, (Italy), no collection date. Accessioned 1938.

Prestige Level 7

Alabasters and onyx (*cotognino, fiorito, a pecorella*, etc.); breccias: *Aleppo, corallina, medicea, Serravezza,* and *Settebasi; verde antico*

Alabastro

Alabaster is highly variable, and small pieces in particular can be very difficult to identify. Unless we are confident of an identification, we list simply "alabastro," while noting conjectures recorded in the accession books. The same uncertainty frustrates alphabetization for the most part. Sources: Nile Valley, Egypt; Pamukkale (ancient Hierapolis), Turkey.

280. KM 6767 **pl. 28**
Description: Paving or veneer, L. 14.1, W. 10.7, D. 0.48 cm, no preserved edges.
Lithotype: *Alabastro.*
Description in Accession: "Alabastro Agatato rosa"; "Alabastro agatino rosso?" (Neither type matches any in the literature.)
Collection: E. B. Van Deman, (Italy), no collection date. Accessioned 1938.

281. KM 2111 **pl. 28**
Description: Unidentifiable object, L. 20.4, W. 8.8, D. 2.4 cm, two pieces glued together, no preserved edges.
Lithotype: *Alabastro*
Description in Accession: A partially illegible note in the accession book reads "bianco Aventr[--]a."
Collection: F. W. Kelsey, purchased from F. Romano, (Rome), (1900–1901). Accessioned 1901.

282. KM 6769 **pl. 28**
Description: Paving or veneer? L. 9.5, W. 11.0, D. 3.7 cm, one preserved edge, small margin shows marks of small pick or point chisel, polished surface probably modern.
Lithotype: *Alabastro.*
Description in Accession: "alabastro di Civitavecchia bruciato."
Collection: E. B. Van Deman, (Civitavecchia, Rome), no collection date. Accessioned 1938.

283. KM 6770 **pl. 28**
Description: Modern sample plaque which Van Deman evidently took as representative of the ancient lithotype, L. 17.4, W. 10.5, D. 1.0 cm, four preserved edges.
Lithotype: *Alabastro.*
Description in Accession: "Alabastro bigio di Sicilia."
Collection: E. B. Van Deman, (Italy), no collection date. Accessioned 1938.

284. KM 6762 **pl. 28**
Description: Paving or veneer, L. 15.9, W. 13.1, D. 1.0, one preserved edge, traces of saw marks on bottom surface.
Lithotype: *Alabastro cotognino.*
Collection: E. B. Van Deman, (Italy), no collection date. Accessioned 1938.

285. KM 6761 **pl. 28**
Description: Paving or veneer, L. 16.5, W. 21.0, D. 0.8 cm, one preserved edge.
Lithotype: *Alabastro cotognino.*
Collection: Van Deman, (Italy), no collection date. Accessioned 1938.

286. KM 6763 **fig. 1.4, pl. 28**
Description: Paving or veneer, L. 17.8, W. 13.5, D. 6.5 cm, one preserved edge.
Lithotype: *Alabastro cotognino.*
Description in Accession: "Alabastro Orientale Ghiacciato."
Collection: E. B. Van Deman, (Italy), no collection date. Accessioned 1938.

287. KM 6841 **pl. 28**
Description: Paving or veneer, L. 3.8, W. 2.6, D. 2.7 cm, two preserved edges, wax on polished surface.
Lithotype: *Alabastro cotognino.*
Collection: F. W. Kelsey, no provenance, no collection date. Accessioned 1938 (posthumous).

288. KM 29626 **pl. 28**
Description: *Opus sectile* element? L. 7.2, W. 3.9, D. 0.63 cm, one preserved edge.
Lithotype: *Alabastro cotognino?*
Description in Accession: "Alabastro giallo"; "almost a stained travertine."
Collection: J. D. Candler, no provenance, (1860s). Accessioned 1964.

289. KM 29518 **pl. 29**
Description: Paving or veneer, L. 13.0, W. 16.4, D. 1.06 cm, one preserved edge, bottom shows line of breakage at end of saw cut.
Lithotype: *Alabastro fiorito.*

Alabastro (cont.)

Collection: E. B. Van Deman, (Italy), no collection date. Accessioned 1938.

290. KM 6765 **pl. 29**
Description: Paving or veneer, L. 10.4, W. 9.9, D. 1.9 cm, one preserved edge, trace of breakage at end of saw cut on bottom.
Lithotype: *Alabastro fiorito*.
Description in Accession: "Alabastro Rosso Fasciato."
Collection: E. B. Van Deman, (Italy), no collection date. Accessioned 1938.

291. KM 29617 **pl. 29**
Description: Paving or veneer, L. 17.2, W. 9.1, D. 3.67 cm, no preserved edges.
Lithotype: *Alabastro fiorito*.
Description in Accession: "alabastro listato."
Collection: J. D. Candler, "from the Great Theater at Ephesus, Sept. 6, 1867." Accessioned 1964.

292. KM 29624 **pl. 29**
Description: Paving or veneer, L. 8.0, W. 3.2, D. 1.39 cm, no preserved edges.
Lithotype: *Alabastro fiorito*?
Description in Accession: "Alabastro fiorito rosso."
Collection: J. D. Candler, no provenance, (1860s?). Accessioned 1964.

293. KM 6771 **pl. 29**
Description: Paving or veneer, L. 18.9, W. 11.1, D. 0.48 cm, no preserved edges.
Lithotype: *Alabastro fiorito*?
Description in Accession: "Alabastro rosso."
Collection: E. B. Van Deman, (Italy), no collection date. Accessioned 1938.

294. KM 6837 **pl. 29**
Description: Paving or veneer, L. 11.3, W. 4.2, D. 2.6 cm, no preserved edges, mortar on back.
Lithotype: *Alabastro fiorito*?
Description in Accession: "Alabastro Amythisto (?)"
Collection: F. W. Kelsey, no provenance, no collection date. Accessioned 1938 (posthumous).

295. KM T2008.58 **pl. 29**
Description: Paving or veneer, L. 15.5, W. 7.2, D. 3.5 cm, no preserved edges. Masking tape affixed to the bottom of the piece reads "Cap du Garde 2."
Lithotype: *Alabastro*.
Collection: (F. W. Kelsey, purchased from Père R. P. Delattre, Carthage, Tunisia, March 2, 1893). Accessioned 2008.
Note: This piece may originally have come from Cap de Garde, Algeria. See note at cat. no. 135.

296. KM 2123 **pl. 29**
Description: Architectural element? L. 22.9, W. 16.3, D. 9.9 cm, wax on polished surface, one preserved edge, pick marks on one side.
Lithotype: *Alabastro*.
Collection: F. W. Kelsey, "Purchased of F. Romano?," (Rome), February 1901. Accessioned 1901.

297. KM 2122 **pls. 29–30**
Description: Internal fragment, L. 14.8, W. 9.7, D. 9.3 cm, no worked surfaces, no preserved edges.
Lithotype: *Alabastro*.
Description in Accession: "fasciato orientale [illegible] di Egitto."
Collection: F. W. Kelsey, purchased from F. Romano, Baths of Diocletian, Rome, March 9, 1901. Accessioned 1901.

298. KM 2108 **fig. 2.36c; pl. 30**
Description: Sculpture fragment, lower part of a bust, L. 22.5, W. 9.5, D. 7.9 cm, dowel hole with iron residue on bottom surface.
Lithotype: *Alabastro*.
Description in Accession: "di S. Paulo"; "Alabastro colognino."
Collection: F. W. Kelsey, "Isola Farnese near Fiumicino," Italy, (1900–1901). Accessioned 1901.
Note: "Isola Farnese" may be a mistake for Isola Sacra.

299. KM 2110 **pl. 30**
Description: Unworked or internal fragment, L. 15.4, W. 7.4, D. 10.8 cm, wax on (modern) polished surface, no preserved edges.
Lithotype: *Alabastro*.
Description in Accession: "di Palombara antica."
Collection: F. W. Kelsey, "Isola Farnese near Fiumicino," Italy, (1900–1901). Accessioned 1901.
Note: "Isola Farnese" may be a mistake for Isola Sacra.

300. KM 2124 **pl. 30**
Description: Unworked or internal fragment, L. 13.2, W. 12.0, D. 8.5 cm, no preserved edges.
Lithotype: *Alabastro a ghiaccione*.
Description in Accession: "Breccia di Cosi, modern" (Orlandi); "Alabastro giaccione".
Collection: F. W. Kelsey, no provenance, (1900–1901). Accessioned 1901.
Note: The *alabastro a ghiaccione* from the La Batteria quarries at Circeo, Italy (Bruno 1998, 2002), is a good match and was much used in the Julio-Claudian period.

301. KM 6838 **pl. 30**
Description: Paving or veneer, L. 7.0, W. 4.7, D. 2.9 cm, two preserved edges.
Lithotype: *Alabastro a ghiaccione*. For source, see note at cat. no. 300.
Description in Accession: "Alabastro a giaccione."
Collection: F. W. Kelsey, no provenance, no collection date. Accessioned 1938 (posthumous).
Note: "Aqua gelata" is penned on the piece but this meaning is unclear.

302. KM 6843 **pl. 30**
Description: Paving or veneer, L. 10.3, W. 6.3, D. 1.6 cm, three preserved edges, irregular shape with notched corner, probable saw marks.
Lithotype: *Alabastro.*
Description in Accession: "Alabastro giallo."
Collection: F. W. Kelsey, no provenance, no collection date. Accessioned 1938 (posthumous).

303. KM 2114 **pl. 30**
Description: Paving or veneer, L. 4.3, W. 8.6, D. 2.5 cm, one preserved edge.
Lithotype: *Alabastro*, very pure white.
Collection: F. W. Kelsey, "Purchased of workman," Volterra, Italy, September 1900. Accessioned 1901.

304. KM 6840 **pl. 30**
Description: Paving or veneer, L. 10.4, W. 5.5, D. 2.5 cm, no preserved edges.
Lithotype: *Alabastro.*
Description in Accession: "Alabastro Occhiale"; "Alabastro cotignino occhiale."
Collection: F. W. Kelsey, no provenance, no collection date. Accessioned 1938 (posthumous).

305. KM 6861 **pl. 30**
Description: Unworked stone, L. 7.6, W. 5.0, D. 2.5 cm, no preserved edges.
Lithotype: *Alabastro.*
Description in Accession: "Alabastro orbiculare."
Collection: F. W. Kelsey, no provenance, no collection date. Accessioned 1938 (posthumous).

306. KM 2118 **pl. 30**
Description: Paving or veneer, L. 14.0, W. 11.3, D. 7.0 cm, two preserved edges.
Lithotype: *Alabastro* ("orientale").
Collection: F. W. Kelsey, Baths of Diocletian, Rome, January 1900 and January 7, 1901. Accessioned 1901.

307. KM 2329 **pl. 31**
Description: Paving or veneer? L. 15.9, W. 11.1, D. 10.6 cm, three preserved edges with a probably modern polish.
Lithotype: *Alabastro.*
Description in Accession: "Oriental."
Collection: F. W. Kelsey, (Italy), (1900–1901). Accessioned 1901.

308. KM 6839 **fig. 2.9b; pl. 31**
Description: *Opus sectile* curved fragment, L. 7.3, W. 3.5, D. 2.1 cm, two preserved edges, some mortar on broken edges, saw marks on straight edge.
Lithotype: *Alabastro.*
Description in Accession: "Alabastro orientale."
Collection: F. W. Kelsey, no provenance, no collection date. Accessioned 1938 (posthumous).

309. KM 6768 **pl. 31**
Description: Paving or veneer, L. 22.4, W. 17.0, D. 2.68 cm, one preserved edge.
Lithotype: *Alabastro di Palombara*, a variety of *alabastro a tartaruga*, q.v.
Collection: E. B. Van Deman, (Italy), no collection date. Accessioned 1938.

310. KM 6766 **pl. 31**
Description: Paving or veneer, L. 15.7, W. 12.0, D. 0.97 cm, no preserved edges, one surface polished.
Lithotype: *Alabastro.*
Description in Accession: "Alabastro Palombaro Attosoni."
Collection: E. B. Van Deman, (Italy), no collection date. Accessioned 1938.

311. KM 29517 **pl. 31**
Description: Paving or veneer, L. 17.0, W. 16.1, D. 1.9 cm, no preserved edges, wax on polished surface, mortar and saw marks on bottom surface.
Lithotype: *Alabastro.*
Description in Accession: "Alabastro palombino."
Collection: E. B. Van Deman, (Italy), no collection date. Accessioned 1938.

312. KM 6760 **pl. 31**
Description: Paving or veneer, L. 19.6, W. 15.6, D. 0.84 cm, one preserved edge.
Lithotype: *Alabastro a pecorella.*
Description in Accession: "Alabastro ametisto(?) dendritico."
Collection: E. B. Van Deman, (Italy), no collection date. Accessioned 1938.

313. KM 6844 **pl. 31**
Description: Paving or veneer, L. 6.4, W. 4.2, D. 1.4 cm, one preserved edge.
Lithotype: *Alabastro.*
Description in Accession: "alabastro sanguigno."
Collection: F. W. Kelsey, no provenance, no collection date. Accessioned 1938 (posthumous).

314. KM 2109 **pl. 31**
Description: Unworked rock, L. 11.9, W. 7.8, D. 4.6 cm, wax on (modern) polished surface, no preserved edges.
Lithotype: *Alabastro.*
Description in Accession: "di S. Paulo"; "Alabastro di Civita Vecchia antica" (Orlandi).
Collection: F. W. Kelsey, (Isola Sacra), Italy, (1900–1901). Accessioned 1901.
Note: Isola Sacra is assumed for this piece because of the similarity of description to two pieces, cat. nos. 298 and 299, collected at "Isola Farnese," though we believe this to be a mistake for Isola Sacra.

315. KM 2112 **pl. 31**
Description: Paving or veneer, L. 13.8, W. 10.0, D. 1.8 cm, no preserved edges.
Lithotype: *Alabastro a tartaruga.*
Description in Accession: "fiorito occhiale."

Alabastro (cont.)

Collection: F. W. Kelsey, (Italy), (1900–1901). Accessioned 1901.

316. KM 2113 **pl. 31**
Description: Paving or veneer, L. 19.6, W. 15.8, D. 4.3 cm, one preserved edge.
Lithotype: *Alabastro a tartaruga.*
Description in Accession: "lumagalo"; "Alabastro di Montauto" (Orlandi).
Collection: F. W. Kelsey, Baths of Diocletian, Rome, December 1900. Accessioned 1901.

317. KM 6842 **pl. 32**
Description: Paving or veneer, L. 8.1, W. 7.8, D. 1.8 cm, no preserved edges, wax on polished surface.
Lithotype: *Alabastro.*
Description in Accession: "Alabastro a Tartaruga giallastro?"
Collection: F. W. Kelsey, no provenance, no collection date. Accessioned 1938 (posthumous).

318. KM 6753 **pl. 32**
Description: Paving or veneer, L. 17.6, W. 12.1, D. 0.7 cm, top and bottom surfaces polished, no preserved edges.
Lithotype: *Alabastro.*
Description in Accession: "Alabastro a tartaruga giallo e rosso."
Collection: E. B. Van Deman, (Italy), no collection date. Accessioned 1938.

319. KM 2117 **pl. 32**
Description: Paving or veneer, L. 16.3, W. 8.3, D. 2.6 cm, no preserved edges.
Lithotype: *Alabastro.*
Collection: F. W. Kelsey, "Purchased of Vecchio," Ostia, (1900–1901). Accessioned 1901.

320. KM 2121 **pl. 32**
Description: Paving or veneer, L. 21.6, W. 6.0, D. 5.6 cm, no preserved edges.
Lithotype: *Alabastro.*
Collection: F. W. Kelsey, "Fiumicino near Ostia," Italy, (1900–1901). Accessioned 1901.

321. KM 6866 **pl. 32**
Description: Paving or veneer, L. 4.2, W. 2.9, D. 2.6 cm, one preserved edge.
Lithotype: *Alabastro*?
Collection: F. W. Kelsey, no provenance, no collection date. Accessioned 1938 (posthumous).

322. KM 6764 **fig. 2.29f; pl. 32**
Description: Small section of a column, L. 10.9, W. 9.9, D. 0.97 cm, one preserved edge, polished modern surface, curvature of unpolished surface is a segment of a circle ca. 38 cm in diameter (see ch. 2).
Lithotype: *Alabastro.*
Description in Accession: "Alabastro fasciato Giallo"; "alabastro listato."
Collection: E. B. Van Deman, (Italy), no collection date. Accessioned 1938.

323. KM 29609 **pl. 32**
Description: Paving or veneer, L. 14.6, W. 9.3, D. 1.64 cm, one preserved edge, two pieces joined.
Lithotype: *Alabastro.*
Description in Accession: "Alabastro rosso fiorito."
Collection: J. D. Candler, no provenance, (1860s?). Accessioned 1964.

324. KM 29640 **pl. 32**
Description: Paving or veneer, L. 6.5, W. 6.4, D. 1.61 cm, no preserved edges.
Lithotype: *Alabastro.*
Collection: J. D. Candler, no provenance, (1860s?). Accessioned 1964.

325. KM 93793 **pl. 32**
Description: Paving or veneer, L. 21.9, W. 12.6, D. 3.62 cm, no preserved edges.
Lithotype: *Alabastro.*
Collection: G. De Criscio, Pozzuoli or vicinity, no collection date. Accessioned 1923.

326. KM 2545 **pl. 32**
Description: Paving or veneer from "corner of wall," L. 9.0, W. 4.8, D. 4.0 cm, two(?) preserved edges.
Lithotype: *Alabastro.*
Collection: E. Jones, "Palatine," Rome, (1885). Accessioned 1912.

327. KM 2349 **pl. 32**
Description: Paving or veneer? L. 16.2, W. 14.2, D. 6.4 cm, one preserved edge, wax on polished surface.
Lithotype: *Alabastro.*
Collection: F. W. Kelsey, "Palestrina, Hadrian's Villa and adjoining sites," March 2, 1901. Accessioned 1901.

328. KM 2396 **pl. 32**
Description: Paving or veneer, L. 21.6, W. 12.4, D. 2.6 cm, no preserved edges, wax on polished surface.
Lithotype: *Alabastro.*
Collection: F. W. Kelsey, (Italy), (1900–1901). Accessioned (1901).

329. KM 2403 **pl. 33**
Description: Paving or veneer, L. 20.2, W. 15.9, D. 8.4 cm, no preserved edges, wax on polished surface.
Lithotype: *Alabastro.*
Collection: F. W. Kelsey, Rome, 1901. Accessioned (1901).

330. KM 2405 **pl. 33**
Description: Paving or veneer? L. 18.3, W. 14.0, D. 9.3 cm, no preserved edges.
Lithotype: *Alabastro.*
Collection: F. W. Kelsey, purchased from F. Romano, Rome, (1900–1901). Accessioned (1901).

331. KM 2555 **pl. 33**
Description: Paving or veneer, L. 8.7, W. 2.6, D. 2.2 cm, four(?) preserved edges, cut and polish appear modern.
Lithotype: Alabastro.
Collection: E. Jones, "Tomb of Lucilla, Via Salaria," Rome, (1885). Accessioned 1912.

332. KM 93788 **pl. 33**
Description: Paving or veneer, L. 2.11, W. 1.41, D. 0.93 cm, one preserved edge, surface polished.
Lithotype: Alabastro of uncertain type.
Description in Accession: "small fragment of polished pink marble."
Collection: E. Jones, (Rome), (1885). Accessioned (1912).

333. KM 1969.5.4 **pl. 33**
Description: Paving or veneer, L. 5.6, W. 4.4, D. 1.07 cm, no preserved edges.
Lithotype: Alabastro or onyx.
Description in Accession: "white and gray marble."
Collection: J. G. Winter, no provenance, no collection date. Accessioned 1969.

334. KM 1969.5.5 **pl. 33**
Description: Paving or veneer, L. 4.96, W. 3.5, D. 1.23 cm, no preserved edges.
Lithotype: Alabastro or onyx.
Collection: J. G. Winter, no provenance, no collection date. Accessioned 1969.

335. KM 29706 **pl. 33**
Description: Paving or veneer, L. 6.2, W. 4.6, D. 3.1 cm, top and bottom polished, other surfaces broken.
Lithotype: Alabastro, white-beige.
Collection: P. O. E. Burton, no provenance, no collection date. Accessioned 1961.

Breccia di Aleppo

Source: Kariès, island of Chios, Greece

336. KM 2101 **fig. 2.29g; pl. 33**
Description: Column fragment, L. 13.1, W. 14.5, D. 3.58 cm; reconstructed diameter is 36 cm (see ch. 2); no preserved edges.
Lithotype: Breccia di Aleppo.
Description in Accession: "Breccia traccagnina" (Orlandi); "Brecciola da latte stimata assai rara."
Collection: F. W. Kelsey, purchased from F. Romano, Rome, March 9, 1901. Accessioned 1901.

337. KM 2093 **pl. 33**
Description: Paving or veneer, L. 9.04, W. 7.5, D. 0.54 cm, no preserved edges.
Lithotype: Breccia di Aleppo?
Description in Accession: "Breccia Traccagnina" (Orlandi).
Collection: F. W. Kelsey, no provenance, (1901). Accessioned 1902.

Breccia corallina

Sources: Bilecik, Vezirhan Province, Turkey; Karaburun Peninsula, Turkey

338. KM 2432 **fig. 2.29h; pls. 33–34**
Description: Column fragment with circular base, L. 12.0, W. 14.1, D. 11.0 cm; reconstructed diameter is 45.5 cm (see ch. 2); one preserved edge, flat surface has modern cut and polish.
Lithotype: Breccia corallina.
Collection: F. W. Kelsey, Baths of Diocletian, Rome, February 1, 1901. Accessioned (1901).

339. KM 2102 **pl. 34**
Description: Paving or veneer, L. 15.5, W. 12.5, D. 3.2 cm, one preserved edge.
Lithotype: Breccia corallina.
Description in Accession: "Corallina Antica."
Collection: F. W. Kelsey, purchased from F. Romano, Rome, (1900–1901). Accessioned 1901.

340. KM 2100 **pl. 34**
Description: Paving or veneer, L. 12.5, W. 18.7, D. 2.8 cm, no preserved edges, quarry pick marks on bottom.
Lithotype: Breccia corallina?
Description in Accession: "Corallina minuta."
Collection: F. W. Kelsey, purchased from F. Romano, (Rome), (1901). Accessioned 1902.

341. KM 2176 **pl. 34**
Description: Paving or veneer, L. 9.7, W. 7.8, D. 2.1 cm, no preserved edges, pick marks on bottom surface.
Lithotype: Breccia corallina.
Collection: F. W. Kelsey, "Isola Farnese near Fiumicino," Italy, 1900–1901. Accessioned 1901.
Note: "Isola Farnese" may be a mistake for Isola Sacra.

342. KM 2103 **pl. 34**
Description: Paving or veneer, L. 19, W. 20, D. 2 cm, no preserved edges.
Lithotype: Breccia corallina.
Description in Accession: "Corallina spaglia" (Orlandi).
Collection: F. W. Kelsey, Rome, (1900–1901). Accessioned 1901.

343. KM 29613 **pl. 34**
Description: Paving or veneer, L. 9.8, W. 8.3, D. 0.8 cm, no preserved edges.
Lithotype: Breccia corallina.
Description in Accession: "Breccia corallina."
Collection: J. D. Candler, no provenance, (1860s?). Accessioned 1964.
Note: The network of lines shading from gold to dark red is similar to the variety of *breccia corallina* in room 64–65 of Villa A at Oplontis.

344. KM 93561 **fig. 2.6g; pl. 34**
Description: Opus sectile triangle, L. 5.22, W. 5.38, D. 1.45 cm, two preserved edges, all corners broken.

Breccia corallina (cont.)

Lithotype: *Breccia corallina?*
Collection: J. D. Candler, no provenance, 1860–1890. Accessioned 1964.

345. KM 93564 **pl. 34**
Description: Paving or veneer, L. 5.37, W. 2.29, D. 1.24 cm, no preserved edges.
Lithotype: *Breccia corallina.*
Collection: J. D. Candler, no provenance, 1860–1890. Accessioned 1964.

346. KM 93851 **pl. 34**
Description: Paving or veneer, L. 8.66, W. 5.50, D. 1.70 cm, two preserved edges.
Lithotype: *Breccia corallina.*
Collection: G. De Criscio, Pozzuoli or vicinity, Italy, no collection date. Accessioned 1923.

347. KM 6759 **pl. 34**
Description: Paving or veneer, L. 11.5, W. 8.7, D. 1.7 cm, one preserved edge.
Lithotype: *Breccia corallina.*
Collection: E. B. Van Deman, (Italy), no collection date. Accessioned 1938.

348. KM 1969.5.2 **pl. 35**
Description: Paving or veneer, L. 10.0, W. 5.71, D. 1.85 cm, no preserved surfaces.
Lithotype: *Breccia corallina.*
Collection: J. G. Winter, no provenance, no collection date. Accessioned 1969.

Breccia medicea and *Serravezza*

Source: Apuan Alps, northern Italy

349. KM 2144 **pl. 35**
Description: Paving or veneer, L. 24, W. 18, D. 2 cm, no preserved edges, polished.
Lithotype: *Breccia medicea*, a subvariety of *breccia di Seravezza*, q.v.
Collection: F. W. Kelsey, purchased from F. Romano, Ostia, March 2, 1901. Accessioned 1901.

350. KM 2260 **pl. 35**
Description: Paving or veneer, L. 12.0, W. 7.7, D. 1.3 cm, no preserved edges, wax on polished surface.
Lithotype: *Breccia medicea*, a subvariety of *breccia di Seravezza*, q.v.
Description in Accession: "macchia di settebasi."
Collection: F. W. Kelsey, Paestum, Italy, (1900–1901). Accessioned (1901).

351. KM 29619 **pl. 35**
Description: Paving or veneer, L. 21.6, W. 12.6, D. 0.9 cm, three mating pieces glued together, two preserved edges, wax on polished surface, saw marks on edge.
Lithotype: *Breccia medicea*, a subvariety of *breccia di Seravezza*, q.v.
Description in Accession: "Macchia di Settebase"; "Africa" [africano?].
Collection: J. D. Candler, no provenance, (1860s?). Accessioned 1964.

Breccia di Settebasi

Source: Renes and nearby inlets, island of Skyros, Greece

352. KM T2008.61 **pl. 35**
Description: Paving or veneer, L. 16.3, W. 11.6, D. 1.3 cm, two preserved edges.
Lithotype: *Breccia di Settebasi* or possibly *pavonazzetto.*
Collection: (F. W. Kelsey, purchased from Père R. P. Delattre, Carthage, Tunisia, March 2, 1893). Accessioned 2008.
Note: This piece may originally have come from Cap de Garde, Algeria. See note at cat. no. 135.

353. KM 2126 **pl. 35**
Description: Paving or veneer, L. 8.0, W. 7.2, D. 1.64 cm, no preserved edges.
Lithotype: *Breccia di Settebasi* or *breccia di Seravezza.*
Collection: F. W. Kelsey, Baths of Diocletian, Rome, October 1900. Accessioned 1901.

354. KM 2099+2368 **pl. 35**
Description: Paving or veneer, L. 18.2, W. 28.2, D. 1.9 cm, two pieces joined with glue.
Lithotype: *Breccia di Settebasi.*
Collection: F. W. Kelsey, KM 2099: "Found in Baths of Diocletian," Rome, March 9, 1901; KM 2368: "Rome," 1901. Accessioned 1901.

355. KM 2105 **pl. 35**
Description: Paving or veneer, L. 14.5, W. 9.4, D. 2.3 cm, one preserved edge.
Lithotype: *Breccia di Settebasi.*
Description in Accession: "Corallina brecciata rossa."
Collection: F. W. Kelsey, purchased of F. Romano, (Rome), (1900–1901). Accessioned 1901.

356. KM 2106 **pl. 35**
Description: Paving or veneer, L. 15.5, W. 10.6, D. 3.5 cm, no preserved edges, possible ancient drill holes on bottom.
Lithotype: *Breccia di Settebasi.*
Collection: F. W. Kelsey, "Purchased of Vecchio," Rome, (1900–1901). Accessioned 1901.

357. KM 2107 **pl. 35**
Description: Paving or veneer, L. 21.7, W. 23.5, D. 3.0 cm, no preserved edges.
Lithotype: *Breccia di Settebasi.*
Collection: F. W. Kelsey, Astura, Italy, (1901). Accessioned 1901.

358. KM 2119 **pl. 35**
Description: Paving or veneer, L. 12.8, W. 10.3, D. 4.7 cm, no preserved edges.

Lithotype: *Breccia di Settebasi.*
Description in Accession: "Alabastro broccatellone brecciato."
Collection: F.W. Kelsey, purchased from F. Romano, (Rome), (1900–1901). Accessioned 1901.

359. KM 2440 **pl. 35**
Description: Paving or veneer, L. 8.0, W. 6.1, D. 2.0 cm, no preserved edges.
Lithotype: *Breccia di Settebasi.*
Collection: F.W. Kelsey, (Italy), (1900–1901). Accessioned (1901).

360. KM 2255 **pl. 36**
Description: Paving or veneer, L. 18.4, W. 11.2, D. 2.5 cm, one preserved edge, wax on polished surface.
Lithotype: *Breccia di Settebasi.*
Collection: F.W. Kelsey, (Italy), (1900–1901). Accessioned 1901.

361. KM 2257 **pl. 36**
Description: Paving or veneer, L. 16.6, W. 11.2, D. 7.9 cm, two preserved edges, wax on polished surface.
Lithotype: *Breccia di Settebasi.*
Collection: F.W. Kelsey, purchased from F. Romano, Baths of Diocletian, Rome, August 10, 1901. Accessioned (1901).

362. KM 25583 **pl. 36**
Description: Paving or veneer, L. 6.5, W. 8.2, D. 1.4 cm, one preserved edge.
Lithotype: Fine-grained (MGS 0.4 mm) white marble with thin and sparse coral-colored veins: *breccia di Settebasi* or possibly *breccia corallina.*
Collection: F.W. Kelsey, "Antium" (Anzio), Italy, (1901). Accession date unknown.

363. KM 2263 **pl. 36**
Description: Paving or veneer, L. 12.0, W. 9.2, D. 1.7 cm, two preserved edges.
Lithotype: *Breccia di Settebasi?*
Description in Accession: "Serravezza"; "Cottonella fiorita settebasi"; "Serravezza rossa" (Orlandi).
Collection: F.W. Kelsey, purchased from F. Romano, (Rome), (1900–1901). Accessioned (1901).

364. KM 2389 **pl. 36**
Description: Paving or veneer, L. 16.2, W. 10.1, D. 2.0 cm, no preserved edges.
Lithotype: *Breccia di Settebasi?*
Description in Accession: "Broccatello (Settebasi a rosa)"; "Seravezza."
Collection: F.W. Kelsey, purchased from F. Romano, (Rome), (1900–1901). Accessioned (1901).

365. KM 2325 **pl. 36**
Description: Paving or veneer, L. 12.3, W. 11.0, D. 2.2 cm, no preserved edges, wax on polished surface.
Lithotype: *Breccia di Settebasi?*
Description in Accession: "Breccia di Montana."
Collection: F.W. Kelsey, (Italy), (1900–1901). Accessioned 1901.

366. KM 2437 **pl. 36**
Description: Paving or veneer, L. 8.6, W. 4.7, D. 3.9 cm, no preserved edges.
Lithotype: *Breccia di Settebasi* or *pavonazzetto.*
Collection: F.W. Kelsey, (Italy), (1900–1901). Accessioned 1901.

367. KM 2259 **fig. 2.29i; pl. 36**
Description: Column fragment, L. 11.1, W. 16.8, D. 4.6 cm; reconstructed diameter is 34.5 cm (see ch. 2); one preserved edge, sawed and polished.
Lithotype: *Breccia di Settebasi.*
Collection: F.W. Kelsey, purchased from F. Romano, (Rome), (1900–1901). Accessioned (1901).

368. KM 6831 **pl. 36**
Description: Paving or veneer? L. 8.6, W. 5.3, D. 2.2 cm, three preserved edges, unusual shape with edges cut at angles not parallel, wax on polished surface.
Lithotype: *Breccia di Settebasi.*
Collection: F.W. Kelsey, no provenance, no collection date. Accessioned 1938 (posthumous).

369. KM 6835 **pl. 36**
Description: Paving or veneer, L. 8.5, W. 7.7, D. 1.3 cm, one preserved edge, wax on polished surface.
Lithotype: *Breccia di Settebasi.*
Description in Accession: "Breccialina Antico di Sicilia."
Collection: F.W. Kelsey, no provenance, no collection date. Accessioned 1938 (posthumous).

370. KM 6832 **pl. 36**
Description: Paving or veneer, L. 12.5, W. 2.6, D. 1.0 cm, one preserved edge, wax on polished surface.
Lithotype: *Breccia di Settebasi.*
Collection: F.W. Kelsey, no provenance, no collection date. Accessioned 1938 (posthumous).

371. KM 6863 **pl. 37**
Description: Paving or veneer, L. 10.9, W. 7.0, D. 0.8 cm, two preserved edges, wax on polished surface.
Lithotype: *Breccia di Settebasi.*
Collection: F.W. Kelsey, no provenance, no collection date. Accessioned 1938 (posthumous).

372. KM 29641 **pl. 37**
Description: Paving or veneer, L. 6.3, W. 5.5, D. 0.5 cm, no preserved edges.
Lithotype: *Breccia di Settebasi.*
Collection: J. D. Candler, no provenance, (1860s?). Accessioned 1964.

373. KM 6786 **fig. 2.32b; pl. 37**
Description: Fragment of a *columna rudentata,* L. 18.1, W. 16.0, D. 2.4 cm; reconstructed diameter is 54.5 cm (see ch. 2); one preserved edge, flutes with rope filling.

Breccia di Settebasi (cont.)

Lithotype: *Breccia di Settebasi* or *breccia di Seravezza.*
Description in Accession: "Serravezza rosso."
Collection: E. B. Van Deman, (Italy), no collection date. Accessioned 1938.

374. KM 29516 **pl. 37**
Description: Paving or veneer, L. 24.0, W. 17.1, D. 1.06 cm, one preserved edge.
Lithotype: *Breccia di Settebasi.*
Collection: E. B. Van Deman, (Italy), no collection date. Accessioned 1938.

375. KM 6731 **pl. 37**
Description: Paving or veneer, L. 17.9, W. 6.7, D. 0.6 cm, three pieces glued together, one preserved edge.
Lithotype: *Breccia di Settebasi.*
Description in Accession: "Giallo antico dorato?"
Collection: E. B. Van Deman, (Italy), no collection date. Accessioned 1938.

376. KM 93852 **pl. 37**
Description: Paving or veneer, L. 18.1, W. 6.0, D. 1.30–1.44 cm, no preserved edges.
Lithotype: *Breccia di Settebasi?* Or possibly *pavonazzetto.*
Collection: G. De Criscio, Pozzuoli or vicinity, no collection date. Accessioned 1923.

377. KM 93854 **pl. 37**
Description: Paving or veneer, L. 13.5, W. 10.1, D. 1.6 cm, two preserved edges.
Lithotype: *Breccia di Settebasi?*
Collection: G. De Criscio, Pozzuoli or vicinity, no collection date. Accessioned 1923.

378. KM 2547 **pl. 37**
Description: Paving or veneer, L. 8.9, W. 6.2, D. 0.4 cm, saw marks on edge.
Lithotype: *Breccia di Settebasi.*
Collection: E. Jones, "St. John Lateran," Rome, (1885). Accessioned 1912.

379. KM 29645 **pl. 37**
Description: Paving or veneer, L. 5.9, W. 9.8, D. 1.4 cm, no preserved edges.
Lithotype: *Breccia di Settebasi?*
Description in Accession: "Bianco e giallo reticulato."
Collection: J. D. Candler, no provenance, (1860s?). Accessioned 1964.

Verde antico

Source: Kasambali near Larissa, Thessaly, Greece

380. KM 25574 **pl. 37**
Description: Paving or veneer, L. 9.8, W. 7.8, D. 1.1 cm, no preserved edges.
Lithotype: *Verde antico, m. Thessallicum.*
Description in Accession: "breccia verde."
Collection: F. W. Kelsey, purchased from Père R. P. Delattre, Carthage, Tunisia, March 2, 1893. Accessioned 1893.

381. KM 2216 **pl. 37**
Description: Paving or veneer, L. 13.0, W. 11.6, D. 1.9 cm, no preserved edges.
Lithotype: *Verde antico, m. Thessalicum.*
Description in Accession: "Verde di Grecia, modern" (Orlandi).
Collection: F. W. Kelsey, Baths of Diocletian, Rome, January and July 1, 1901. Accessioned (1901).

382. KM 2223 **pl. 38**
Description: Paving or veneer, L. 10.0, W. 7.7, D. 1.5 cm, one preserved edge.
Lithotype: *Verde antico, m. Thessalicum.*
Description in Accession: "Verde antico (di seconda) (antico) smaralde" (Orlandi).
Collection: F. W. Kelsey, purchased from F. Romano, (Rome), (1900–1901). Accessioned (1901).

383. KM 2320 **pl. 38**
Description: Paving or veneer, L. 16.0, W. 11.0, D. 4.3 cm, no preserved edges.
Lithotype: *Verde antico, m. Thessalicum.*
Collection: F. W. Kelsey, purchased from F. Romano, (Rome), (1900–1901). Accessioned 1901.

384. KM 25571 **pl. 38**
Description: Paving or veneer, L. 5.6, W. 5.0, D. 1.5 cm, no preserved edges.
Lithotype: *Verde antico, m. Thessalicum.*
Collection: F. W. Kelsey, "Purchased in Rome in 1901." Accessioned 1901.

385. KM 2365 **fig. 2.21d; pl. 38**
Description: Fillet, L. 15, W. 5, D. 3 cm, two preserved edges.
Lithotype: *Verde antico, m. Thessalicum.*
Collection: F. W. Kelsey, "FWK purchase 1901," Rome. Accessioned 1901.

386. KM 2225 **pl. 38**
Description: Paving or veneer, L. 19.0, W. 13.8, D. 1.9 cm, one preserved edge, wax on polished surface.
Lithotype: *Verde antico, m. Thessalicum.*
Collection: F. W. Kelsey, purchased from F. Romano, (Rome), February 1901. Accessioned 1901.

387. KM 2247 **pl. 38**
Description: Paving or veneer, L. 18.8, W. 17.1, D. 7.3 cm, no preserved edges.
Lithotype: *Verde antico, m. Thessalicum?*
Collection: F. W. Kelsey, purchased from F. Romano, (Rome), (1900–1901). Accessioned (1901).

388. KM 2218 **pl. 38**
Description: Paving or veneer, interior fragment? L. 19.0, W. 15.0, D. 6.6 cm, no preserved edges.
Lithotype: *Verde antico, m. Thessalicum.*
Description in Accession: "Verde antico."

Collection: F.W. Kelsey, "Pherae (Velestino) Acropolis, Greek Trip, 1900–01." Accessioned 1901.
Note: Velestino (ancient Pherae) is near Volos, in Magnesia. Kelsey's group was there in late April 1901, according to Pedley 2012, 72, though it is not explicitly mentioned.

389. KM 2220 **pl. 38**
Description: Paving or veneer, L. 14.1, W. 8.0, D. 5.8 cm, two preserved edges.
Lithotype: *Verde antico, m. Thessalicum.*
Collection: F.W. Kelsey, no provenance (Greek trip?), (1900–1901). Accessioned (1901).

390. KM 2221 **pl. 38**
Description: Paving or veneer, L. 13.1, W. 9.0, D. 1.5 cm, no preserved edges.
Lithotype: *Verde antico, m. Thessalicum.*
Collection: F.W. Kelsey, no provenance (Greek trip?), (1900–1901). Accessioned (1901).

391. KM 6735 **pl. 38**
Description: Paving or veneer, L. 22.7, W. 11.4, D. 0.6 cm, two preserved edges, mortar preserved on sides and back.
Lithotype: *Verde antico, m. Thessalicum.*
Description in Accession: "Verde antico or Breccia Verde."
Collection: E.B. Van Deman, (Italy), no collection date. Accessioned 1938.

392. KM 6736 **pl. 38**
Description: Paving or veneer, L. 28.3, W. 13.5, D. 1.0 cm, one preserved edge.
Lithotype: *Verde antico, m. Thessalicum.*
Description in Accession: "Verde antico."
Collection: E.B. Van Deman, (Italy), no collection date. Accessioned 1938.

393. KM 93855 **pl. 39**
Description: Paving or veneer, L. 15.8, W. 8.1, D. 1.63 cm, no preserved edges.
Lithotype: *Verde antico, m. Thessalicum.*
Collection: G. De Criscio, Pozzuoli or vicinity, no collection date. Accessioned 1923.

394. KM 2717 **pl. 39**
Description: Paving or veneer, L. 16.4, W. 6.3, D. 0.9 cm, no preserved edges.
Lithotype: *Verde antico, m. Thessalicum.*
Collection: G. De Criscio, Pozzuoli or vicinity, no collection date. Accessioned 1923.

Prestige Level 6

Broccatello, broccatellone, cipollino rosso, fior di pesco, lumachella including *occhio di Pavone, rosso antico*, other breccias and limestones

Broccatello

Source: vicinity of Tortosa (ancient Dertosa), northeastern Spain

395. KM 2194 **pl. 39**
Description: Paving or veneer, L. 13.8, 5.3, D. 1.6 cm, two preserved edges.
Lithotype: *Broccatello.*
Collection: F.W. Kelsey, purchased from F. Romano, (Rome), (1900–1901). Accessioned 1901.

396. KM 2317 **pl. 39**
Description: Paving or veneer, L. 8.3, W. 8.3, D. 2.6 cm, no preserved edges.
Lithotype: *Broccatello.*
Collection: F.W. Kelsey, "Bòt. from Vecchio," (Italy), (1900–1901). Accessioned 1901.

397. KM 2351 **pl. 39**
Description: Paving or veneer, L. 12.4, W. 8.0, D. 2.9 cm, no preserved edges, mortar on back, wax on polished surface.
Lithotype: *Broccatello.*
Collection: F.W. Kelsey, no provenance, 1900–1901. Accessioned 1901.

398. KM 2402 **pl. 39**
Description: Paving or veneer, L. 18.9, W. 14.8, D. 11.0 cm, no preserved edges.
Lithotype: *Broccatello?*
Collection: F.W. Kelsey, Rome, 1901. Accessioned (1901).

399. KM 2443 **pl. 39**
Description: Paving or veneer, L. 8.3, W. 7.1, D. 1.9 cm, one preserved edge, mortar on edges.
Lithotype: *Broccatello.*
Collection: F.W. Kelsey, "Hadrian's Villa and adjoining sites, Palestrina," March 2, 1901. Accessioned (1901 or 1902).

400. KM 6836 **pl. 39**
Description: Paving or veneer, L. 9.0, W. 5.3, D. 1.3 cm, one(?) preserved edge, wax on polished surface.
Lithotype: *Broccatello.*
Description in Accession: "Broccatellone"; "Lumachella di Spagna."
Collection: F.W. Kelsey, no provenance, no collection date. Accessioned 1938 (posthumous).

401. KM 6867 **pl. 39**
Description: Paving or veneer, L. 5.9, W. 5.8, D. 0.9 cm, one preserved edge.

Broccatello (cont.)

Lithotype: *Broccatello.*
Description in Accession: "Breccia dorata pavonazza?"; "Lumachella di Spagna."
Collection: F. W. Kelsey, no provenance, no collection date. Accessioned 1938 (posthumous).

402. KM 6782 **pl. 39**
Description: Paving or veneer, L. 16.4, W. 8.5, D. 0.7 cm, no preserved edges.
Lithotype: *Broccatello.*
Collection: E. B. Van Deman, (Italy), no collection date. Accessioned 1938.

403. KM 29608 **pl. 39**
Description: Paving or veneer, L. 5.6, W. 3.6, D. 1.3 cm, one preserved edge, wax on polished surface.
Lithotype: *Broccatello.*
Description in Accession: "Lumachella di Spagna. Red and yellow, w/ white shells, w/ quartz deposits" (the latter being not quartz but calcite infill).
Collection: J. D. Candler, no provenance, no collection date. Accessioned 1964.

404. KM 1969.5.7 **pl. 39**
Description: *Opus sectile* fragment, L. 3.18, W. 2.3, D. 0.68 cm, no preserved edges.
Lithotype: *Broccatello.*
Collection: J. G. Winter, no provenance, no collection date. Accessioned 1969.

Broccatellone

Source: Bilecik, Vezirhan Province, Turkey

405. KM 2324 **pl. 40**
Description: Paving or veneer, L. 9.6, W. 8.9, D. 1.9 cm, no preserved edges, wax on polished surface.
Lithotype: *Broccatellone?*
Description in Accession: "Brocatellone antico."
Collection: F. W. Kelsey, (Italy), (1900–1901). Accessioned 1901.

406. KM 2166 **pl. 40**
Description: Paving or veneer, L. 13.0, W. 10.9, D. 1.9 cm, one preserved edge.
Lithotype: *Broccatellone?*
Description in Accession: "Agata"; "Broccatellone"; "Breccia opaca gialla" (Orlandi).
Collection: F. W. Kelsey, purchased from F. Romano, (Rome), (1900–1901). Accessioned 1901.

407. KM 6757 **pl. 40**
Description: Paving or veneer, L. 10.4, W. 9.0, D. 0.5 cm, no preserved edges, wax on polished surface.
Lithotype: *Broccatellone?*
Description in Accession: "Pink Breccia(?)"; "Brocatellone antico."
Collection: E. B. Van Deman, (Italy), no collection date. Accessioned 1938.

408. KM 6758 **pl. 40**
Description: Paving or veneer, L. 14.2, W. 9.3, W. 0.7 cm, no preserved edges.
Lithotype: *Broccatellone.*
Description in Accession: "Pink marble, rough(?)"; "Probably Brocatellone antico (close to 2324 [cat. no. 405])."
Collection: E. B. Van Deman, (Italy), no collection date. Accessioned 1938.

Cipollino rosso

Source: Kıyıkışlacık (ancient Iasos), western Turkey

409. KM 2173 **fig. 2.24d; pl. 40**
Description: Molding of composite type, L. 29.0, W. 10.3, D. 5.0 cm, unusual recurving profile as if to create a channel in the middle.
Lithotype: *Cipollino rosso.*
Description in Accession: "not Rosso Antico" (Ward-Perkins).
Collection: F. W. Kelsey, Baths of Diocletian, Rome, December 1900. Accessioned 1901.
Note: John Ward-Perkins gave the Jerome Lectures at the University of Michigan in the mid-1970s and looked at the marble samples then.

410. KM 2136 **pl. 40**
Description: Paving or veneer, L. 20.8, W. 9.2, D. 2.58 cm, no preserved edges.
Lithotype: *Cipollino rosso.*
Description in Accession: "Fiore di Persico"; "Cipollino ventresco[?] rosso" (Orlandi).
Collection: F. W. Kelsey, purchased from F. Romano, (Rome), (1900–1901). Accessioned 1901.

411. KM 2143 **pl. 40**
Description: Paving or veneer, L. 22.4, W. 18.0, D. 2.02 cm, no preserved edges.
Lithotype: *Cipollino rosso.*
Collection: F. W. Kelsey, Baths of Diocletian, Rome, January 1901. Accessioned 1901.

412. KM 2137 **pl. 40**
Description: Paving or veneer, L. 21.1, W. 14.8, D. 2.7 cm, no preserved edges.
Lithotype: *Cipollino rosso.*
Description in Accession: "Fiore di Persico fasciato."
Collection: F. W. Kelsey, Baths of Diocletian, Rome, February 8, 1901. Accessioned 1901.

413. KM 2142 **pl. 40**
Description: Paving or veneer, L. 20.6, W. 10.4, D. 1.7 cm, no preserved edges.
Lithotype: *Cipollino rosso.*
Collection: F. W. Kelsey, Baths of Diocletian, Rome, February 8, 1901. Accessioned 1901.

414. KM 2140 **pl. 40**
Description: Paving or veneer, L. 13.5, W. 11.4, D. 2.12 cm, no preserved edges.
Lithotype: *Cipollino rosso.*
Description in Accession: "Fiore di Persico fasciato (modern)"; "Cipollino fasciato pavonazzo" (Orlandi).
Collection: F. W. Kelsey, purchased from F. Romano, (Rome), (1900–1901). Accessioned 1901.

415. KM 2242 **fig. 2.29j; pl. 40**
Description: Column fragment, unfluted, L. 10.0, W. 7.5, D. 21.0 cm; reconstructed diameter is 36 cm (see ch. 2); one preserved edge, polished.
Lithotype: *Cipollino rosso.*
Description in Accession: "Broccatello rosso"; "Rosso brecciato minuto."
Collection: F. W. Kelsey, purchased from F. Romano, (Rome), (1900–1901). Accessioned (1901).

416. KM 2331 **pl. 41**
Description: Architectural element, not further identifiable, D. 25.3, W. 8.5, D. 5.4 cm, polished surface angles downward 10 degrees.
Lithotype: *Cipollino rosso.*
Collection: F. W. Kelsey, (Italy), (1900–1901). Accessioned 1901.

417. KM 2334 **pl. 41**
Description: Fragment of thick paving or possibly a larger artifact, L. 24.8, W. 6.1, D. 12.2 cm, one preserved edge.
Lithotype: *Cipollino rosso.*
Description in Accession: "Serravezza rossa"; "A note on the slip says 1 pc. sawed in two."
Collection: F. W. Kelsey, "Tor Paterno," Italy, 1900–1901. Accessioned 1901.

418. KM 2390 **pl. 41**
Description: Paving or veneer, L. 13.3, W. 12.2, D. 1.3 cm, one preserved edge.
Lithotype: *Cipollino rosso* or possibly *fior di pesco.*
Collection: F. W. Kelsey, Ostia, Italy, March 2, 1901. Accessioned (1901).

419. KM 29629 **pl. 41**
Description: Paving or veneer, L. 6.2, W. 4.4, D. 2.1 cm, one preserved edge, wax on polished surface.
Lithotype: *Cipollino rosso.*
Description in Accession: "Brecciato di Pirenei."
Collection: J. D. Candler, no provenance, no collection date. Accessioned 1964.

420. KM 6755 **fig. 2.9d; pl. 41**
Description: *Opus sectile* semicircle, L. 21.6, W. 23.0, D. 0.4 cm, seven pieces glued together, wax on polished surface.
Lithotype: *Cipollino rosso.*
Description in Accession: "Rosso brecciato."
Collection: E. B. Van Deman, (Italy), no collection date. Accessioned 1938.

Fior di pesco

Source: Eretria, Euboea, Greece

421. KM 2391 **pl. 41**
Description: Paving or veneer, L. 24.7, W. 13.5, D. 2.1 cm, no preserved edges.
Lithotype: *Fior di pesco, m. Chalcidicum.*
Collection: F. W. Kelsey, purchased from F. Romano, (Rome), (1900–1901). Accessioned (1901).

422. KM 2139 **pl. 41**
Description: Paving or veneer, L. 15.0, W. 8.5, D. 2.5 cm, no preserved edges, wax on polished surface.
Lithotype: *Fior di pesco, m. Chalcidicum* (Lazzarini).
Description in Accession: "Fior di Persico lumacato."
Collection: F. W. Kelsey, "Bought of old man," Rome, March 13, 1901. Accessioned 1901.

423. KM 2248 **pl. 41**
Description: Paving or veneer, L. 25, W. 20, D. 2 cm, no preserved edges.
Lithotype: *Fior di pesco, m. Chalcidicum?*
Description in Accession: "Porta Venere rosso"; "Rosso di Levanto modern" (Boni).
Collection: F. W. Kelsey, purchased from F. Romano, (Rome), (1900–1901). Accessioned (1901).

424. KM 2141 **pl. 41**
Description: Paving or veneer, L. 13.8, W. 18.5, D. 1.8 cm, no preserved edges; accession note: "Has been near fire."
Lithotype: *Fior di pesco, m. Chalcidicum* (Lazzarini).
Description in Accession: "Fior di Persico fiorito"; "Cottonella fasciata" (Orlandi).
Collection: F. W. Kelsey, Baths of Diocletian, Rome, February 8, 1901. Accessioned 1901.

425. KM 2318 **pl. 41**
Description: Paving or veneer, L. 7.4, W. 5.8, D. 1.3 cm, no preserved edges.
Lithotype: *Fior di pesco, m. Chalcidicum?*
Collection: F. W. Kelsey, "Theatre, Sparta," "Greek Trip, 1900–01." Accessioned 1901.
Note: The group was at Sparta the first week of April 1901, according to Pedley 2012, 69–70.

426. KM 2145 **pl. 42**
Description: Paving or veneer, L. 23.4, W. 20.5, D. 2.2 cm, no preserved edges.
Lithotype: *Fior di pesco, m. Chalcidicum.*
Description in Accession: "Fiore di Persico"; Pavonazzetta brecciata verde"; "Serravezza rossa" (Orlandi).
Collection: F. W. Kelsey, purchased from F. Romano, (Rome), February 1901. Accessioned 1901.

427. KM 6847 **pl. 42**
Description: Paving or veneer? L. 10.1, W. 3.6, D. 1.3 cm, three preserved edges, wax on polished surface, saw marks on edge.
Lithotype: *Fior di pesco, m. Chalcidicum.*

Fior di pesco **(cont.)**

Collection: F.W. Kelsey, no provenance, no collection date. Accessioned 1938 (posthumous).

428. KM 6828 **pl. 42**
Description: Paving or veneer, L. 7.5, W. 7.5, D. 2.2 cm, no preserved edges, two lines incised on top surface.
Lithotype: Fior di pesco, m. Chalcidicum.
Description in Accession: "Fiore di Persico."
Collection: F.W. Kelsey, no provenance, no collection date. Accessioned 1938 (posthumous).

429. KM 6754 **pl. 42**
Description: Paving or veneer, L. 15.4, W. 13.8, D. 1.05 cm, one preserved edge.
Lithotype: Fior di pesco, m. Chalcidicum.
Description in Accession: "Cottanello?"; "Fiore di Persico bruniccio?"
Collection: E. B. Van Deman, (Italy), no collection date. Accessioned 1938.

Lumachella

Sources: Henchir el Kasbat (ancient Thuburbo Maius), Tunisia; Verona, Italy; quarries across the Mediterranean

430. KM 2234+2436 **pl. 42**
Description: Paving or veneer, L. 19.2, W. 12.8, D. 1.9 cm, two pieces glued together, no preserved edges.
Lithotype: Gray *lumachella.*
Description in Accession: "Occhio di Pavone."
Collection: F.W. Kelsey, Baths of Diocletian, Rome, December 1900. Accessioned (1901).

431. KM 2160 **pl. 42**
Description: Paving or veneer, L. 12.4, W. 12.0, D. 1.6 cm, two preserved edges.
Lithotype: Gray *lumachella.*
Description in Accession: "Bigio lumacato"; "dorato greco."
Collection: F.W. Kelsey, "Tor Paterno, Castel Fusano," Italy, (1900–1901). Accessioned (1901).

432. KM 2172 **pl. 42**
Description: Paving or veneer, L. 11.6, W. 10.9, D. 3.6 cm, one preserved edge, saw mark on one edge.
Lithotype: Lumachella.
Description in Accession: "Lumachella mandolata."
Collection: F.W. Kelsey, purchased from F. Romano, Baths of Diocletian, Rome, August 10, 1901. Accessioned 1901.

433. KM 2356 **pl. 42**
Description: Quarry debris, L. 19.8, W. 15.5, D. 6.22 cm, no preserved edges, polished and waxed surface appears modern.
Lithotype: Black *lumachella.*
Description in Accession: "Granite."
Collection: F.W. Kelsey, Rome, 1901. Accessioned (1901).

434. KM 2171 **fig. 2.41b; pl. 42**
Description: Paving or veneer, L. 18.9, W. 10.1, D. 3.8 cm, three preserved edges, saw marks on edges and top surface.
Lithotype: Gray-red *lumachella?*
Description in Accession: "Ostracana lumacata."
Collection: F.W. Kelsey, purchased from F. Romano, "Al Porto Antico, Ostia," Italy, (1900–1901). Accessioned (1901).

435. KM 2155 **pl. 42**
Description: Paving or veneer, L. 15.1, W. 9.5, D. 1.8 cm, one preserved edge.
Lithotype: Gray-red *lumachella?*
Description in Accession: "Bigio fasciato lumacato rosso."
Collection: F.W. Kelsey, Baths of Diocletian, Rome, January 1901. Accessioned 1901.

436. KM 2097 **pl. 43**
Description: Paving or veneer, L. 9.5, W. 8.9, D. 4.2 cm, two preserved edges.
Lithotype: Lumachella with fossils set in a mauve groundmass
Description in Accession: "Breccia tigrata lumacata"; "Cottanello chiara"; "Verona Mandolato" (Boni).
Collection: F.W. Kelsey, purchased from F. Romano, Rome, (1900–1901). Accessioned 1902.

437. KM 6773 **pl. 43**
Description: Paving or veneer, L. 20.4, W. 9.2, D. 2.1 cm, one preserved edge.
Lithotype: Lumachella.
Description in Accession: "Lumachella di Calabria or Lum. degli Abruzzi."
Collection: E. B. Van Deman, (Italy), no collection date. Accessioned 1938.
Note: "Lumachella di Calabria or Lum. degli Abruzzi," is probably Van Deman's identification. This seems to be a case where she attributed an origin in the mode of Italian artisanal nomenclature, but neither *lumachella* from Calabria or the Abruzzi is given as a distinct lithotype in the standard sources.

438. KM 6756 **pl. 43**
Description: Paving or veneer, L. 12.5, W. 7.3, D. 1.4 cm, one preserved edge.
Lithotype: Red *lumachella*, possibly *lumachella pavonazza.*
Description in Accession: "EBVD Portasanta lumacata."
Collection: E. B. Van Deman, (Italy), no collection date. Accessioned 1938.

439. KM 6774+6775 **pl. 43**
Description: Paving or veneer, L. 16.7, W. 14.1, D. 0.8 cm, one preserved edge, two pieces joined.
Lithotype: Lumachella.
Description in Accession: "Lumacella or Lumachellone" (but not clear what Van Deman's threshold size for the fossils was to distinguish "snail" from "big snail").
Collection: E. B. Van Deman, (Italy), no collection date. Accessioned 1938.

440. KM 6779 **pl. 43**
Description: Paving or veneer, L. 9.6, W. 17.0, D. 1.7 cm, one or possibly two preserved edges.
Lithotype: *Lumachella orientale.* (See Lazzarini and Mariottini 2012, 446–47.)
Description in Accession: "occhiale."
Collection: E. B. Van Deman, (Italy), no collection date. Accessioned 1938.

441. KM 6777 **pl. 43**
Description: Paving or veneer, L. 14.0, W. 9.2, D. 1.1 cm, one preserved edge, wax on polished surface.
Lithotype: *Lumachella*, probably *orientale.*
Description in Accession: "EBVD Lumachella degli Abruzzi."
Collection: E. B. Van Deman, (Italy), no collection date. Accessioned 1938.

442. KM 29611 **pl. 43**
Description: Paving or veneer, L. 7.0, W. 4.9, D. 2.8 cm, three preserved edges, saw marks preserved on edges, wax on polished surfaces.
Lithotype: *Lumachellone antica* or *Lumachella di San Vitale.*
Description in Accession: "Lumachella grande."
Collection: J. D. Candler, no provenance, (1860s?). Accessioned 1964.

Occhio di pavone rosso and *pavonazzo*

Source: Kutluca and the Sakarya (ancient Sangarius) Valley, north central Turkey

443. KM 93843 **pl. 43**
Description: Paving or veneer, L. 5.6, W. 3.7, D. 1.5 cm, no preserved edges.
Lithotype: *Occhio di pavone pavonazzo.*
Collection: F. W. Kelsey, Mérida, Spain, January 10–12, 1893. Accessioned 1893.

444. KM 6784 **pl. 43**
Description: Paving or veneer, L. 18.7, W. 8.5, D. 1.9 cm, one preserved edge.
Lithotype: *Occhio di pavone pavonazzo.*
Collection: E. B. Van Deman, (Italy), no collection date. Accessioned 1938.

445. KM 29616 **pl. 43**
Description: Paving or veneer, L. 19.0, W. 13.1, D. 1.1 cm, two pieces joined, no preserved edges, mortar on bottom.
Lithotype: *Occhio di pavone pavonazzo.*
Description in Accession: "Lumachella rosso."
Collection: J. D. Candler, no provenance, (1860s?). Accessioned 1964.

446. KM 29618 **pl. 43**
Description: Paving or veneer, L. 17.4, W. 12.9, D. 1.7 cm, no preserved edges.
Lithotype: *Occhio di pavone pavonazzo.*
Description in Accession: "Lumachella rosso (Almost red-brown ground)."
Collection: J. D. Candler, no provenance, (1860s?). Accessioned 1964.

Rosso antico

Source: Mani Peninsula, Greece

447. KM 93783 **fig. 2.26a; pl. 43**
Description: Cornice molding, L. 9.2, W. 3.1, D. 2.2 cm, two preserved edges.
Lithotype: *Rosso antico.*
Collection: F. W. Kelsey, Rome, 1893. Accessioned 1893.

448. KM 592 **fig. 2.26b; pl. 44**
Description: Cornice molding, L. 12.0, W. 3.5, D. 3.0 cm, three preserved edges.
Lithotype: *Rosso antico.*
Collection: F. W. Kelsey, Rome, (1893?). Accessioned (1893?).

449. KM 2506 **fig. 2.10; pl. 44**
Description: *Opus interrasile* fragment, L. 6.16, W. 4.08, D. 0.86 cm, only a small tendril of the inlay is in situ.
Lithotype: Gray limestone with *rosso antico* inlay.
Collection: F. W. Kelsey, Rome, 1901. Accessioned (1901).

450. KM 2046a **pl. 44**
Description: Paving or veneer, L. 11.5, W. 9.4, D. 2.4 cm, no preserved edges.
Lithotype: *Rosso antico.*
Collection: F. W. Kelsey, "Isola Farnese near Fiumicino," Italy, January 1901. Accessioned 1902.
Note: "Isola Farnese" may be a mistake for Isola Sacra.

451. KM 2073 **fig. 2.26c; pl. 44**
Description: Cornice molding, L. 12, W. 5, D. 5 cm, three preserved edges.
Lithotype: *Rosso antico.*
Collection: F. W. Kelsey, "in front of 38, Via Firenze," Rome, "Found Jan. 21, 1901"; "July 1901" is also given. Accessioned 1902.

452. KM 2074 **fig. 2.26d; pl. 44**
Description: Cornice molding, L. 17.2, W. 3.8, D. 3.1 cm; profile similar to that of cat. no. 462.
Lithotype: *Rosso antico.*
Collection: F. W. Kelsey, "Place marked Rudei [Ruderi] on staff map 145-III (Cassoli) S.E. of Roviano & E of Anticoli Corrado near Mola Nuova," province of L'Aquila, Italy, February 15, 1901. Accessioned 1902.
Note: Same provenance and collection date as cat. nos. 219 and 600.

453. KM 2075 **fig. 2.27b; pl. 44**
Description: Cornice molding, L. 16.0, W. 7.4, D. 8.4 cm, three preserved edges, pick dressed on back.
Lithotype: *Rosso antico.*
Collection: F. W. Kelsey, Rome, 1901. Accessioned 1902.

Rosso antico **(cont.)**

454. KM 2076 pl. 44
Description: Paving or veneer, L. 10.4, W. 7.5, D. 2.9 cm, one preserved edge, wax on polished edge.
Lithotype: *Rosso antico.*
Collection: F.W. Kelsey, Rome, 1901. Accessioned 1902.

455. KM 2387 pl. 44
Description: Paving or veneer, L. 24.3, W. 20.1, D. 1.9 cm, no preserved edges.
Lithotype: *Rosso antico.*
Description in Accession: Boni and Orlandi both called it modern, but it appears to be ancient.
Collection: F.W. Kelsey, "Al Porto Antico, Ostia," March 2, 1901. Accessioned (1901).
Note: Based on the date and place of acquisition, possibly purchased from F. Romano.

456. KM 2421 figs. 2.19, 2.39d; pl. 44
Description: Fragment of a Corinthian pilaster capital, L. 10.2, W. 12.4, D. 2.4 cm, no preserved edges, back roughly chiseled. Lower right portion preserves five-petal rosette tethered to branch, surrounded by scrolling vines and vegetal motifs.
Lithotype: *Rosso antico.*
Collection: F.W. Kelsey, Rome, 1901. Accessioned (1901).

457. KM 25593 pl. 44
Description: Paving or veneer, L. 10.2, W. 4.4, D. 1.4 cm, two preserved edges.
Lithotype: *Rosso antico.*
Collection: F.W. Kelsey, "Hadrian's Villa and adjoining sites, Palestrina," March 2, 1901. Accessioned (1901 or 1902).

458. KM 6823 fig. 2.27c; pl. 44
Description: Cornice molding, L. 10.0, W. 3.0, D. 2.9 cm, no preserved edges, profiles on both sides.
Lithotype: *Rosso antico.*
Collection: F.W. Kelsey, no provenance, no collection date. Accessioned 1938 (posthumous).

459. KM 6824 pls. 44–45
Description: Paving or veneer, L. 5.2, W. 1.9, D. 2.1 cm, one preserved edge, wax on polished surfaces.
Lithotype: *Rosso antico.*
Collection: F.W. Kelsey, no provenance, no collection date. Accessioned 1938 (posthumous).

460. KM 6825 pl. 45
Description: Paving or veneer, L. 7.2, W. 3.8, D. 0.8 cm, no preserved edges, wax on polished surface.
Lithotype: *Rosso antico.*
Collection: F.W. Kelsey, no provenance, no collection date. Accessioned 1938 (posthumous).

461. KM 6826 pl. 45
Description: Paving or veneer, L. 4.7, W. 2.8, D. 1.1 cm, one preserved edge, mortar on broken edges, wax on polished surface.
Lithotype: *Rosso antico.*
Collection: F.W. Kelsey, no provenance, no collection date. Accessioned 1938 (posthumous).

462. KM 6751 fig. 2.26e; pl. 45
Description: Cornice molding, L. 16.1, W. 3.6, D. 2.3 cm, one preserved edge; profile similar to that of cat. no. 452.
Lithotype: *Rosso antico.*
Collection: E.B. Van Deman, (Italy), no collection date. Accessioned 1938.

463. KM 29654 pl. 45
Description: Paving or veneer, L. 9.6, W. 6.1, D. 1.4 cm, three preserved edges, wax patch on bottom.
Lithotype: *Rosso antico.*
Collection: J.D. Candler, no provenance, (1860s?). Accessioned 1964.

464. KM 93559 fig. 2.11a; pl. 45
Description: *Opus sectile* figural inlay piece, L. 7.75, W. 1.73, D. 0.47 cm, three preserved edges; piece forms a nearly complete "S" shape. May not pair with cat. no. 465 because of difference in thickness.
Lithotype: *Rosso antico.*
Collection: J.D. Candler, no provenance, 1860–1890. Accessioned 1964.

465. KM 93560 fig. 2.11b; pl. 45
Description: *Opus sectile* figural inlay piece, L. 6.28, W. 1.69, D. 0.75 cm, three preserved edges (piece intact), sides beveled to allow easy insertion. May not pair with cat. no. 464 because of difference in thickness.
Lithotype: *Rosso antico.*
Collection: J.D. Candler, no provenance, 1860–1890. Accessioned 1964.

466. KM 93562 fig. 2.7a; pl. 45
Description: *Opus sectile* rectangle, L. 9.07, W. 3.87, D. 1.21 cm, two preserved edges.
Lithotype: *Rosso antico.*
Collection: J.D. Candler, no provenance, 1860–1890. Accessioned 1964.

467. KM 93563 pl. 45
Description: *Opus sectile* element? L. 3.7, W. 5.05, D. 7.8 cm, no preserved edges.
Lithotype: *Rosso antico.*
Collection: J.D. Candler, no provenance, 1860–1890. Accessioned 1964.

468. KM 2973 fig. 2.28a; pl. 45
Description: Cornice molding, L. 9.5, W. 4.8, D. 2.6 cm, no preserved edges, drill hole on bottom surface.
Lithotype: *Rosso antico.*

Collection: G. De Criscio, Pozzuoli or vicinity, no collection date. Accessioned 1923.

469. KM 29730 **fig. 2.28d; pl. 45**
Description: Cornice molding, L. 10.1, W. 2.7, D. 2.4 cm, one preserved edge.
Lithotype: Rosso antico.
Collection: G. De Criscio, Pozzuoli or vicinity, no collection date. Accessioned 1923.

470. KM 29737 **fig. 2.28b; pls. 45–46**
Description: Cornice molding, L. 21.2, W. 4.1, D. 3.7 cm, one preserved edge, one end dressed at 45 degrees to form an inside join.
Lithotype: Rosso antico.
Collection: G. De Criscio, Pozzuoli or vicinity, no collection date. Accessioned 1923.

471. KM 93780 **fig. 2.28e; pl. 46**
Description: Cornice molding, L. 10.6, W. 8.1, D. 4.5 cm, one preserved edge, four pieces joining.
Lithotype: Rosso antico.
Collection: G. De Criscio, Pozzuoli or vicinity, no collection date. Accessioned 1923.

472. KM 93781 **fig. 2.28f; pl. 46**
Description: Cornice molding, L. 30.2, W. 5.4, D. 4.1 cm, one preserved edge, badly weathered.
Lithotype: Rosso antico.
Collection: G. De Criscio, Pozzuoli or vicinity, no collection date. Accessioned 1923.

473. KM 93782 **fig. 2.28g; pl. 46**
Description: Cornice molding, L. 19.9, W. 5.0, D. 3.3, two preserved edges.
Lithotype: Rosso antico.
Collection: G. De Criscio, Pozzuoli or vicinity, no collection date. Accessioned 1923.

474. KM 93859 **pl. 46**
Description: Small beach cobble, evenly rounded and pitted, L. 5.1, W. 2.38, D. 1.66 cm, no preserved (or original?) edges.
Lithotype: Rosso antico.
Collection: G. De Criscio, Pozzuoli or vicinity, no date. Accessioned 1923.

Other breccias and limestones

475. KM 2230 **fig. 2.8c; pl. 46**
Description: Opus sectile lozenge, L. 12.0, W. 17.6, D. 3.8 cm, four preserved edges.
Lithotype: Breccia cenerina.
Collection: F. W. Kelsey, "Purchased of vecchio," (Italy), (1900–1901). Accessioned (1901).

476. KM 2152 **pl. 46**
Description: Paving or veneer, L. 11.5, W. 18.0, D. 1.8 cm, no preserved edges.
Lithotype: Breccia cenerina?
Collection: F. W. Kelsey, (Rome), "Said to have come from Forum," March 9, 1901. Accessioned 1901.
Note: Based on the date and place of acquisition, it is likely Kelsey bought this piece from F. Romano.

477. KM 2262 **pl. 46**
Description: Paving or veneer, L. 20.2, W. 10.2, D. 8.2 cm, no preserved edges.
Lithotype: Breccia rossa Appenninica.
Description in Accession: "Breccia di Simone."
Collection: F. W. Kelsey, "Purchased," (Rome), February 7, 1901. Accessioned (1901).
Note: This piece is probably part of a group with cat. nos. 478 and 479, noted as purchased on the following day.

478. KM 2132 **fig. 2.5b; pl. 46**
Description: Paving or veneer, L. 23.6, W. 16.7, D. 4.0 cm, three preserved edges, drill hole, chisel marks on bottom.
Lithotype: Breccia rossa Appenninica.
Collection: F. W. Kelsey, Baths of Diocletian, Rome, February 8, 1901. Accessioned 1901.
Note: This piece is probably part of a group with cat. nos. 477 and 479. Note the similar features in cat. nos. 478 and 479.

479. KM 2195 **fig. 2.5a; pl. 46**
Description: Paving or veneer? L. 15.0, W. 9.6, D. 4.0 cm, two preserved edges, drill hole, chisel marks on bottom.
Lithotype: Breccia rossa Appenninica?
Collection: F. W. Kelsey, Baths of Diocletian, Rome, February 8, 1901. Accessioned 1901.
Note: This piece is probably part of a group with cat. nos. 477 and 478. Note the similar features in cat. nos. 478 and 479.

480. KM 2265 **pl. 46**
Description: Paving or veneer, L. 20.3, W. 12.0, D. 5.0 cm, no preserved edges, wax on polished surface.
Lithotype: Breccia rossa Appenninica.
Description in Accession: "Breccia gregoriana antica."
Collection: F. W. Kelsey, (Italy), (1900–1901). Accessioned 1901.

481. KM 2261 **pl. 46**
Description: Paving or veneer, L. 20.7, W. 15.6, D. 5.8 cm, no preserved edges, wax on surface.
Lithotype: Brecciated pink limestone.
Description in Accession: "Breccia di Simone"; "Carnagione, antico" (Orlandi).
Collection: F. W. Kelsey, (Rome?), (1900–1901). Accessioned (1901).

482. KM 2252 **pl. 47**
Description: River or beach cobble, L. 12.2, W. 6.0, D. 3.3 cm, no preserved edges.
Lithotype: Conglomerate with light groundmass and light quartz pebbles, not similar to the Roman *breccia frutti-*

Other breccias and limestones (cont.)

colosa with clam gouges in the pebbles and dark green groundmass.
Collection: F.W. Kelsey, "Purchased from vecchio," (Italy), (1900–1901). Accessioned (1901).

483. KM 2395 **pl. 47**
Description: Paving or veneer, L. 15.5, W. 14.8, D. 1.7 cm, no preserved edges.
Lithotype: *Breccia di Seravezza*? The clast boundaries are very sharp.
Collection: F.W. Kelsey, (Rome?), (1900–1901). Accessioned (1901).

484. KM 2096 **pl. 47**
Description: Paving or veneer, L. 17.0, W. 12.5, D. 1.1 cm, one preserved edge.
Lithotype: *Breccia di Seravezza*, resembling *africano*, which, however, does not have the purple blotches present here.
Collection: F.W. Kelsey, purchased from F. Romano, Rome, (1900–1901). Accessioned 1902.

485. KM 2241 **pl. 47**
Description: Original object unidentifiable, L. 13.7, W. 8.8, D. 7.6 cm, two preserved edges.
Lithotype: *Breccia traccagnina*?
Description in Accession: "Broccatello pelle di Tigre (valuable)."
Collection: F.W. Kelsey, purchased from F. Romano, (Rome), (1900–1901). Accessioned (1901).

486. KM 6732 **fig. 2.22; pl. 47**
Description: Fragment of a *tondino*, circular molding, L. 20.0, W. 3.8, D. 3.4 cm, three preserved edges.
Lithotype: *Breccia gialla.*
Collection: E. B. Van Deman, (Italy), no collection date. Accessioned 1938.

Prestige Level 5

Greek and Anatolian white marbles, *greco scritto, nero antico*

Greek and Anatolian white marbles

487. KM 93867 **fig. 2.25a; pl. 47**
Description: Cornice molding fragment, L. 12, W. 7, D. 6 cm, four preserved edges.
Lithotype: Dokimeion white marble? (visual; the provenance and MGS of 0.06 mm support this identification). See s.v. *Pavonazzetto* in Lithotype section.
Collection: F.W. Kelsey (posthumous), "Material brought back from the U-M expedition to Antioch in Pisidia," directed by D. M. Robinson, 1924. Accession date unknown.
Note: Cat. nos. 487, 488, and 489 are from the same run of cornice although they do not join precisely.

488. KM 93865 **fig. 2.25b; pl. 47**
Description: Cornice molding fragment, L. 12, W. 7, D. 6 cm, four preserved edges.
Lithotype: Dokimeion white marble? (visual; MGS 0.50 mm).
Collection: F.W. Kelsey (posthumous), "Material brought back from the U-M expedition to Antioch in Pisidia," directed by D. M. Robinson, 1924. Accession date unknown.
Note: Cat. nos. 487, 488, and 489 are from the same run of cornice although they do not join precisely.

489. KM 93866 **fig. 2.25c; pl. 47**
Description: Cornice molding fragment, L. 16, W. 7, D. 6 cm, four preserved edges.
Lithotype: Dokimeion white marble? (visual; MGS 0.40 mm).
Collection: F.W. Kelsey (posthumous), "Material brought back from the U-M expedition to Antioch in Pisidia," directed by D. M. Robinson, 1924. Accession date unknown.
Note: Cat. nos. 487, 488, and 489 are from the same run of cornice although they do not join precisely.

490. KM 2550 **fig. 2.6h; pl. 47**
Description: *Opus sectile* triangle, L. 8.2, W. 5.2, D. 1.6 cm, all edges preserved.
Lithotype: Dokimeion? White marble with very fine grain (MGS 0.40 mm) and creamy cast.
Description in Accession: "Breccia di Sette Basse."
Collection: E. Jones, "Caelian," Rome, (1885). Accessioned 1912.

491. KM 2015 **fig. 2.30d; pl. 47**
Description: Fragment of fluted column, L. 17.4, W. 12.1, D. 8.4 cm, two preserved edges, modern(?) surface cut and polished.
Lithotype: Doliana? White marble with fine grain (MGS 0.40 mm); grain size and context are consistent with origin from Doliana.
Description in Accession: "Greek?"
Collection: F.W. Kelsey, (Greece?), 1900–1901. Accessioned 1901.

492. KM 2001 **table 3.1; fig. 3.3; pl. 47**
Description: Paving or veneer, L. 16.1, W. 5.7, D. 3.3 cm, one preserved edge.
Lithotype: Doliana (MGS 1.10 mm), according to isotopic and petrographic analysis (see ch. 3).
Description in Accession: "Bit of Carrara marble."
Collection: F.W. Kelsey, "Theatre, Sparta," 1901. Accessioned 1901.

Note: Kelsey's group was at Sparta the first week of April 1901 (Pedley 2012, 69–70).

493. KM 2019 table 3.1; fig. 3.3; pl. 48
Description: Paving or veneer, L. 8.0, W. 6.5, D. 1.4 cm, no preserved edges.
Lithotype: Doliana (MGS 01.10 mm), according to isotopic and petrographic analysis (see ch. 3).
Collection: F.W. Kelsey, "Theatre, Sparta," 1901. Accessioned 1901.
Note: Kelsey's group was at Sparta the first week of April 1901 (Pedley 2012, 69–70).

494. KM 2007 table 3.1; figs. 2.27a, 3.3; pl. 48
Description: Cornice molding, L. 16.1, W. 5.6, D. 7.1 cm.
Lithotype: White marble with fine grain (MGS 0.40 mm). Isotopic and petrofabric analysis indicates either Doliana or Pentelikon as a source (see ch. 3).
Description in Accession: "Bit of Pentelic."
Collection: F.W. Kelsey, "Olympia Hill of Kronos," 1901. Accessioned 1901.
Note: Kelsey's group was at Olympia on March 20–21, 1901 (Pedley 2012, 68).

495. KM 2002 table 3.1; fig. 3.3; pl. 48
Description: Architectural fragment, L. 14.4, W. 11.7, D. 3.8 cm, no preserved edges, modern surface sawn and polished.
Lithotype: White-gray marble with medium grain (MGS 1.00 mm). Isotopic and petrofabric analysis indicates Doliana as a source (see ch. 3).
Description in Accession: "Bit of Carrara marble."
Collection: F.W. Kelsey, "Temple of Athena Alea, Tegea," "Greek Trip, 1900–01." Accessioned 1901.
Note: Kelsey's group was at Tegea the first week of April 1901 (Pedley 2012, 70).

496. KM 2020 table 3.1; fig. 3.3; pl. 48
Description: Paving or veneer? L. 11.4, W. 10.3, D. 2.8 cm, no preserved edges.
Lithotype: Grayish-white marble with medium-coarse grain (MGS 1.20). Isotopic and petrofabric analysis indicates Doliana as a possible source (see ch. 3).
Description in Accession: "Hymettian(?) marble."
Collection: F.W. Kelsey, "Velestino Pherae Acropolis," "Greek Trip, 1900–01." Accessioned 1901.
Note: Velestino (ancient Pherae) is near Volos, in Magnesia. Kelsey's group was at Pherae in late April 1901, according to Pedley 2012, 72, though it is not explicitly mentioned.

497. KM 2017 table 3.1; fig. 3.3; pl. 48
Description: Paving or veneer, L. 14.0, W. 10.4, D. 4.8 cm, no preserved edges.
Lithotype: White marble (MGS 1.30 mm). Isotopic and petrofabric analysis indicates Hymettus or Mani(?) as source (see ch. 3).
Description in Accession: "Marmo greco"; "Pentelic."
Collection: F.W. Kelsey, "Odeum, Patras," Greece, 1901. Accessioned 1901.
Note: Kelsey's group was at Patras on March 19, 1901 (Pedley 2012, 67).

498. KM 93472 table 3.1; fig. 3.2c; pl. 48
Description: Paving or veneer, L. 16.5, W. 12.2, D. 6.3 cm, no preserved edges.
Lithotype: *Pentelic.* White marble with fine grain (MGS 0.95 mm). Isotopic and petrographic analysis indicates Pentelikon as source (see ch. 3).
Description in Accession: "Pentelic marble."
Collection: F.W. Kelsey, Baths of Diocletian, Rome, "found October 1900". Accessioned 1901.
Note: Accession book notes, "from bit of concrete," hence debris from an earlier building reused as aggregate.

499. KM 2012 table 3.1; fig. 3.2b; pl. 48
Description: Paving or veneer, L. 12.1, W. 11.8, D. 6.3 cm, two preserved edges.
Lithotype: *Pentelic.* White marble with medium grain (MGS 1.15 mm). Isotopic and petrographic analysis indicates Pentelikon as source (see ch. 3).
Description in Accession: "Pentelic marble."
Collection: Kelsey, Baths of Diocletian, Rome, October 1900. Accessioned 1901.
Note: Accession book notes, "from bit of concrete," hence debris from an earlier building reused as aggregate.

500. KM T2018.1 pl. 48
Description: Quarry or sculptural debris, L. 16.4, W. 13.4, max. D. 6.2 cm, no worked edges.
Lithotype: *Pentelic* (contextual).
Collection: F.W. Kelsey, "Piece from block cut for monument of Von Moltke, Berlin, North quarry, Pentelicus, Arthur Hill" (on which see ch. 2), (1901). Accessioned 2018.
Note: Kelsey's group was at Athens in mid-April 1901, according to Pedley 2012, 70.

501. KM 2340 pl. 48
Description: Quarry debris, L. 14.2, W. 7.3, D. 7.5 cm, no preserved edges.
Lithotype: *Pentelic* (contextual).
Collection: F.W. Kelsey, "new quarry, north side," Mt. Pentelikon, Athens, (1901). Accessioned 1901.
Note: Kelsey's group was at Athens in mid-April 1901, according to Pedley 2012, 70.

502. KM 2013 pl. 48
Description: Quarry debris, L. 18.9, W. 8.1, D. 1.9 cm, no preserved edges.
Lithotype: *Pentelic* (contextual).
Collection: F.W. Kelsey, "new quarry, above monastery, collected on trip[?]," Mt. Pentelikon, Athens, "Greek Trip, 1900–01." Accessioned 1901.
Note: Kelsey's group was at Athens in mid-April 1901, according to Pedley 2012, 70.

Greek and Anatolian white marbles (cont.)

503. KM 2014 **pl. 48**

Description: Architectural fragment with curved lip, L. 11.3, W. 8.4, D. 3.1 cm, cut and polished surface seems modern.

Lithotype: *Pentelic* (contextual).

Collection: F. W. Kelsey, "old quarry, w. of entrance to cave," Mt. Pentelikon, Athens, "Greek Trip, 1900–01." Accessioned 1901.

Note: Kelsey's group was at Athens in mid-April 1901, according to Pedley 2012, 70.

504. KM 2018 **fig. 2.35a; pl. 49**

Description: Candelabrum fragment, height of column fragment 21.0 cm, diameter 11.3 cm; twisting double helix shaft, edge of band of twisting filets turns into acanthus leaves.

Lithotype: *Pentelic* (visual).

Collection: F. W. Kelsey, purchased from F. Romano, (Rome), 1900. Accessioned 1901.

505. KM 2008 **pl. 49**

Description: Paving or veneer, L. 6.6, W. 13.5, D. 3.8 cm, no preserved edges, marks of point chisel and quarry pick on bottom.

Lithotype: *Pentelic* or Hymettan marble, streaks of beige-pink, muscovite and schist (visual; MGS 0.40 mm).

Collection: F. W. Kelsey, Eleusis, Greece, 1901. Accessioned 1901.

Note: Kelsey's group was at Eleusis in mid-April 1901, according to Pedley 2012, 70.

506. KM 6747 **pl. 49**

Description: Paving, veneer, or *opus sectile*, L. 14.7, W. 7.5, D. 1.9 cm, both edges preserved. Lozenge shape is apparently modern cutting.

Lithotype: White marble with fine grain (MGS 0.40 mm), possibly *Pentelic*.

Collection: E. B. Van Deman, (Italy), no collection date. Accessioned 1938.

Note: Since Van Deman did not collect stone not used in antiquity, this must have been intended as a specimen of an ancient variety.

507. KM 6819 **pl. 49**

Description: Paving or veneer, L. 10.2, W. 2.4, D. 1.0 cm, three preserved edges.

Lithotype: White marble, medium grain size

Description in Accession: "Parian Marble."

Collection: F. W. Kelsey, no provenance, no collection date. Accessioned 1938 (posthumous).

508. KM 2371 **pl. 49**

Description: Unworked or internal fragment, L. 8.8, W. 7.1, D. 2.8 cm, no preserved edges, sawn and polished surface is modern. "CORINTO" is inscribed in block letters 0.4 cm tall on this surface by the same hand that inscribed cat. no. 551. Since this is the Italian form of the toponym, it seems likely that these samples were prepared for Kelsey by a *marmista*, probably F. Romano (who is mentioned as cutting and polishing cat. no. 272), after Kelsey's return from the Greek trip of the spring of 1901.

Lithotype: White-grayish marble with fine grain (MGS 0.50 mm).

Description in Accession: "Pentelic" (but this is excluded by the small MGS).

Collection: F. W. Kelsey, Corinth, Greece, 1901. Accessioned (1901).

Note: May 4, 1901, is the only date on which Kelsey is known to have been at Corinth; see Pedley 2012, 73.

509. KM 25596 **pl. 49**

Description: Paving or veneer, L. 9.5, W. 11.2, D. 2.7 cm, no preserved edges.

Lithotype: White marble with gray bedding planes and coarse grain, possibly *m. Proconnesium*.

Collection: F. W. Kelsey, Astura, Italy, March 2, 1901. Accessioned (1901).

Note: See note at cat. no. 176.

510. KM 2006 **table 3.1; fig. 3.2a; pl. 49**

Description: Paving or veneer, L. 13.0, W. 12.5, D. 4.8 cm, no preserved edges.

Lithotype: Coarse-grained (MGS 2.25 mm) dolomitic marble, probably *m. Thasium*. Isotopic and petrographic analysis indicates Cape Vathy on the island of Thasos as the source (see ch. 3).

Collection: F. W. Kelsey, "Found Jan. 23, 1901 . . . In front of 38, Via Firenze," Rome (although a few columns over in the accession book the find date is January 21, 1901). Accessioned 1901.

511. KM 2005 **pl. 49**

Description: Paving or veneer, L. 12.5, W. 17.7, D. 1.5 cm.

Lithotype: White-gray marble with coarse grain (MGS 2.80 mm). Probably *m. Thasium*, dolomitic marble from Cape Vathy on the island of Thasos.

Description in Accession: "Carrara Palombino"; "Pentelic" (Vecchiatto); "Greek marble badly stained by earth" (Boni).

Collection: F. W. Kelsey, "Tor Paterno," Italy, 1901. Accessioned 1901.

512. KM 93794 **pl. 49**

Description: Paving or veneer, L. 17.6, W. 15.3, D. 3.68 cm, one preserved edge.

Lithotype: White marble with large grain (MGS not readable), probably *m. Thasium*, dolomitic marble from Cape Vathy on the island of Thasos.

Collection: G. De Criscio, Pozzuoli or vicinity, no collection date. Accessioned 1923.

513. KM 2025 **pl. 49**

Description: Paving, veneer, or *opus sectile*, L. 5.0, W. 2.5, D. 2.1 cm, one preserved edge.

Lithotype: White marble with medium-coarse grain (MGS 1.50 mm).

Collection: G. De Criscio, Pozzuoli or vicinity, no collection date. No accession date.

514. KM 29642 **pl. 49**
Description: Paving or veneer, L. 9.4, W. 8.7, D. 1.7 cm, two preserved edges.
Lithotype: White marble with coarse grain (MGS 3.20 mm) and short gray streaks characteristic of *greco scritto*, but MGS rules out both sources.
Collection: J. D. Candler, "Olympia," Greece, (1867 or 1868?). Accessioned 1964.

515. KM 2016 **fig. 2.38d; pl. 50**
Description: Paving or veneer, L. 17.8, W. 11.5, D. 4.4 cm, no preserved edges, pick marks on bottom.
Lithotype: White-beige marble with coarse grain (MGS 3.30 mm).
Description in Accession: "Parian marble"; "Polished Rome."
Collection: F. W. Kelsey, Eleusis, Greece, "Greek Trip, 1900–01." Accessioned 1901.
Note: Kelsey's group was at Eleusis in mid-April 1901, according to Pedley 2012, 70.

Greco scritto

Sources: Cap de Garde and Mt. Filfila, Algeria; Hasançavuşlar, western Turkey

516. KM 2168 **pl. 50**
Description: Paving or veneer, L. 13.2, W. 11.0, D. 0.98 cm, two preserved edges.
Lithotype: *Greco scritto.*
Description in Accession: "Bigio scritto" (Orlandi); written on object: "bianca e nero."
Collection: F. W. Kelsey, "Bought of F. Romano," (Rome), (1900–1901). Accessioned 1901.

517. KM 2170 **fig. 2.7b; pl. 50**
Description: *Opus sectile* rectangle, L. 18.0, W. 6.8, D. 2.36 cm, four preserved edges.
Lithotype: *Greco scritto.*
Description in Accession: "marmor scritto."
Collection: F. W. Kelsey, "Bought of old man," Rome, March 13, (1901). Accessioned 1901.

518. KM 2344 **pl. 50**
Description: Paving or veneer, or possibly a very worn *opus sectile* square, L. 12.8, W. 12.8, D. 1.61 cm, three preserved edges.
Lithotype: *Greco scritto.*
Description in Accession: "marmor scritto."
Collection: F. W. Kelsey, "Purchased of old man," (Rome), March 13, 1901. Accessioned 1901.

519. KM 2441 **pl. 50**
Description: Paving or veneer, L. 10.4, W. 8.3, D. 1.33 cm, no preserved edges.
Lithotype: *Greco scritto.*
Collection: F. W. Kelsey, no provenance, no collection date. Accession date unknown.

520. KM T2008.63 **pl. 50**
Description: Paving or veneer? L. 7.4, W. 4.6, D. 1.8 cm, one preserved edge.
Lithotype: *Greco scritto.*
Collection: (F. W. Kelsey, purchased from Père R. P. Delattre, Carthage, Tunisia, March 2, 1893). Accessioned 2008.
Note: This piece may originally have come from Cap de Garde, Algeria. See note at cat. no. 135.

521. KM 25584 **fig. 2.21e; pl. 50**
Description: Fillet, L. 9.1, W. 2.6, D. 1.45 cm, one preserved edge, rounded edge, molding.
Lithotype: *Greco scritto.*
Collection: F. W. Kelsey, "Antium" (modern Anzio), Italy, 1901. Accessioned (1901).

522. KM 2150 **pl. 50**
Description: Paving or veneer, L. 13.3, W. 10.2, D. 1.38 cm, two preserved edges.
Lithotype: *Greco scritto.*
Description in Accession: "marmor moschiato" (= fly-speck marble).
Collection: F. W. Kelsey, (Italy), (1900–1901). Accessioned 1901.

523. KM 2169 **pl. 50**
Description: Paving, veneer, or *opus sectile*, L. 11.4, W. 8.5, D. 1.6 cm, three preserved edges.
Lithotype: *Greco scritto* (MGS ca. 1.7 mm).
Description in Accession: "Marmor scritto."
Collection: F. W. Kelsey, (Italy), (1900–1901). Accessioned (1901).

524. KM 6742 **fig. 2.20d; pl. 50**
Description: Fragment of dado or orthostat, L. 18.0, W. 10.7, D. 1.33 cm, upper and lower edges preserved.
Lithotype: *Greco scritto.*
Description in Accession: "Bigio greco fiorito."
Collection: E. B. Van Deman, (Italy), no collection date. Accessioned 1938.

525. KM 6745 **pl. 51**
Description: Paving or veneer, L. 17.3, W. 12.3, D. 1.3 cm, no preserved edges, two pieces glued together.
Lithotype: *Greco scritto.*
Description in Accession: Written on object: "EBVD Bigio dislocato."
Collection: E. B. Van Deman, (Italy), no collection date. Accessioned 1938.

526. KM 29637 **pl. 51**
Description: Paving or veneer, L. 4.8, W. 5.1, D. 0.86 cm, no preserved edges.
Lithotype: *Greco scritto.*
Collection: J. D. Candler, no provenance, (1860s?). Accessioned 1964.

Nero antico

Sources: Ain el Ksir and Jebel Aziza/Jebel Oust, Tunisia; Mani Peninsula and island of Chios, Greece; Göktepe, near Aphrodisias, western Turkey

527. KM 2161 **pl. 51**
Description: Paving or veneer, L. 20.31, W. 12.40, D. 2.25 cm, one preserved edge, bottom surface sawn.
Lithotype: *Nero antico.*
Description in Accession: "Nero antico"; "nero di Como, modern" (Orlandi); "modern" (Boni) (the provenance and architectural function make it unlikely that the piece is modern).
Collection: F.W. Kelsey, Baths of Diocletian, Rome, October 1900. Accessioned 1901.
Bibliography: Gazda and Clarke 2016, 226, no. 124.

528. KM 2162 **pl. 51**
Description: Modillion? L. 11.0, W. 11.8, D. 7.6 cm, four preserved edges.
Lithotype: *Nero antico* or very fine-grained black limestone.
Description in Accession: "modern carving."
Collection: F.W. Kelsey, purchased from F. Romano, (Rome), (1900–1901). Accessioned 1901.

529. KM 6748 **fig. 2.6i; pl. 51**
Description: *Opus sectile* equilateral triangle, L. 22.0, W. 22.0 (originally 22.0 cm all sides), D. 2.05 cm, two preserved edges.
Lithotype: *Nero antico.*
Description in Accession: "Nero antico."
Collection: E.B. Van Deman, (Italy), no collection date. Accessioned 1938.
Bibliography: Gazda and Clarke 2016, 226, no. 125.

Prestige Level 4

Luna bardiglio, bigio antico, and other gray marbles

Luna bardiglio, bigio antico, and other gray marbles

Sources: Carrara (ancient Luna), Italy; Moria, Lesbos, Greece; Belevi, Sığacık (ancient Teos), and Göktepe, western Turkey; Mani Peninsula, island of Chios, island of Rhodes, Greece; other sources are numerous

530. KM T2008.60 **pl. 51**
Description: Paving or veneer? L. 6.8, W. 6.3, D. 1.4 cm, no preserved edges.
Lithotype: *m. Lunense: Luna bardiglio* or *bigio antico?*
Collection: (F.W. Kelsey, purchased from Père R.P. Delattre, Carthage, Tunisia, March 2, 1893). Accessioned 2008.
Note: This piece may originally have come from Cap de Garde, Algeria. See note at cat. no. 135.

531. KM 2153 **pl. 51**
Description: Paving or veneer, L. 18.2, W. 11.9, D. 5.3 cm, no preserved edges, mortar adhering to bottom, saw marks on same surface.
Lithotype: *m. Lunense: Luna bardiglio* (visual).
Description in Accession: "Bigio scritto (variato Orlandi)"; "Bardiglio moschiato"; "Lunese" (Boni).
Collection: F.W. Kelsey, Baths of Diocletian, Rome, October 1900. Accessioned 1901.

532. KM 2342 **pl. 51**
Description: Paving or veneer, L. 16.4, W. 12.3, D. 6.2 cm, four preserved edges, surface polished.
Lithotype: *m. Lunense: Luna bardiglio* (visual).
Description in Accession: "Bardiglio"; "Bigio (morato) venato."
Collection: F.W. Kelsey, (Italy), (1900–1901). Accessioned 1901.

533. KM 2154 **pl. 51**
Description: Paving or veneer, L. 16.5, W. 11.2, D. 1.58 cm, no preserved edges.
Lithotype: *m. Lunense: Luna bardiglio* (visual).
Description in Accession: "Bigio rigato."
Collection: F.W. Kelsey, (Italy), (1900–1901). Accessioned 1901.

534. KM 2158 **pl. 51**
Description: Paving or veneer, L. 22.0, W. 15.0, D. 2.14 cm, no preserved edges.
Lithotype: *m. Lunense: Luna bardiglio* (visual).
Description in Accession: "bigio fasciato (striato)."
Collection: F.W. Kelsey, Ostia, Italy, March 2, 1901. Accessioned (1901).

535. KM 2159 **fig. 2.24b; pl. 51**
Description: Molding of composite type, L. 17.1, W. 10.5, D. 5.0 cm, four preserved edges.
Lithotype: Gray marble.
Description in Accession: "Bigio fasciato"; "Bigio fiorito"; "Bardiglio fasciato"; "Bigio greco" (Orlandi).
Collection: F.W. Kelsey, Baths of Diocletian, Rome, (1900–1901). Accessioned (1901).

536. KM 2251 **pl. 51**
Description: Paving or veneer, L. 25.6, W. 19.0, D. 3.8 cm, no preserved edges, surface polished.
Lithotype: *m. Lunense: Luna bardiglio* (visual).
Collection: F.W. Kelsey, "Purchased from vecchio," (Italy), 1900–1901. Accessioned 1901.

537. KM 2249 **pl. 52**
Description: Paving or veneer, L. 15.0, W. 11.2, D. 1.26 cm, two preserved edges, surface polished and waxed.
Lithotype: *m. Lunense: Luna bardiglio* (visual).
Description in Accession: "Bardiglio lumacato."

Collection: F.W. Kelsey, "Purchased from vecchio," (Italy), (1900–1901). Accessioned (1901).

538. KM 2357 **pl. 52**
Description: Paving or veneer? L. 14.1, W. 11.5, D. 6.5 cm, no preserved edges.
Lithotype: *m. Lunense: Luna bardiglio*; less likely *bigio antico* or *m. Lesbium*.
Description in Accession: "Bigio."
Collection: F.W. Kelsey, Rome, 1901. Accessioned (1901).

539. KM 2361 **pl. 52**
Description: Polished quarry sample or other formless chunk, L. 11.0, W. 6.5, D. 3.5 cm.
Lithotype: White marble with roughly aligned gray veins, possibly *Luna bardiglio* (s.v. *m. Lunense*).
Description in Accession: "Bardiglio fasciato."
Collection: F.W. Kelsey, Rome, 1901. Accessioned (1901).

540. KM 25598 **pl. 52**
Description: Paving, veneer, or *opus sectile* triangle or broken rectangle, L. ca. 7.5, W. ca. 5, D. ca. 1 cm, two preserved edges, beveled.
Lithotype: *m. Lunense: Luna bardiglio* (visual).
Collection: F.W. Kelsey, Astura, Italy, March 2, 1901. Accessioned (1901).
Note: See note at cat. no. 176.

541. KM 6822 **pl. 52**
Description: Paving or veneer, L. 8.4, W. 6.3, D. 1.1 cm, two preserved edges, both surfaces polished and waxed.
Lithotype: *m. Lunense: Luna bardiglio* (visual).
Collection: F.W. Kelsey, no provenance, no collection date. Accessioned 1938.

542. KM 93844 **pl. 52**
Description: Paving or veneer, L. 7.0, W. 5.6, D. 2.4 cm, no preserved edges, mortar on back.
Lithotype: Luna white marble (visual).
Collection: F.W. Kelsey, Mérida, Spain, January 10–12, 1893. Accessioned 1893.

543. KM 93840 **pl. 52**
Description: Paving or veneer, L. 11.1, W. 7.1, D. 2.3 cm, no preserved edges, two pieces glued together (modern break?).
Lithotype: White-gray marble with coarse grain (MGS ca. 4 mm).
Collection: F.W. Kelsey, Mérida, Spain, January 10–12, 1893. Accessioned 1893.

544. KM 93837 **pl. 52**
Description: Paving or veneer, L. 8.4, W. 6.0, D. 1.9 cm, no preserved edges.
Lithotype: White marble with gray-purple streaks, lack of polished surface impedes identification; *pavonazzetto* (*m. Phrygium*) is unlikely given the provenance from western Spain.
Collection: F.W. Kelsey, Mérida, Spain, January 10–12, 1893. Accessioned 1893.

545. KM 2951 **figs. 2.33c, 2.37b; pl. 52**
Description: Fragment of a *labrum*, L. 28.0, W. 23.0, D. 2.6 cm. Preservation allows reconstruction of original dimensions as 57.5 cm diameter and 15 cm depth.
Lithotype: *m. Lunense: Luna bardiglio* (visual).
Collection: G. De Criscio, Pozzuoli or vicinity, no collection date. Accessioned 1923.

546. KM 93850 **fig. 2.7c; pl. 52**
Description: *Opus sectile* rectangle, L. 14.8, W. 9.1, D. 0.89–1.2 cm, three preserved edges.
Lithotype: *m. Lunense: Luna bardiglio* (visual).
Collection: G. De Criscio, Pozzuoli or vicinity, no collection date. Accessioned 1923.

547. KM 93849 **pl. 52**
Description: Paving or veneer, L. 12.6, W. 10.1, D. 2.05 cm, no preserved edges.
Lithotype: Gray marble, probably *Luna bardiglio* (s.v. *m. Lunense*).
Collection: G. De Criscio, Pozzuoli or vicinity, no collection date. Accessioned 1923.

548. KM 6743 **pl. 52**
Description: Paving or veneer, L. 22.5, W. 17.0, D. 1.86 cm, one preserved edge.
Lithotype: Gray marble with sinuous white and darker gray areas, probably *Luna bardiglio* (s.v. *m. Lunense*).
Description in Accession: "Bigio intrecciato."
Collection: E. B. Van Deman, (Italy), no collection date. Accessioned 1938.

549. KM 29628 **pl. 53**
Description: Paving or veneer, L. 9.5, W. 7.6, D. 1.2 cm, no preserved edges.
Lithotype: *m. Lunense*: *Luna bardiglio* (visual) or possibly *cipollino grigio*.
Collection: J. D. Candler, no provenance, (1860s?). Accessioned 1964.

550. KM 2156 **fig. 2.29k; pl. 53**
Description: Column fragment, L. 20.6, W. 14.8, D. 4.4 cm, reconstructed diameter is 41 cm; modern surface polished from internal surface.
Lithotype: *Bigio antico*? *Bardiglio* is excluded by the MGS of 1.2 mm.
Description in Accession: "Bigio chiaro (fiorito)"; "Balthium" (Boni).
Collection: F.W. Kelsey, purchased from F. Romano, "Al Porto Antico, Ostia," 1900–1901. Accessioned 1901.

551. KM 2370 **pl. 53**
Description: Paving or veneer, L. 10.1, W. 6.2, D. 2.08 cm, no preserved edges. "CORINTO" is incised on the polished surface in block letters cut with a fine-point chisel, by the same hand as the same word on cat. no. 508.

Luna bardiglio and other gray marbles (cont.)

Since this is the Italian form of the toponym, it seems likely that these samples were prepared for Kelsey by a *marmista*, probably F. Romano, after his return from the Greek trip of the spring of 1901.
Lithotype: Dark gray marble, possibly the *bigio antico* from Siğacık, Turkey (ancient Teos) since it is mottled like *africano* but lacks red or green patches.
Collection: F.W. Kelsey, Corinth, Greece, 1901. Accessioned 1901.
Note: May 4, 1901, is the only date on which Kelsey is known to have been at Corinth (Pedley 2012, 73).

552. KM 2348 **pl. 53**
Description: Small column fragment or drill core, L. 4.13, diameter 11.0 cm, one preserved edge. (For another possible drill core, see cat. no. 3.)
Lithotype: Gray marble not further identifiable, or white marble with gray streaks under dingy surface.
Collection: F.W. Kelsey, Corinth, Greece, 1901. Accessioned 1901.
Note: May 4, 1901, is the only date on which Kelsey is known to have been at Corinth (Pedley 2012, 73).

553. KM 2363 **pl. 53**
Description: Paving or veneer, L. 19.8, W. 10.0, D. 4.5 cm, no preserved edges, marks of round chisel and small pick on bottom surface, top surface polished.
Lithotype: *Bigio antico* or possibly *Luna bardiglio*.
Description in Accession: "Sette Basi."
Collection: F.W. Kelsey, Rome, 1901. Accessioned 1901.

554. KM 2157 **pl. 53**
Description: Paving or veneer, L. 15.0, W. 10.6, D. 4.11 cm, three edges preserved in outline but badly abraded; surface given modern polish but probably is original.
Lithotype: Gray marble with algae blooms and light orange swirls; fine grain (MGS 0.90 mm).
Collection: F.W. Kelsey, "Tunnel under Quirinal," Rome, February 14, 1901. Accessioned 1901.
Note: Construction on the Umberto I tunnel under the Quirinal was begun just as Kelsey arrived in Rome for the 1900–1901 academic year and was dedicated October 20, 1902; see ch. 2.

555. KM 25585 **pl. 53**
Description: Paving or veneer, L. 11.6, W. 6.8, D. 1.88 cm, one preserved edge.
Lithotype: Gray mottled marble, possibly from Lesbos. Compare to cat. no. 196.
Collection: F.W. Kelsey, "Antium" (modern Anzio), Italy, 1901. Accessioned (1901).

556. KM 6741 **pl. 53**
Description: Paving or veneer, L. 19.4, W. 10.5, D. 1.14 cm, one preserved edge.
Lithotype: Gray marble, probably not *Luna bardiglio* because of the orange blooms of iron oxide.
Description in Accession: "Bigio morata Radicellato ~~giallo~~ dorato"; "Bigio, at any rate."
Collection: E. B. Van Deman, (Italy), no collection date. Accessioned 1938.

Prestige Level 3

m. Lunense, unidentified white marble, *palombino*

m. Lunense (white)

Source: Carrara (ancient Luna), Tuscany, Italy

557. KM 1999 **pl. 53**
Description: Internal fragment or unworked quarry debris, L. 13.8, W. 5.6, D. 5.0 cm, no preserved edges.
Lithotype: *m. Lunense: Luna ordinario* (visual and contextual).
Description in Accession: "Marmo Greco?"
Collection: F.W. Kelsey, Carrara, Italy, September 1900. Accessioned 1901.

558. KM 2000 **pl. 53**
Description: Internal fragment? L. 13.5, W. 4.4, D. 3.8 cm, no preserved edges.
Lithotype: *m. Lunense: Luna ordinario* (visual and contextual).
Description in Accession: "Marmo Greco?"
Collection: F.W. Kelsey, Carrara, Italy, September 1900. Accessioned 1901.

559. KM 1996 **pl. 53**
Description: Debris or internal fragment, L. 10.0, W. 13.8, D. 9.3 cm, no preserved edges.
Lithotype: *m. Lunense: Luna ordinario* (visual and contextual).
Description in Accession: "Bit of Carrara marble."
Collection: F.W. Kelsey, Carrara, Italy, September 1900. Accessioned 1901.

560. KM 2022 **table 3.1; fig. 3.2d; pl. 54**
Description: Sarcophagus fragment, L. 14.3, W. of sarcophagus walls 4.7, D. 5.9 cm, no preserved edges.
Lithotype: White marble with fine grain (MGS 0.65 mm). Isotopic and petrographic analysis indicates *Luna ordinario* from Carrara (see ch. 3).
Description in Accession: "Carrara antica" (Orlandi).
Collection: F.W. Kelsey, Baths of Diocletian, Rome, October 1900. Accessioned 1901.

561. KM 1997 **pl. 54**
Description: Paving or veneer, L. 14.0, W. 14.7, D. 2.6 cm, two preserved edges.
Lithotype: *m. Lunense: Luna ordinario* (visual).
Collection: F. W. Kelsey, "Found In front of 38, Via Firenze," Rome, January 21, 1901. Accessioned 1901.

562. KM 2023 **table 3.1; figs. 2.20e, 3.2e; pl. 54**
Description: Dado or orthostat with molding, L. 10.2, W. 18.5, D. 3.6 cm, three preserved edges.
Lithotype: White marble with fine grain (MGS 0.75 mm). Isotopic and petrographic analysis indicates *Luna ordinario* from Carrara (see ch. 3).
Description in Accession: "white marble Pentellic."
Collection: F. W. Kelsey, "Tunnel under Quirinal," Rome, February 14, 1901. Accessioned 1901.
Note: For the findspot, see note to cat. no. 554.

563. KM 2439 **pl. 54**
Description: Paving or veneer, L. 10.8, W. 6.0, D. 2.1 cm, no preserved edges.
Lithotype: White marble with gray tones, medium to coarse grain.
Collection: F. W. Kelsey, (Italy), (1900–1901). Accessioned 1901.

564. KM 6820 **pl. 54**
Description: Paving or veneer, L. 6.1, W. 8.4, D 1.9 cm, one preserved edge.
Lithotype: *m. Lunense: Luna ordinario?* (visual).
Collection: F. W. Kelsey, no provenance, no collection date. Accessioned 1938.

565. KM 6821 **pl. 54**
Description: Paving or veneer, L. 7.2, W. 8.3, D 2.0 cm, no preserved edges, worn by water?
Lithotype: *m. Lunense: Luna ordinario?* (visual).
Collection: F. W. Kelsey, no provenance, no collection date. Accessioned 1938.

566. KM 93853 **pl. 54**
Description: Paving or veneer, L. 23.3, W. 12.0, D. 3.4 cm, no preserved edges.
Lithotype: White marble with coarse grain (MGS 2.10 mm), possibly *m. Thasium*, dolomitic marble from Cape Vathy on the island of Thasos.
Collection: G. De Criscio, Pozzuoli or vicinity, no collection date. Accessioned 1923.

Unidentified white marbles

567. KM 93532 **fig. 2.20f; pl. 54**
Description: Dado or orthostat with molding, L. 13.2, W. 12.7, D. 2.7 cm, no preserved edges.
Lithotype: White marble.
Collection: J. D. Candler, "From Church of St. John, Ephesus," Turkey, (September 1867). Accessioned 1964.
Note: Collection date based on cat. no. 291.

568. KM 93539 **fig. 2.13b; pl. 54**
Description: *Opus sectile* figural piece in form of a leaf, L. 9.59, W. 3.75, D. 1.11–1.39 cm, tip of leaf broken, incised vein lines with no paint, edges beveled to aid insertion.
Lithotype: White marble with fine grain (MGS ca. 0.5 mm).
Collection: J. D. Candler, no provenance, 1860–1890. Accessioned 1964.
Note: Probably part of a set with cat. nos. 569 and 570; see ch. 2.

569. KM 93541 **fig. 2.13c; pl. 54**
Description: *Opus sectile* figural piece in form of a leaf, L. 9.52, W. 3.68, D. 0.96–1.11 cm, one edge broken, incised vein lines with no paint.
Lithotype: White marble with fine grain.
Collection: J. D. Candler, no provenance, 1860–1890. Accessioned 1964.
Note: Probably part of a set with cat. nos. 568 and 570; see ch. 2.

570. KM 93542 **fig. 2.13d; pl. 54**
Description: *Opus sectile* figural piece in form of a leaf, L. 8.40, W. 4.45, D. 0.89–1.13 cm, upper edge of leaf broken, incised vein lines with no paint.
Lithotype: White marble with fine grain (MGS not readable).
Collection: J. D. Candler, no provenance, 1860–1890. Accessioned 1964.
Note: Probably part of a set with cat. nos. 568 and 569; see ch. 2.

571. KM 29651 **pl. 54**
Description: Paving, L. 5.9, W. 5.7, D. 1.9 cm, no preserved edges.
Lithotype: White marble.
Collection: J. D. Candler, "Portion of the floor of St. Peter's, Rome," (1867?). Accessioned 1964.
Note: 1867 was the year of Candler's Grand Tour.

572. KM 93546 **fig. 2.13g; pl. 55**
Description: *Opus sectile* figural piece, L. 5.61, W. 2.92, D. 0.44 cm, S-curve of tapering width, incised lines following curved edges.
Lithotype: White marble, or possibly pale *giallo antico*.
Collection: J. D. Candler, no provenance, 1860–1890. Accessioned 1964.
Note: Too thin to be a match with cat. nos. 568 and 569; see ch. 2.

Palombino

Source: Tolfa, Lazio, Italy

573. KM 93540 **fig. 2.13e; pl. 55**
Description: *Opus sectile* figural piece in form of a leaf, L. 7.67, W. 2.92, D. 0.6 cm, one edge broken, incised border lines emphasized with paint.
Lithotype: *Palombino?*
Collection: J. D. Candler, no provenance, 1860–1890. Accessioned 1964.
Note: Part of a set with cat. no. 574; see ch. 2.

Palombino (cont.)

574. KM 93545 **fig. 2.13f; pl. 55**
Description: *Opus sectile* figural piece in form of a leaf, L. 7.2–8.4, W. 3.4, D. 0.8 cm, tip of leaf broken.
Lithotype: *Palombino?*
Collection: J. D. Candler, no provenance, 1860–1890. Accessioned 1964.
Note: Part of a set with cat. no. 573; see ch. 2

575. KM 93547 **fig. 2.13h; pl. 55**
Description: *Opus sectile* figural piece, L. 6.68, W. 4.46, D. 0.70–0.84 cm, incised border lines (no paint) and vegetal design within.
Lithotype: *Palombino?*
Collection: J. D. Candler, no provenance, 1860–1890. Accessioned 1964.

576. KM 93548 **fig. 2.13i; pl. 55**
Description: *Opus sectile* figural piece, L. 5.72, W. 3.55, D. 0.46 cm, two pieces glued together; incised border lines, no design within.
Lithotype: *Palombino?*
Collection: J. D. Candler, no provenance, 1860–1890. Accessioned 1964.
Note: Matches with cat. nos. 577, 578, and 579; see ch. 2.

577. KM 93549 **fig. 2.13j; pl. 55**
Description: *Opus sectile* figural piece, L. 4.03, W. 2.57, D. 0.51 cm, edges sharply beveled to aid insertion; incised border lines, no design within; curling lip of leaf?
Lithotype: *Palombino?*
Collection: J. D. Candler, no provenance, 1860–1890. Accessioned 1964.
Note: Matches with cat. nos. 576, 578, and 579; see ch. 2.

578. KM 93550 **fig. 2.13k; pl. 55**
Description: *Opus sectile* figural piece, L. 4.09, W. 3.28, D. 0.63 cm, incised border lines; flower bud?
Lithotype: *Palombino?*
Collection: J. D. Candler, no provenance, 1860–1890. Accessioned 1964.
Note: Matches with cat. nos. 576, 577, and 579; see ch. 2.

579. KM 93551 **fig. 2.13l; pl. 55**
Description: *Opus sectile* figural piece, L. 2.77, W. 3.04, D. 0.61 cm, incised border lines, no design within; section of body of leaf?
Lithotype: *Palombino?*
Collection: J. D. Candler, no provenance, 1860–1890. Accessioned 1964.
Note: Matches with cat. nos. 576, 577, and 578; see ch. 2.

580. KM 93569 **pl. 55**
Description: Paving or veneer, L. 5.58, W. 6.16, D. 1.1 cm, one preserved edge.
Lithotype: *Palombino* or white marble.
Collection: J. D. Candler, no provenance, 1860–1890. Accessioned 1964.

581. KM 2167 **pl. 55**
Description: Paving or veneer, L. 13.2, W. 8.4, D. 1.31 cm, no preserved edges.
Lithotype: White marble with roughly aligned gray streaks somewhat resembling *greco scritto* (MGS 2.0 mm).
Collection: F. W. Kelsey, (Italy), (1900–1901). Accessioned (1901).

582. KM 2315 **fig. 2.8d; pl. 55**
Description: *Opus sectile* lozenge, L. 14.0, W. 8.0, D. 2.5, four preserved edges.
Lithotype: White marble.
Collection: F. W. Kelsey, "bought of farmer about two miles north of Arricia," Italy, February 18, (1901). Accessioned 1901.
Note: The entry in the accession book is difficult to read. The collection place could also be read as "Arezzo," but since Kelsey was still in Rome as of February 15, Arricia in the Alban hills seems more likely than Arezzo in Tuscany.

583. KM 2003 **fig. 2.8e; pl. 55**
Description: *Opus sectile* hexagon fragment, L. 14.0, W. 7.1, D. 2.0 cm, most of three sides preserved.
Lithotype: Beige limestone, possibly *palombino*.
Description in Accession: "Bit of palombino."
Collection: F. W. Kelsey, (Rome?), 1901. Accessioned 1901.

584. KM 25590 **fig. 2.24a; pl. 56**
Description: Composite molding, ogee or *cyma reversa*, L. 10.3, W. 2.4, D. 2.5 cm, top and bottom edges preserved, both sides broken.
Lithotype: *Palombino* (visual).
Collection: F. W. Kelsey, "Hadrian's Villa and adjoining sites, Palestrina," March 2, 1901. Accessioned (1901 or 1902).

585. KM 25594 **pl. 56**
Description: Paving or veneer, L. 7.7, W. 6.4, D. 1.5 cm, no preserved edges.
Lithotype: White marble.
Collection: F. W. Kelsey, Astura, Italy, March 2, 1901. Accessioned (1901).
Note: See note at cat. no. 176.

586. KM 23397 **fig. 2.21f; pl. 56**
Description: Fillet, L. 14.6, W. 3.4, D. 1.3 cm, four preserved edges, one long edge beveled at 60 degrees.
Lithotype: White marble with rose streaks (MGS 0.6 mm).
Collection: F. W. Kelsey, Ostia, no collection date. Accessioned 1924.
Note: For possible use as a teaching tool, see note at cat. no. 141.

587. KM 29688 **pl. 56**
Description: Mosaic tessera, L. 1.1, W. 1.1, D. 1.0 cm, four preserved edges.
Lithotype: White limestone.
Collection: P. O. E. Burton, Delos, Greece, no collection date. Accessioned 1961.

Note: The accession book notes, "Mosaic piece from a Private House on the Island of Delos, Greece." This artifact is unique among Burton donations in having a recorded provenance other than printed commercial labels; this one, written on the container bag, was presumably transcribed from an original note. It is also the most suggestive evidence that Miss Burton made one or more collecting trips in Europe, which we would in any case be tempted to infer from several pieces which were not much used or known outside their native locales (cat. nos. 665, 670).

588. KM 29714 **pl. 56**
Description: Mosaic tessera, L. 1.98, W. 2.00, D. 1.19 cm, four preserved edges.
Lithotype: White marble or limestone.
Collection: P. O. E. Burton, no provenance, no collection date. Accessioned 1961.

589. KM 6746 **pl. 56**
Description: Paving, veneer, or *opus sectile*, L. 16.8, W. 8.8, D. 3.1 cm, four preserved edges, three preserved corners.
Lithotype: Creamy limestone, *palombino*?
Description in Accession: "Palombino?"; "Worked as floor tile?"
Collection: E. B. Van Deman, (Italy), no collection date. Accessioned 1938.
Note: The thickness of the piece, unusual for an *opus sectile* element, may have prompted the suggestion of reworking in the accession note; see ch. 2.

590. KM 6807 **fig. 2.13a; pl. 56**
Description: *Opus sectile* figural piece in the shape of an acanthus leaf, L. 11.6, W. 7.6, D. 1.6 cm, four preserved edges, incised line representing the leaf's middle vein.
Lithotype: White marble.
Collection: E. B. Van Deman, (Italy), no collection date. Accessioned 1938.

591. KM 774 **fig. 2.34c; pl. 56**
Description: Fragment of a krater rim, L. 6.8, W. 13.8, D. 1.8 cm, profile is vertical for only 1.3 cm below the top, then begins a convex curve; three preserved edges.
Lithotype: White marble.
Collection: C. L. Meader, "Purchased for the University . . . in the spring of 1898, (Rome). Accessioned 1898.
Note: Rome is presumed because Meader also bought lamps there in the same year.

592. KM 1969.5.3d **pl. 56**
Description: Paving or veneer, L. 5.6, W. 4.1, D. 1.36 cm, no preserved edges.
Lithotype: Microcrystalline white limestone, possibly *palombino* (dirt on surface impedes identification).
Collection: J. G. Winter, no provenance, no collection date. Accessioned 1969.
Note: Inventoried together with cat. no. 160 = three fragments of *giallo antico*.

Prestige Level 2

Basalt for paving, limestones, metamorphic clay, sandstones, stones chiefly associated with late Republican decoration

Basalto

Source: Alban Hills south of Rome, Italy

593. KM 2212 **pl. 56**
Description: Quarry debris, L. 10, W. 7, D. 6 cm, no preserved edges.
Lithotype: *Basalto.*
Description in Accession: "Basalt"; "Salce" (i.e., *selce* or basalt typically used for road metal).
Collection: F. W. Kelsey, Frascati, Italy, October 1900. Accessioned (1901).
Note: Kelsey evidently made several expeditions to quarries of selce in the Alban Hills in the fall of 1900.

594. KM 2213 **pl. 56**
Description: Quarry debris, L. 10.0, W. 8.0, D. 0.68–1.91 cm, one preserved edge.
Lithotype: *Basalto.*
Collection: F. W. Kelsey, "from rock near station Monte Comprati [*sic*] . . . Hammered at side of railway cutting," December 1900. Accessioned 1901.
Note: The town of Monte Compatri is on the southeastern outskirts of Rome.

Sperone

Source: Gabii, Italy

595. KM 2291 **pl. 57**
Description: Edge or corner of an architectural element, L. 6.9, W. 6.2, D. 3.7 cm, no preserved edges, mortar on two sides.
Lithotype: *Sperone, l. Gabinus.*
Collection: F. W. Kelsey, "Bottom of trench for foundations of new building," Baths of Diocletian, Rome, October 1900. Accessioned 1901.

Limestones

596. KM 2267 **fig. 2.40; pl. 57**
Description: Internal fragment of an architectural block, L. 19.0, W. 16.5, D. 6.3 cm, three partial edges with claw chisel marks, quarry pick marks on bottom.
Lithotype: White limestone.

Limestones (cont.)

Collection: F.W. Kelsey, "from ancient block used in rebuilding of s. wall of Olympieum," Athens, (1901). Accessioned 1901.
Note: Kelsey's group was at Athens in mid-April 1901, according to Pedley 2012, 70.

597. KM 25572 **pl. 57**
Description: Paving or veneer, L. 11.7, W. 6.2, D. 1.3 cm, no preserved edges.
Lithotype: *Travertino*, *l. Tiburtinus.*
Collection: F.W. Kelsey, "Collected in Rome, purchased in 1901." No accession date.

598. KM 2021 **pl. 57**
Description: Paving or veneer, L. 12.8, W. 9.0, D. 0.9 cm, no preserved edges.
Lithotype: Fossiliferous creamy limestone, probably *palombino*. It cannot be travertine, which is a depositional rock formed by precipitation from hot water and therefore contains no fossils.
Description in Accession: "Travertine"; "Quarry near Subiaco"; "Palombino Latiato" (Orlandi).
Collection: F.W. Kelsey, Subiaco, Italy, 1900. Accessioned 1901.

599. KM 2542 **pl. 57**
Description: Paving or veneer, L. 6.9, W. 5.6, D. 2.1 cm, two preserved edges.
Lithotype: *Travertino*, *l. Tiburtinus.*
Collection: E. Jones, (Rome), (1885). Accessioned (1912).

600. KM 2004 **pl. 57**
Description: Paving or veneer, L. 13.6, W. 6.4, D. 2.5 cm, mortar on edges.
Lithotype: Cream-colored limestone, *palombino*?
Collection: F.W. Kelsey, "Place marked Rudei [Ruderi] on staff map 145-III (Cassoli) S.E. of Roviano & E of Anticoli Corrado near Mola Nuova," province of L'Aquila, Italy, February 15, 1901. Accessioned 1901.
Note: Same provenance and collection date as cat. nos. 219 and 452.

601. KM 2116 **pl. 57**
Description: Paving or veneer? L. 13.4, W. 10.9, D. 2.8 cm, one preserved edge, not polished.
Lithotype: Sandstone.
Collection: F.W. Kelsey, purchased from F. Romano, (Rome), (1900–1901). Accessioned 1901.

602. KM 93795 **pl. 57**
Description: Paving or veneer, L. 13.3, W. 8.57, D. 2.05 cm, no preserved edges.
Lithotype: Gray limestone.
Collection: G. De Criscio, Pozzuoli or vicinity, no collection date. Accessioned 1923.

603. KM 29620 **pl. 57**
Description: Fragment of casing block? L. 21.2, W. 9.4, D. 2.5 cm, no preserved edges, surface is pitted from exposure.
Lithotype: Light gray limestone.
Collection: J. D. Candler, Cairo, "From the Great Pyramid, 'Cheops,'" October 4, 1867. Accessioned 1964.

604. KM 29699 **pl. 57**
Description: Paving or veneer (if in fact from Italy)? L. 12.9, W. 7.8, D. 3.7 cm, one preserved edge.
Lithotype: Black limestone.
Description in Accession: "Nero, improperly called Serpentino."
Collection: P. O. E. Burton, no provenance, no collection date. Accessioned 1961.

Prestige Level 1

Casual collection, surface finds, unworked samples

605. KM 2134 **pl. 58**
Description: Unworked, L. 9.7, W. 7.1, D. 6.6 cm, no preserved edges.
Lithotype: Obsidian?
Description in Accession: "Cipollino venato nero"; "A note on slip says 'Arenaria antica Volcano da Vesuvio come una cinere ma dura come il diaspro.'" (The suggested identifications run the geological gamut: veined marble, sandstone, pumice, jasper.)
Collection: F.W. Kelsey, Baths of Diocletian, Rome, December 1900. Accessioned 1901.

606. KM 25581 **pl. 58**
Description: Beach cobble or wave-eroded artifact, L. 10.8, W. 9.2, D. 2.26 cm, no preserved edges.
Lithotype: Gray marble(?) with white striations.
Collection: F.W. Kelsey, "Antium," (modern Anzio), Italy, 1901. Accessioned (1901).
Note: The imperial villa at Anzio (generally called "Nero's" although it spans five centuries) has been overwhelmed by the sea (Coarelli 1982, 295–96); thus Kelsey's piece may well have come from a visit to the remnants of the villa rather than a day on the beach.

607. KM 2011 **pl. 58**
Description: Quarry debris, L. 10.2, W. 7.5, D. 5.6 cm, no worked edges.
Lithotype: Quartz.
Description in Accession: "Crystal"; "alabastro bianco."

Collection: F. W. Kelsey, "mouth of cave, Mt. Pentelicus," Mt. Pentelikon, Athens, 1901. Accessioned 1901.
Note: Kelsey's group was at Athens in early April 1901, according to Pedley 2012, 70.

608. KM 2321 **pl. 58**
Description: Unworked surface sample or internal fragment, L. 12.2, W. 11.5, D. 2.5 cm.
Lithotype: Unidentifiable.
Collection: F. W. Kelsey, Langada Gorge, Greece, "Greek Trip, 1900–01." Accessioned 1901.
Note: Kelsey and party crossed the Langada Gorge in late March 1901 (Pedley 2012, 69).

609. KM 2322 **pl. 58**
Description: Unworked surface sample, L. 17.8, W. 11.2, D. 1.52 cm.
Lithotype: Slate.
Description in Accession: "Schistose rock."
Collection: F. W. Kelsey, Langada Gorge, Greece, "Greek Trip, 1900–01." Accessioned 1901.
Note: Kelsey and party crossed the Langada Gorge in late March 1901 (Pedley 2012, 69).
Bibliography: Gazda and Clarke 2016, 226, no. 127.

610. KM 2237 **pl. 58**
Description: Unworked cobble, L. 12.5, W. 8.6, D. 5.9 cm, rounded from water action, one surface cut and polished.
Lithotype: Black microcrystalline limestone with quartz veining.
Description in Accession: "Bianco e nero antica."
Collection: F. W. Kelsey, "Tor Paterno, Castel Fusano," Italy, (1900–1901). Accessioned 1901.

611. KM 2149 **pl. 58**
Description: Architectural element or paving with beveled edges, L. 11.5, W. 9.3, D. 2.8 cm, one preserved curving and beveled edge.
Lithotype: Green slate or siltstone.
Description in Accession: "Malachite bastarda"; "Verde Argilla"; "Verde Prato." (These proposed identifications are geologically unrelated to each other.)
Collection: F. W. Kelsey, purchased from F. Romano, (Rome), (1901). Accessioned 1901.
Bibliography: Gazda and Clarke 2016, 225, no. 122.

612. KM 29717 **pl. 58**
Description: Unworked fragment, L. 3.0, W. 3.7, D. 2.5 cm.
Lithotype: Obsidian.
Collection: P. O. E. Burton, no provenance, no collection date. Accessioned 1961.

Modern and/or non-Mediterranean Samples

Alabaster

613. KM 29661 **pl. 58**
Description: Modern sample? L. 5.9, W. 3.0, D. 1.2 cm, one preserved edge.
Lithotype: *Alabastro.*
Description in Accession: "Alabastro."
Collection: P. O. E. Burton, no provenance, no collection date. Accessioned 1961.

614. KM 29687 **pl. 58**
Description: Modern molding sample, L. 10.6, W. 8.0, D. 1.5 cm, four preserved edges.
Lithotype: *Alabastro di Volterra* is the closest visual match (see OUMNH nos. 334–35), if the sample is in fact from Italy; cf. cat. no. 598).
Description in Accession: "Alabastro? (Called Italian Onyx, but is very soft. Gypsum?)"
Collection: P. O. E. Burton, no provenance, no collection date. Accessioned 1961.

Botticino

615. KM 29669 **pl. 59**
Description: Modern(?) veneer or molding, L. 10.9, W. 6.7, D. 2.8 cm, three preserved edges, dowel hole in one edge.
Lithotype: *Botticino.*
Description in Accession: "Marble: Botticino (unpolished) Light dusky gray. Bigio."
Collection: P. O. E. Burton, no provenance, no collection date. Accessioned 1961.
Note: Burton's three *botticino* samples (cat. nos. 615–617) are consistent in small fossils (although abundance varies), stylolites, color, and micro-fine grain.

616. KM 29670 **pl. 59**
Description: Paving or veneer? L. 10.2, W. 10.1, D. 1.8 cm, no preserved edges.
Lithotype: *Botticino.*
Description in Accession: "Marble: Botticino – polished (Nomenclature appears confused, as this piece has little relation to the preceding specimen [cat. no. 615])."
Collection: P. O. E. Burton, no provenance, no collection date. Accessioned 1961.
Note: See note at cat. no. 615.

617. KM 29671 **pl. 59**
Description: Modern sample? L. 10.5, W. 6.0, D. 2.2 cm, three preserved edges.
Lithotype: *Botticino.*
Description in Accession: "Marble: Botticino – polished."

Botticino (cont.)

Collection: P. O. E. Burton, no provenance, no collection date. Accessioned 1961.
Note: See note at cat. no. 615.

Breccia

618. KM 2146 **pl. 59**
Description: Modern sample as recut. L. 10.5, W. 8.1, D. 4.0 cm, six modern edges preserved.
Lithotype: Breccia di Hereke, or more accurately, pudding-stone of Hereke since the clasts here are rounded. This stone was much used in Constantinople/Istanbul from the early Byzantine to the Ottoman periods and likely caught Kelsey's eye from its use in Ottoman buildings in Greece.
Collection: F. W. Kelsey, "bought in Athens," "Athens, Acropolis, said to be rock of Areopagus" (dealer's attribution?)," Greek Trip, 1900–01." Accessioned 1901.
Note: Kelsey made several visits to the Acropolis between April 7 and 17, 1901 (Pedley 2012, 70).

619. KM 29674 **pl. 59**
Description: Modern sample plaque, L. 11.9, W. 9.2, D. 2.6 cm, one preserved edge.
Lithotype: Macchia Vecchia di Arzo.
Description in Accession: "Brocatello rosso, called Breche Vecchia Meghia."
Collection: P. O. E. Burton, no provenance, no collection date. Accessioned 1961.
Note: Arzo is in the province of Lugano, Switzerland. This stone was widely used in the northern Italian baroque and is prominent in the cathedral of Milan (see Bugini and Folli 2013, reference thanks to L. Lazzarini).

620. KM 29673 **pl. 59**
Description: Modern sample plaque, L. 7.7, W. 7.6, D. 1.1 cm, four preserved edges.
Lithotype: Brèche Nouvelle, which is still quarried at the fishing village of Port-la-Nouvelle, Languedoc-Roussillon, southern France (Price 2007, 144), sometimes subsumed under the name *brèche Rouge* (see https://www.christies.com/lot/lot-a-french-breche-rose-marble-wall-fountain-5127894/?).
Description in Accession: "Breccia Rosastro, called Breche Rose."
Collection: P. O. E. Burton, from a Chicago marble importer. Accessioned 1961.
Note: The label affixed to the piece is torn. What is left is hard to decipher: "[...]se Marble (Impo[...][rters and Contractors, established 1871, Wentworth Ave. Chicago."

621. KM 29703 **pl. 59**
Description: Rectangular strip, not evidently molding, probably modern, L. 17.3, W. 4.4, D. 2.6 cm, two preserved edges, both ends broken.
Lithotype: Light beige breccia.
Description in Accession: Written on the object's back surface: "lumachella" in ink with accession number, "Tavernelle" in pencil. However, this is not a *lumachella* since there are no fossils. Such breccias are very common in Italy.
Collection: P. O. E. Burton, no provenance, no collection date. Accessioned 1961.
Note: Tavernelle is a village in the commune of Licciana Nardi, in the Lunigiana region of the province of Massa and Carrara, Italy. It is at the foot of the Apuan Alps, which may be the provenance of this sample, although no lithotype is associated with this Tavernelle in the literature.

622. KM 29692 **pl. 59**
Description: Paving or veneer, probably not ancient, L. 9.2, W. 8.2, D. 3.3 cm, no preserved edges.
Lithotype: Escalettes.
Description in Accession: "Escalette (Pink and yellow w/ green veining – rather similar to Giallo Antico)."
Collection: P. O. E. Burton, no provenance, no collection date. Accessioned 1961.
Note: Escalettes was quarried in the French Pyrenees near the Pass of Escalettes, in the department of Ariège, apparently entirely in the early modern period into the twentieth century (https://quarriesandbeyond.org/, accessed October 12, 2017).

623. KM 29702 **pl. 59**
Description: Modern sample plaque, L.15.1, W. 10.1, D. 2.3 cm.
Lithotype: Beige breccia resembling *breccia gialla*, a subvariety of *giallo di Siena*, from which its commercial name, Sienna, was drawn.
Collection: P. O. E. Burton, Gray Knox Marble Company, Knoxville, Tennessee, no collection date. Accessioned 1961.
Note: The Gray Knox Marble Company owned at least three quarries, including that of Gray Knox in the suburbs of Knoxville; it was in operation from 1917 to 1965 (https://news.utk.edu/2018/11/01/new-exhibit-honors-marble-history-of-facilities-services-site/).

Granites, serpentinites, and similar

624. KM 29682 **pl. 59**
Description: Modern sample plaque, L. 7.0, W. 5.0, D. 0.5 cm, four preserved edges; the closely spaced and parallel saw marks on the back are not ancient.
Lithotype: Fine-grained gray granite similar to that of Kozak Dağ.
Collection: P. O. E. Burton, no provenance, no collection date. Accessioned 1961.

625. KM 29716 **pl. 59**
Description: Modern industrial quarry artifact or possibly fragment of half-round molding, if ancient, L. 11.7, D. 3.2 cm, cylinder split lengthwise exactly in two, both ends broken.
Lithotype: Red granite.

Collection: P. O. E. Burton, no provenance, no collection date. Accessioned 1961.

626. KM 29678 pl. 60

Description: Modern sample plaque, L. 13.0, W. 7.5, D. 0.8 cm, four preserved edges.

Lithotype: Red granite. Identified by the label affixed to the piece as "American Beauty Supreme."

Collection: P. O. E. Burton, North Star Granite Corporation, St. Cloud, Minnesota, no collection date. Accessioned 1961.

Note: A roadside marker erected by the North Star Granite Corporation in 1930 commemorates the origins of the granite business in 1868 and its contemporary importance ("St. Cloud Historical Marker," at *Minnesota Historic Roadside Structures and Historic Places* (website). http://roadsidestructures.blogspot.com/2013/10/st-cloud-historical-marker.html, accessed August 27, 2017).

627. KM 29709 pl. 60

Description: Modern sample plaque, L. 114.8, W. 5.64, D. 2.04 cm, one preserved edge.

Lithotype: Granite.

Description in Accession: "Granite (Granito rosso) w/ quartz vein. Tennessee."

Collection: P. O. E. Burton, Tennessee, no collection date. Accessioned 1961.

628. KM 29680 pl. 60

Description: Modern sample plaque, L. 7.8, W. 7.0, D. 1.1 cm, four preserved edges.

Lithotype: Granite resembling Aswan but not Egyptian.

Description in Accession: "Granito Rossastro or Corallino."

Collection: P. O. E. Burton, no provenance, no collection date. Accessioned 1961.

629. KM 29684 pl. 60

Description: Object not identifiable, L. 13.8, W. 9.7, D. 3.1 cm, finished and polished on one narrow edge.

Lithotype: Granite resembling Troad *m. troadense, granito violetto.*

Collection: P. O. E. Burton, no provenance, no collection date. Accessioned 1961.

630. KM 29679 pl. 60

Description: Paving or veneer? L. 11.6, W. 4.4, D. 4.3 cm, no preserved edges, large dowel hole in one corner.

Lithotype: Red Peterhead granite?

Description in Accession: "Granito Rosso Grande."

Collection: P. O. E. Burton, no provenance, no collection date. Accessioned 1961.

Note: Price 2007, 223.

631. KM 2246 pl. 60

Description: Paving or veneer, L. 7.9, W. 13.2, D. 3.5 cm, two preserved edges.

Lithotype: *Rosso di Levanto.*

Description in Accession: "Verde fiorito rosso Greco" (Orlandi).

Collection: F. W. Kelsey, Baths of Diocletian, Rome, January and July 1901. Accessioned (1901).

Note: If quarrying of *rosso di Levanto* began only in the sixteenth century as asserted by Price (2007, 179), Romano or a similar scout may have taken advantage of Kelsey's enthusiasm for buying pieces from the excavations of the Baths of Diocletian.

632. KM 29691 pl. 60

Description: Modern sample? L. 13.3, W. 10.0, D. 2.5 cm, one preserved edge.

Lithotype: *Rosso di Levanto*, if in fact the sample is from Italy.

Description in Accession: "Marble: Rosso e verde, called Levantine."

Collection: P. O. E. Burton, no provenance, no collection date. Accessioned 1961.

Note: See MUSNAF no. 1873; Price 2007, 179.

633. KM 29662 pl. 60

Description: Paving or veneer?, L. 10.5, W. 3.6, D. 7.6 cm, one preserved edge.

Lithotype: *Verde Alpi*, Alps Green or Green Alps.

Description in Accession: "Marble Verde (Called Alps Green)."

Collection: P. O. E. Burton, no provenance, no collection date. Accessioned 1961.

Note: Quarried in the Val d'Aosta from the early twentieth century, still continuing at a diminished level. See Price 2007, 185; although this specific name is not listed, it is consistent with the descriptions and illustrations.

634. KM 29664 pl. 60

Description: Modern sample, L. 12.6, W. 9.7, D. 2.1 cm, two preserved edges.

Lithotype: *Verde Alpi*?

Description in Accession: "Verde Antico (Called Antico Verde)." (This identification seems inaccurate since the characteristic black inclusions are absent here.)

Collection: P. O. E. Burton, no provenance, no collection date. Accessioned 1961.

Note: For source information, see note at cat. no. 633.

635. KM 29700 pl. 60

Description: Modern sample plaque, L. 20.3, W. 10.8, D. 1.4 cm, four preserved edges.

Lithotype: *Verde antique* marble from Vermont.

Description in Accession: "Verde Antico, improperly called Serpentino, from Roxbury, Vermont." (The website of the Vermont Department of Environmental Conservation states that "*verde antique* marble" is "actually a metamorphosed ultramafic hydrous magnesium silicate containing minerals similar to chlorite and talc"; http://dec.vermont.gov/geological-survey/resources-energy/minres/marble, accessed September 7, 2017.)

Collection: P. O. E. Burton, no provenance, no collection date. Accessioned 1961.

Note: The Vermont Geological Survey states that the quarries at Rochester were opened in the 1920s (a photo of the *verde antique* quarry at Roxbury in the Hood Museum at Dartmouth College dates to 1925; https://hoodmuseum.dartmouth.edu/objects/2015.72.1, accessed September 7, 2017), but the Roxbury Verd-Antique

Granites, serpentinites, and similar (cont.)

Marble Company was founded in 1857 and supplied stone for the Capital Extension project in Washington, D.C. It is still in operation.

636. KM 29701 **pl. 60**

Description: Modern sample plaque, L. 15.2, W. 6.2, D. 2.1 cm, three preserved edges.

Lithotype: *Verde antique* marble.

Description in Accession: "Verde Antico, improperly called Serpentino."

Collection: P. O. E. Burton, no provenance, no collection date. Accessioned 1961.

Note: For source information, see note at cat. no. 635.

637. KM 6785 **pl. 61**

Description: Modern sample, L. 13.6, W. 13.6, D. 1.2 cm, no preserved edges.

Lithotype: Orbicular granite.

Description in Accession: "EB Van Deman Orbicular gabbro"; "Lumacella?"

Collection: E. B. Van Deman, (Italy), no collection date. Accessioned 1938.

Note: Price (2007, 229) notes that orbicular granite with its deceptively organic figure was discovered in 1785 on Corsica and was subsequently popular in Paris. It is not hard to imagine that the market of Rome also attracted some samples, one of which caught Van Deman's eye, although she must have known that it was atypical.

638. KM 29715 **pl. 61**

Description: Paving or veneer, modern? L. 7.6, W. 6.2, D. 3.2 cm, bottom surface broken, two preserved edges.

Lithotype: Green quartz, prasiolite?

Collection: P. O. E. Burton, no provenance, no collection date. Accessioned 1961.

639. KM 2219 **pl. 61**

Description: Paving or veneer? L. 18.5, W. 10.8, D. 6.0 cm, no preserved edges.

Lithotype: *Verde di Prato*, not used in antiquity.

Collection: F. W. Kelsey, purchased from F. Romano, (Rome), January or February 1901. Accessioned 1901.

Note: Cf. OUMNH nos. 179–80; Price 2007, 178, although the photo there is atypical of the dark spots of enstatite.

640. KM 6739 **pl. 61**

Description: Modern sample? L. 12.8, W. 10.7, D. 1.9 cm, no preserved edges, pick or chisel marks on back.

Lithotype: *Verde di Prato.*

Description in Accession: "Verde di Prato (modern)."

Collection: E. B. Van Deman, (Italy), no collection date. Accessioned 1938.

Note: Quarried at Prato, near Florence, Italy. Cf. OUMNH nos. 579–81.

641. KM 2201 **pl. 61**

Description: Paving or veneer, L. 16.0, W. 14.1, D. 3.6 cm, one preserved edge with saw marks, wax on polished surface.

Lithotype: Black marble with the characteristic dark inclusions of *verde di Prato*, not granite.

Description in Accession: "Granito verde"; "Granitello cristallino" (Orlandi); "Granito al Erbetta, Granito nero. difficile a trovare."

Collection: F. W. Kelsey, purchased from F. Romano, (Rome), 1900–1901. Accessioned 1901.

642. KM 6734 **pl. 61**

Description: Paving or veneer, not clear if ancient, L. 18.6, W. 10.9, D. 1.3 cm, one preserved edge (noted in accession book, "EDGED ONE SIDE"), wax on polished surface and notation on that surface in carefully incised letters with serifs, "WAX" with a left-facing arrow below pointing toward a surface void at the seam between the two pieces of a (recent?) break that constitute the piece.

Lithotype: Serpentinite.

Description in Accession: "Verde Tarquinio."

Collection: E. B. Van Deman, (Italy), no collection date. Accessioned 1938.

Note: The closest visual match is Corsi's *verde ranocchia* (OUMNH no. 568), which is not the well-known Egyptian *ophites* (*Granito verde della Sedia di San Lorenzo*) and is of unknown origin. "Verde Tarquino" from the accession book may perhaps point toward Tarquinia, but there are generally similar serpentinites from Varallo near Novara (OUMNH nos. 569–70), the Val di Susa north of Torino (OUMNH nos. 571–72), and Calabria (OUMNH no. 573).

Limestones and travertines

643. KM 29693 **pl. 61**

Description: Modern sample plaque, L. 19.5, W. 13.5, D. 2.1 cm, four preserved edges.

Lithotype: Limestone with fine grain, no fossils or calcite infills, resembles Indiana limestone.

Description in Accession: Label on object: "Sample of Cartha[g?]e (Clear) Marble, from the Lloyd Brothers Co."

Collection: P. O. E. Burton, Lloyd Brothers Company, Sylvania, Ohio, no collection date. Accessioned 1961.

Note: Lloyd Brothers was founded in 1846 and went out of business in the 1990s. It was active in the Detroit area, where it may have come to Miss Burton's attention.

644. KM 6781 **pl. 61**

Description: Paving or veneer, L. 12.8, W. 9.6, D. 2.1 cm, one preserved edge, mortar on back, wax on polished surface.

Lithotype: *Lumachella di San Vitale,* from quarries near Roveré, north of Verona (Lazzarini 2014; OUMNH no. 921; Price 2007, 168; Antonelli et al. 2003).

Collection: E. B. Van Deman, (Italy), no collection date. Accessioned 1938.

645. KM 6778 **pl. 61**
Description: Modern sample plaque? L. 13.9, W. 8.5, D. 1.1 cm, three preserved edges.
Lithotype: Fossiliferous fine limestone.
Description in Accession: "Lumacella [*sic*] perlifera gialla."
Collection: E. B. Van Deman, (Italy), no collection date. Accessioned 1938.

646. KM 29694 **fig. 2.21g; pl. 61**
Description: Fillet (modern?), L. 11.5, W. 10.7, D. 3.6 cm, one preserved edge, triangular profile on front face.
Lithotype: Beige marble in limestone matrix.
Description in Accession: "Pentelic, although this is doubtful" (unfortunately the author of this skeptical note is not known).
Collection: P. O. E. Burton, no provenance, no collection date. Accessioned 1961.

647. KM 29705 **pl. 61**
Description: Paving or veneer, L. 6.34, W. 6.87, D. 2.52 cm, one preserved edge.
Lithotype: *Travertino, l. Tiburtinus?*
Collection: P. O. E. Burton, no provenance, no collection date. Accessioned 1961.

648. KM 29707 **pl. 61**
Description: Modern sample from a molding(?) with a square profile, L. 8.44, W. 2.63, D. 1.72 cm, two preserved edges, ends of strip broken off.
Lithotype: Travertine. The tan color and small, scarce fissures are not consistent with ancient *l. Tiburtinus*.
Description in Accession: "Marble: Travertine."
Collection: P. O. E. Burton, (Gray Knox Marble Company, Knoxville, Tennessee), no collection date. Accessioned 1961.
Note: Probably from the Gray Knox Marble Company of Knoxville, Tennessee, although label is absent; cf. cat. no. 648. For more on the Gray Knox company, see note at cat. no. 623.

649. KM 29708 **pl. 62**
Description: Modern sample plaque, L. 11.15, W. 15.21, D. 2.83 cm, four preserved edges.
Lithotype: Travertine.
Description in Accession: Printed label on back: "Roman Travertine from Gray Knox Marble Company, Knoxville, Tennessee. [...] Submitted in connection our quotation November 25th."
Collection: P. O. E. Burton, Gray Knox Marble Company, Knoxville, Tennessee, no collection date. Accessioned 1961.
Note: For more on the Gray Knox company, see note at cat. no. 623.

650. KM 29704 **pl. 62**
Description: Fragment of modern strip, L. 15.2, W. 7.1, W. 2.2 cm, two preserved edges.
Lithotype: Beige travertine (not the Roman travertine from Tivoli).
Collection: P. O. E. Burton, no provenance, no collection date. Accessioned 1961.

Marbles and fine-grained limestones

651. KM 2239 **pl. 62**
Description: Modern sample from quarry debris, L. 13.1, W. 9.2, D. 6.14 cm, one surface polished.
Lithotype: *Bianco e nero antico, m. Celticum.*
Description in Accession: "Paesina fasciata"; "modern (Diaspro di Sicilia)."
Collection: F. W. Kelsey, Ostia, (1900–1901). Accessioned (1901).

652. KM 29686 **pl. 62**
Description: Modern sample plaque, L. 15.2, W. 10.1, D. 0.7 cm, four worked edges.
Lithotype: Blue-gray marble.
Description in Accession: "Verde cippolino (called cippolino)" (almost certainly erroneously).
Collection: P. O. E. Burton, no provenance, no collection date. Accessioned 1961.

653. KM 29712 **pl. 62**
Description: Modern sample plaque, L. 15.32, W. 11.25, D. 1.84 cm, four preserved edges.
Lithotype: Carrara white marble.
Description in Accession: "Marble: Bigio"; written on object: "White Italian."
Collection: P. O. E. Burton, no provenance, no collection date. Accessioned 1961.

654. KM 29675 **pl. 62**
Description: Modern sample plaque, L. 15.3, W. 10.3, D. 0.9 cm, four preserved edges.
Lithotype: Carrara white marble.
Description in Accession: "Bianco di Carrara."
Collection: P. O. E. Burton, no provenance, no collection date. Accessioned 1961.

655. KM 29676 **pl. 62**
Description: Modern sample? L. 9.1, W. 6.7, D. 2.3 cm, three preserved edges.
Lithotype: Carrara white marble.
Description in Accession: "Bianco di Carrara."
Collection: P. O. E. Burton, no provenance, no collection date. Accessioned 1961.

656. KM 29690 **pl. 62**
Description: Modern sample plaque? L. 10.3, W. 7.5, D. 2.2 cm, three preserved edges, dowel hole in one edge.
Lithotype: White marble with beige veins.
Description in Accession: "Bianco e Giallo, called Creamo."
Collection: P. O. E. Burton, no provenance, no collection date. Accessioned 1961.
Note: Such marbles are very common worldwide; for example, Crema Venato from India and Crema Marfil from Spain are on the market today.

Marbles and fine-grained limestones (cont.)

657. KM 2250 **pl. 62**
Description: Modern sample, L. 9.6, W. 6.6, D. 1.9 cm, four preserved edges.
Lithotype: *Luna bardiglio* (visual).
Description in Accession: "Bardiglio Chiaro (Blue-Gray)"; "modern, no value."
Collection: F. W. Kelsey, purchased from F. Romano, (Rome), (1900–1901). Accessioned 1901.

658. KM 29574 **pl. 62**
Description: Modern plaster plaque painted as *giallo di Siena*, L. 15.5, W. 11.8, D. 1.5 cm.
Lithotype: None
Collection: "Piece of imitation Siena Marble or Giallo Antico, presented by a Miss Wanere on January 10, 1896," no provenance. No accession date.

659. KM 6776 **pl. 62**
Description: Modern sample? L. 10.9, W. 7.5, D. 0.3 cm, four preserved edges.
Lithotype: *Lumachella nera.*
Description in Accession: "Lumacella nera."
Collection: E. B. Van Deman, (Italy), no collection date. Accessioned 1938.

660. KM 6780 **pl. 63**
Description: Modern sample plaque, L. 14.3, W. 8.4, D. 1.4 cm, four preserved edges.
Lithotype: *Astracane* or *lumachella.*
Description in Accession: Written on back of object: "Astracane or Lumachella EBVD."
Collection: E. B. Van Deman, (Italy), no collection date. Accessioned 1938.

661. KM 6733 **pl. 63**
Description: Modern sample plaque? L. 22.4, W. 11.3, D. 1.7 cm, two preserved edges.
Lithotype: *Astracane di Verona* or *Astracane dorata.*
Description in Accession: "Yellow marble"; "Lumachella gialla bigiastra minuta?"
Collection: E. B. Van Deman, (Italy), no collection date. Accessioned 1938.
Note: Known sources of this fossiliferous variety of *lumachella* are Verona, Italy, and Henchir el Kasbat (ancient Thuburbo Maius), in Tunisia. Both types were quarried in antiquity as well as later, but Van Deman's interest is likely to have been taken by the ancient use.

662. KM 29667 **pl. 63**
Description: Paving or veneer, modern, L. 10.1, W. 15.1, D. 2.5 cm, two preserved edges.
Lithotype: Belgian black marble.
Description in Accession: "Nero (Called Blue Belge, from Belgium)."
Collection: P. O. E. Burton, no provenance, no collection date. Accessioned 1961.
Note: From the well-known quarries around Namur in southern Belgium (cf. Price 2007, 78).

663. KM 2163 **pl. 63**
Description: Modern sample plaque, L. 8.4, W. 6.7, D. 0.77 cm, four preserved edges, evenly beveled.
Lithotype: Very fine-grained black marble or limestone with no inclusions.
Description in Accession: "Said to be from Eleusis."
Collection: F. W. Kelsey, "Purchased on Acropolis, Athens," "Greek Trip, 1900–'01." Accessioned (1901).
Note: Kelsey made several visits to the Acropolis between April 7 and 17, 1901 (Pedley 2012, 70).

664. KM 1933 **pl. 63**
Description: Sample formatted as a plaque, certainly modern in light of the provenance, L. 19.0, W. 11.8, D. 3.0 cm, two pieces glued together, three preserved edges.
Lithotype: Fossiliferous limestone.
Collection: Given to F. W. Kelsey by C. P. Sellards ("Mr. Sellards' father rc'd it from J. P. Morgan"), "Persia," March 16, 1923. Accessioned 1923.

665. KM 29689 **pl. 63**
Description: Modern sample plaque, L. 15.4, W. 10.4, D. 0.9 cm, four preserved edges.
Lithotype: White marble with gray-purple streaks, probably *pavonazzetto Toscano* from the Apuan Alps (Istituto commercio estero 1982, 28, fig. 24).
Description in Accession: Label on the back in pencil: "Italian Pavonazzo, PAVONAZETTO"; accession note: "Bigio & Bianca, called Italian Pavonazzetto."
Collection: P. O. E. Burton, no provenance, no collection date. Accessioned 1961.
Note: From Carrara south to Seravezza in the Apuan Alps. Many colors appear alongside white, offering local counterparts to many of the well-known Roman lithotypes.

666. KM 2238 **pl. 63**
Description: Quarry debris, L. 13.0, W. 12.8, D. 4.32 cm, one modern polished surface. Tool marks on back.
Lithotype: *Portoro.*
Description in Accession: "modern" (Orlandi and Boni); "Giallo e nero antica (Porta Venere)."
Collection: F. W. Kelsey, purchased from F. Romano, (Rome), (1900–1901). Accessioned (1901).

667. KM 29663 **pl. 63**
Description: Modern sample plaque, L. 8.1, W. 6.1, D. 3.2 cm, two preserved edges.
Lithotype: *Portoro.*
Description in Accession: "Giallo & nero (Black and Gold) from Italy."
Collection: P. O. E. Burton, no provenance, no collection date. Accessioned 1961.
Note: For stone source, see note at cat. no. 665.

668. KM 29639 **pl. 63**
Description: Modern(?) sample plaque, L. 6.6, W. 4.8, D. 2.0 cm, three preserved edges.
Lithotype: *Portoro.*
Description in Accession: "Marble: Black veined w/ gold. Nero e giallo (?) marked 'Portoso.'"
Collection: J. D. Candler, no provenance, (1860s?). Accessioned 1964.
Note: For stone source, see note at cat. no. 665.

669. KM 2115 **pl. 63**
Description: Paving or veneer, L. 5.1, W. 3.9, D. 4.5 cm, no preserved edges.
Lithotype: *Portoro?*
Description in Accession: "Alabastro nero."
Collection: F. W. Kelsey, (Italy), (1900–1901). Accessioned 1901.
Note: For stone source, see note at cat. no. 665.

670. KM 29666 **pl. 63**
Description: Modern sample plaque, L. 15.0, W. 10.0, D. 2.3 cm, four preserved edges.
Lithotype: Black and gold marble.
Description in Accession: Paper label on object: "Black & Gold, from Gray Knox Marble Company"; accession note: "gold veining not so pronounced as in Italian spec.)" (well diagnosed).
Collection: P. O. E. Burton, Gray Knox Marble Co., Knoxville, Tennessee, no collection date. Accessioned 1961.
Note: For more on the Gray Knox company, see note at cat. no. 623.

671. KM 1089 **pl. 64**
Description: Modern sample plaque, L. 15.1, W. 15.1, D. 2.2 cm, four preserved edges.
Lithotype: *Rose de Numidie.* The ancient quarries of *Rose de Numidie*, near the village of Kleber (now Sidi Benyebka), Oran province, Algeria, were rediscovered in 1880 by Lambert Playfair. The quarries were revived and "were worked in the latter years of the 19th century and the first part of the twentieth … to decorate churches and public buildings in Europe and the United States" (Price 2007, 113). "Numidian" is not geographically inaccurate to describe this marble but does invite confusion with the famous Roman Numidian (*giallo antico*, q.v.). *Rose de Numidie* was used in the Roman city of Cuicul (modern Djémila, northern Algeria) in the province of Africa, but it is not known to have been exported.
Collection: "Presented [to F. W. Kelsey] by Cyrus L. W. Eidlitz, architect, 1123 Broadway, New York, in June, 1898. Specimens of Numidian marble used by him in the decoration of the entrance hallway of the Washington Life Insurance Company's building, 145 Broadway." Accessioned 1899.
Note: See also cat. nos. 672 and 673.

672. KM 1090 **pl. 64**
Description: Modern sample plaque, L. 18.5, W. 16.6, D. 1.9 cm, four preserved edges.
Lithotype: *Rose de Numidie.*
Collection: "Presented [to F. W. Kelsey] by Cyrus L. W. Eidlitz, architect, 1123 Broadway, New York, in June, 1898. Specimens of Numidian marble used by him in the decoration of the entrance hallway of the Washington Life Insurance Company's building, 145 Broadway." Accessioned 1899.
Note: See also cat. nos. 671 and 673.

673. KM 1091 **pl. 64**
Description: Modern sample plaque, L. 17.7, W. 16.3, D. 2.1 cm, four preserved edges.
Lithotype: *Rose de Numidie.*
Collection: Given to F. W. Kelsey by Cyrus Eidlitz, June 1898. Accessioned 1899.
Note: See also cat. nos. 671 and 672.

674. KM 6752 **pl. 64**
Description: Modern sample plaque, L. 13.0, W. 9.4, D. 0.6 cm, two preserved edges. Broken into two pieces.
Lithotype: *Rosso antico.*
Collection: E. B. Van Deman, unknown German stone merchant, no collection date. Accessioned 1938.
Note: The upper part of the paper label affixed to the back of the sample has been aggressively rubbed away so that the name of the stone merchant is illegible. All that can be read is, "aus ge[…] 1795 … Besitzer von Marmorbrüchen, Ateliers für Architectur u. höhere Plastik." "1795" is possibly the the founding date of the firm. It is not clear how Van Deman obtained this sample from a German quarry that was presumably on the Mani Peninsula.

675. KM 2386 **pl. 64**
Description: Paving or veneer, L. 15.7, W. 19.2, D. 1.9 cm, no preserved edges.
Lithotype: *Rosso Verona*, or less likely *breccia di Mastro Simone.* (cf. OUMNH no. 442).
Description in Accession: "Lumacella"; "Breccia di Simone, modern; Brecciola a occhi, modern."
Collection: F. W. Kelsey, purchased from F. Romano, Ostia, March 2, 1901. Accessioned (1901).
Note: Ancient use of *rosso Verona* seems restricted to Verona itself, while it has been popular in the modern period and is widely available today (Price 2007, 115; OUMNH no. 497).

676. KM 29697 **pl. 64**
Description: Modern sample plaque, L. 15.2, W. 10.1, D. 0.8 cm, four pieces reassembled, four preserved edges.
Lithotype: Probably *rosso Verona.*
Description in Accession: "Marble: Roseato."
Collection: P. O. E. Burton, no provenance, no collection date. Accessioned 1961.
Note: Ancient use of *rosso Verona* seems restricted to Verona itself, while it has been popular in the modern period and is widely available today (Price 2007, 115; OUMNH no. 497). Thus it is unlikely that Burton's sample was ancient, but not impossible since she probably collected cat. no. 619, *Macchia Vecchia di Arzo*, from Lugano, Italy,

Marbles and fine-grained limestones (cont.)

since it was widely used in the norther Italian baroque but seems an unlikely candidate to have made its way to the U.S.

677. KM 29698 **pl. 64**

Description: Modern sample plaque, L. 15.2, W. 5.3, D. 2.2 cm, three preserved edges.

Lithotype: Pink marble, not further identifiable.

Description in Accession: "Rouge d'Aautumn/Autonne (Contemporary French building stone?)." (Commercial marble aggregators do not list any similar names today.)

Collection: P. O. E. Burton, no provenance, no collection date. Accessioned 1961.

678. KM 29668 **pl. 64**

Description: Modern sample plaque, L. 11.4, W. 8.8, D. 1.0 cm, four preserved edges.

Lithotype: Blue Vermont marble.

Description in Accession: "Marble: Bigio (Blue Vermont – from Vermont Marble Co. Proctor, Vt.)."

Collection: P. O. E. Burton, Vermont Marble Company, Proctor, Vermont, no collection date. Accessioned 1961.

Note: The Vermont Marble Company, established in 1880 consolidating several local firms, furnished the stone for the Jefferson Memorial and new Supreme Court Building in Washington, D.C. Of the many varieties of marble in the deposit, the best known was Imperial Danby (see Price 2007, 68). Today's visitor to Vermont is well advised to visit the Vermont Marble Museum in Proctor.

679. KM 29710 **pl. 64**

Description: Modern sample plaque, L. 11.49, W. 7.18, D. 1.14 cm, four preserved edges.

Lithotype: Gray Vermont marble.

Collection: P. O. E. Burton, Vermont Marble Company, Proctor, Vermont, no collection date. Accessioned 1961.

Note: For the Vermont Marble Company, see note at cat. no. 678.

680. KM 29711 **pl. 64**

Description: Modern sample plaque, L. 15.32, W. 12.52, D. 0.94 cm, four preserved edges.

Lithotype: Gray Vermont marble. Paper label affixed to object: "#2 Brandon High St." (possibly a reference to quarries in Brandon, Vermont).

Collection: P. O. E. Burton, Vermont Marble Company, Proctor, Vermont, no collection date. Accessioned 1961.

Note: For the Vermont Marble Company, see note at cat. no. 678.

681. KM 29713 **pl. 64**

Description: Modern sample plaque, L. 14.38, W. 11.55, D. 2.14 cm, three preserved edges.

Lithotype: White Vermont marble. Probably Imperial Danby since it is fine grained, of warm color, and free of impurities.

Description in Accession: "Vermont" written in pencil on side of plaque.

Collection: P. O. E. Burton, no provenance, no collection date. Accessioned 1961.

682. KM 29718 **pl. 65**

Description: Top of a modern marker, purpose unknown, L. 7.4, W. 8.0, D. 5.4 cm; the number "75" is incised on the top in 2.2 cm high letters, carefully centered.

Lithotype: White marble with fine grain, possibly Danby marble from Vermont (see note at cat. no. 678).

Collection: P. O. E. Burton, no provenance, no collection date. Accessioned 1961.

683. KM 2098 **pl. 65**

Description: Modern paving or veneer, L. 17.5, W. 13.6, D. 2.1 cm, one preserved edge.

Lithotype: *Mandolato di Verona* (see OUMNH no. 516).

Description in Accession: "Verona mandolato" (Boni); "Modern."

Collection: F. W. Kelsey, purchased from F. Romano, (Rome), March 9, 1901. Accessioned 1902.

Note: There are a number of similar limestones from the Verona area; there seems to have been no Roman use, making this another case of Romano sweeping up interesting stones for Kelsey without regard for chronology.

684. KM 2222+93466 **pl. 65**

Description: Modern industrial artifact? L. 24.1, W. 5.4, D. 5.6 cm, two preserved edges.

Lithotype: *Verde antico, m. Thessalicum.*

Collection: F. W. Kelsey, no provenance, no collection date. Accessioned (1901).

Note: Dimensions and profile are not consistent with ancient moldings.

685. KM 2226 **pl. 65**

Description: Modern quarry debris, L. 22.7, W. 19.4, D. 11.4 cm, saw marks on flat surface.

Lithotype: *Verde antico, m. Thessalicum.*

Description in Accession: "Verde antico."

Collection: F. W. Kelsey, "New Quarry" at Mt. Pentelikon, "quarry whence columns for Westminster Cathedral are being taken," "Greek Trip," (1901). Accessioned (1901).

Note: The collection data is in part extrapolated from the list of objects inserted at p. 80 in the accession book.

686. KM 2227 **pl. 65**

Description: Modern quarry debris, L. 27.6, W. 10.4, D. 11.1 cm.

Lithotype: *Verde antico, m. Thessalicum.*

Description in Accession: "Verde antico."

Collection: F. W. Kelsey, "New Quarry" at Mt. Pentelikon, "quarry whence columns for Westminster Cathedral are being taken," "Greek Trip," (1901). Accessioned (1901).

Note: The collection data is in part extrapolated from the list of objects inserted at p. 80 in the accession book.

687. KM 2228 **pl. 65**
Description: Modern quarry debris, L. 19.3, W. 13.9, D. 7.1 cm, one preserved edge.
Lithotype: *Verde antico, m. Thessalicum.*
Description in Accession: "Verde antico."
Collection: F.W. Kelsey, "New Quarry" at Mt. Pentelikon, "quarry whence columns for Westminster Cathedral are being taken," "Greek Trip," (1901). Accessioned (1901).
Note: The collection data is in part extrapolated from the list of objects inserted at p. 80 in the accession book.

688. KM 29665 **pl. 65**
Description: Modern sample plaque? L. 4.9, W. 1.7, D. 1.1 cm, one preserved edge.
Lithotype: *Verde antico, m. Thesallcum?*
Description in Accession: "Verde Antico."
Collection: P. O. E. Burton, no provenance, no collection date. Accessioned 1961.

Other

689. KM 6738 **pl. 65**
Description: Modern plaque? L. 11.1, W. 11.2, D. 0.7 cm, four preserved edges.
Lithotype: *Argilla*, clay.
Description in Accession: "Verde argilla"; "Verde di Firenze?"
Collection: E. B. Van Deman, (Italy), no collection date. Accessioned 1938.

Unidentified Lithotypes

690. KM 2393 **pl. 65**
Description: Paving or veneer, L. 26, W. 11.8, D. 2.8 cm, one preserved edge. The piece may have been in a fire.
Lithotype: Unidentified.
Collection: F.W. Kelsey, (Italy), (1900–1901). Accessioned (1901).

691. KM 6862 **fig. 2.41c; pl. 65**
Description: Paving or veneer, L. 11.3, W. 5.2, D. 1.5 cm, four preserved edges, evidence of sawing.
Lithotype: Unidentified.
Description in Accession: "Fiore di Persico?"; "fiorito di Spagna"; written on object's edge: "Stillettite di Spagna."
Collection: F.W. Kelsey, no provenance, no collection date. Accessioned 1938 (posthumous).

692. KM 2327 **pl. 66**
Description: Paving or veneer, L. 17, W. 13.9, D. 3.3 cm, no preserved edges, pick marks on back.
Lithotype: Unidentified. Dark red marble with very fine clasts but not *cipollino rosso, rosso antico*, or *portasanta*.
Collection: F.W. Kelsey, (Italy), (1900–1901). Accessioned 1901.

693. KM 2323 **pl. 66**
Description: Unidentified object, L. 17.1, W. 9.0, D. 1.6 cm, no preserved edges.
Lithotype: Green limestone, not further identifiable.
Collection: F.W. Kelsey, Langada Gorge, "Greek Trip, 1900–01." Accessioned 1901.
Note: Kelsey and party crossed the Langada Gorge in late March 1901, according to Pedley 2012, 69.

694. KM 2191 **pl. 66**
Description: Paving or veneer, L. 22, W. 21.5, D. 4.9 cm, three edges partially preserved.
Lithotype: Pink limestone, not further identifiable.
Description in Accession: "Giallo rosso"; "Broccatellone italiano modern."
Collection: F.W. Kelsey, "Purchased of Vecchio," Ostia, (1900–1901). Accessioned 1901.

695. KM 2148 **pl. 66**
Description: Paving or veneer? L. 11.7, W. 9.7, D. 5.1 cm, no preserved edges.
Lithotype: Unidentified green non-calcareous stone.
Description in Accession: "Malachite verde argilla (very rare)"; "Verde antico orifico" (Orlandi); "One of the light stones of the Abruzzi artificially stained" (Boni).
Collection: F.W. Kelsey, purchased from F. Romano, (Rome), (1901). Accessioned 1901.

696. KM 6749 **pl. 66**
Description: Modern sample plaque mounted on slate(?) backing, L. 18.5, W. 11.0, D. 0.56 cm (inclusive of backing, 1.37 cm), two long edges and one short edge preserved.
Lithotype: Unidentified breccia, unknown if ancient. Maroon groundmass with clasts of gray and white marble, calcite infill.
Description in Accession: Handwritten label affixed to object: "Breccia di Villadriana rossa" (Van Deman).
Collection: E. B. Van Deman, (Italy), no collection date. Accessioned 1938.

697. KM 93838 **pl. 66**
Description: Paving or veneer, L. 11.0, W. 6.4, D. 1.3 cm, one preserved edge.
Lithotype: Unidentified.
Collection: F.W. Kelsey, Mérida, Spain, January 10–12, 1893. Accessioned 1893.

698. KM 93845 **pl. 66**
Description: Paving or veneer, L. 5.7, W. 21.9, D. 4.9 cm, no preserved edges.
Lithotype: Unidentified.
Collection: F.W. Kelsey, Mérida, Spain, January 10–12, 1893. Accessioned 1893.

Unidentified lithotypes (cont.)

699. KM 29653 **fig. 2.7d; pl. 66**
Description: *Opus sectile* rectangle, L. 7.2, W. 6.5, D. 0.76 cm, four preserved edges.
Lithotype: Unidentified marble, superficially resembling the dark variety of Aswan granite.
Collection: J. D. Candler, no provenance, (1867). Accessioned 1964.

700. KM 29644 **pl. 66**
Description: Paving or veneer, L. 7.5, W. 3.3, D. 2.4 cm, three preserved edges.
Lithotype: Unidentified ("Marble: Breccia rosso, called or from Brulo di Francia (Triangular prism)").
Collection: J. D. Candler, no provenance, (1860s?). Accessioned 1964.

Lot Accessions

701. KM 595 **pl. 67**
Description: Paving and veneer, thirteen fragments collected and inventoried together.
Lithotypes:
- Four fragments *africano,* ranging in size: L. 3.7–10.2, W. 2.6–3.6, D. 1–2.1 cm (one has evidence of saw on edges).
- One fragment *portasanta, m. Chium*: L. 12.3, W. 6.7, D. 2.5 cm, two preserved edges.
- One fragment *alabastro* (accession book: "alabastro rosso"): L. 7.3, W. 6.5, D. 1.8 cm, one preserved edge.
- Seven fragments *breccia di Settebasi*? (the pieces have not been cleaned and identification is consequently difficult).

Collection: F. W. Kelsey, Timgad, Algeria, February 8, 1893. Accessioned (1893).

702. KM 2950 **pl. 68**
Description: Bag of 38 small marble/granite fragments, not cleaned.
Lithotype: Most appear to be *porfido verde*, consistent with the accession book note, "1 lot small bits of marble (106 pieces) … All but 37 Verde Antico and Serpentino frags. renumbered 93912" (see cat. no. 703).
Collection: G. De Criscio, Pozzuoli and vicinity, no collection date. Accessioned 1923.
Note: KM 2950 was originally a larger lot of 107 (accession book says 106) pieces of "small bits of marble." At some point it was divided in two. These 38 objects remained as KM 2950; the other 69 pieces were bagged and given the new accession number KM 93912 (see cat. no. 703).

703. KM 93912 **pl. 69**
Description: Bag of 69 miscellaneous small marble/granite. No pieces larger than 5 cm, none appear to have any preserved edges.
Lithotypes: Very wide range of varieties, including one small piece each of *granito verde della sedia di San Lorenzo* (*ophites*) and *granito del foro*.
Collection: G. De Criscio, Pozzuoli and vicinity, no collection date. Accessioned 1923.
Note: See note to cat. no. 702.

704. KM 3055 **pl. 70**
Description: Eighteen very small fragments of marble/granite including three mosaic tesserae, received and accessioned as one lot.
Lithotypes: Varieties restricted to white and gray marble with exception of one piece each of *serpentina* and *granito del foro*.
Collection: G. De Criscio, Pozzuoli and vicinity, no collection date. Accessioned 1923.

Appendix A

GAZETTEER OF QUARRIES

Followed by figures A.1–4, Maps of Quarry Locations.

Italian Name(s)	*Ancient Name(s)*	*Geographic Location*	*Coordinates*
Quarries where attestations of objects are identified as ancient			
Africano	*m. Luculleum*	Quarries around Sığacık (ancient Teos), western Turkey	38.178056, 26.787500
Alabastro cotognino, alabastro Egiziano	*l. alabastrites*	Locations throughout Nile Valley, Egypt (esp. Hatnub)	27.5501, 31.02317
Alabastro fiorito	—	Pamukkale (ancient Hierapolis), western Turkey	38.000344, 29.016703
Alabastro a ghiaccione	—	Pamukkale (ancient Hierapolis), western Turkey La Batteria quarries near Monte Circeo, Lazio, Italy	38.000344, 29.016703 41.232500, 13.052778 and 42.993271, 11.515564
Alabastro a pecorella	—	Aïn Tekbalet and Bou Hanifia, western Algeria	35.133186, -1.200086 35.190178, 0.023105979
Variety A: *Alabastro a tartaruga, marmo di Carrara, "Luna ordinario"* Variety B: *Alabastro di Palombara*	— —	Jano di Montaione, Tuscany, Italy	43.514089, 10.877528
Basalto, selce di Roma, basalt	—	Alban Hills, south of Rome, Italy	41.769722, 12.658889
Variety A: *Basanite, basalto verde, pietra bekhen* Variety B: *Breccia verde d'Egitto*	*l. basanites* *l. hexatontolithos, hecatontalithos*	Both: Wâdi Hammâmât, Eastern Desert, Egypt	25.5866, 33.3349
Bigio antico	*m. Lesbium* (one variety)	Moria, Lesbos, Greece Belevi, western Turkey Sığacık (ancient Teos), western Turkey Göktepe, western Turkey Mani Peninsula, Greece Island of Chios, Greece Island of Rhodes, Greece	39.133333, 26.516389 38.01960, 27.44749 38.178056, 26.787500 37.515601, 28.794195 36.386111, 22.483056 38.384987, 26.135160 —
Breccia di Aleppo	—	Kariès, island of Chios, Greece	38.390350, 26.097967
Breccia cenerina	—	Location unknown	—
Breccia corallina	*m. Sa(n)garium*	Bilecik, Vezirhan, Turkey Karaburun Peninsula, Turkey	40.236111, 30.014722 38.429461, 26.568605
Breccia gialla	—	Montagnola Senese, Siena, Italy	43.144706, 11.364681
Breccia di Hereke	—	Hereke, northwest Turkey	40.789170, 29.614982
Breccia rossa Appenninica	—	Coregna, La Spezia, Italy	44.109167, 9.814722
Variety A: *Breccia di Seravezza* Variety B: *Breccia di medicea*	— —	Seravezza and Stazzema, Apuan Alps, Italy	43.994722, 10.227222
Breccia di Settebasi, *Breccia di Sciro*	*m. Scyrium* —	Renes and nearby inlets, Skyros, Greece	38.777222, 24.636944
Breccia traccagnina	—	Source unknown	—

Italian Name(s)	Ancient Name(s)	Geographic Location	Coordinates
Broccatello, broccatello di Spagna, jaspi de la Cinta de Tortosa	—	Barranc de la Llet, La Cinta, Els Valencians, Tortosa (ancient Dertosa), northeastern Spain	38.808415, 27.554482
Broccatellone	—	Çakmakli, Manisa, Turkey	40.236111, 30.014722
Cipollino mandolato, cipollino dei Pirenei, campan (vert), griotte	—	Haute Pyrenees and Campan, France	43.016944, 0.178333
Variety A: *Cipollino rosso* Variety B: *Rosso brecciato*	*m. Carium, m. Iassense*	Kıyıkışlacık (ancient Iasos), western Turkey	37.279444, 27.582500
Cipollino verde, cipollino bigio, cipollino grigio	*m. Carystium, m. Styrium*	Karystos and Styra, Euboea, Greece	38.050000, 24.465833
Diaspro di Corsica	—	San Pietro di Rostino, Orezza, Corsica	42.362159, 9.403413
Diaspro di Sicilia	—	Multiple locations throughout western Sicily	38.016667, 12.538889
Doliana, marmo di Doliana, marmo di Tegea	—	Ano Doliana, Peloponnese, Greece	37.454167, 22.421111
Fior di pesco, fior di persico	*m. Chalcidicum*	Eretria, Euboea, Greece	38.403889, 23.780833
Gabbro eufotide	—	Wâdi Maghrabîya, Eastern Desert, Egypt	26.1874, 33.2378
Giallo antico	*m. Numidicum*	Chemtou (ancient Simitthus), Tunisia	36.492500, 8.576667
Variety A: *Granito rosso* (Aswan granite), *granito di Siene* Variety B: *Granito nero di Siene*	*l. Syenites* or *m. Syeniten, l. pyrrhopoikilos, l. Aethiopicus, l. Thebaicus*	Numerous locations between Aswan (ancient Syene) and Shellal District, Nile River Valley, Egypt	24.370, 32.5370
Granito bianco e nero	*m. Tibereum*	Wâdi Barud, Eastern Desert, Egypt	26.784167, 33.516944
Granito della colonna	—	Wâdi Umm Shegilat, Eastern Desert, Egypt	26.5662, 33.1491
Granito del foro	*m. Claudianum*	Gebel Fatireh, Wâdi Umm Huyut, Eastern Desert, Egypt	26.809167, 33.285556
Granito verde fiorito di bigio	—	Wâdi Umm Balad, Eastern Desert, Egypt	27.132739, 33.278910
Variety A: *Granito verde della sedia di San Lorenzo* Variety B: *Granito verde della sedia di San Pietro*	Both: *ophites, ofite*	Both: Wâdi Semna and Wâdi Umm Wikala, Eastern Desert, Egypt	26.2584, 33.3982
Greco scritto	—	Cap de Garde, Algeria Mt. Filfila, Algeria Hasançavuşlar, western Turkey	36.963056, 7.781667 36.895833, 7.103889 38.033816, 27.313000
Lapis lazuli	—	Sar-i-Sang, Badakhshan, Afghanistan and Tajikistan	36.115407, 70.475594
Variety A: *Astracane / castracane dorata* Variety B: *Lumachella* Variety C: *Lumachella orientale* Variety D: *Lumachella di San Vitale*	— — — —	Henchir el Kasbat (ancient Thuburbo Maius), Tunisia Numerous quarries from Spain to Asia Minor Val Pantena near Verona, Italy	36.398611, 9.906389 — 45.32934, 11.04495
Variety A: *Luna statuario, Luna ordinario, marmo di Carrara* Variety B: *Luna bardiglio, bardiglio di Carrara* (gray)	*m. Lunense* *m. Lunense*	Both: Carrara (ancient Luna), Apuan Alps, Liguria and Tuscany, Italy	44.092500, 10.126667
Nero antico	—	Ain el Ksir, Tunisia Gebel Aziza/Gebel Oust, Tunisia Mani Peninsula, Greece Island of Chios, Greece Göktepe, western Turkey	36.492500, 8.576667 36.498333, 9.885000 36.483056, 22.433611 38.384987, 26.135160 37.515601, 28.794195

Italian Name(s)	Ancient Name(s)	Geographic Location	Coordinates
Occhio di pavone rosso and *pavonazzo*	*m. Triponticum*	Kutluca and Sakarya (ancient Sangarius) Valley, north central Turkey	40.828670, 30.448822
Palombino, palombino bianco	—	Tolfa, Lazio, Italy	42.17527, 12.12813
Variety A: *Pavonazzetto* Variety B: Dokimeion white marble	*m. Phrygium, m. Synnadicum* *l. Dokimenos* (local usage denoted primarily the white marble)	Isçehisar (ancient Dokimeion or Docimium), Turkey Upper Tembris Valley, Turkey	38.837222, 30.783889 39.057500, 30.110000
Pentelico	*m. Pentelicum*	Mt. Pentelikon, Attica, Greece	38.073889, 23.881944
Porfido bigio, granito bigio morviglione Subvariety: *Granito pediculare* or *pidocchioso*		Dramont, Estérel, southern France	43.415000, 6.853611
Variety A: *Porfido rosso* Variety B: *Porfido verde Egiziano* Variety C: *Porfido nero*	*m. porphyriticum , l. porphyrites, leptopsephos* *l. hieracites* —	Gebel Dokhân (Mons Porphyrites), Eastern Desert, Egypt	27.150299, 33.180580
Porfido verde, serpentino, porfido verde antico, porfido verde di Grecia	*l. Krokeatis, l. Lacedaemonius, m. Lacedaemonium*	Krokees (ancient Krokeai), Peloponnese, Greece	36.885000, 22.548333
Portasanta	*m. Chium*	Latomi, Chios, Greece	38.384987, 26.135160
Marmo di Proconneso	*m. Proconnesium/Proconnense, m. Cyzicenum*	Marmara Island (ancient Prokonnesos), Sea of Marmara (ancient Propontis), Turkey	40.650833, 27.656944
Variety A: *Rosso antico* Variety B: White marble	*m. Taenarium*	Mani Peninsula, Greece	36.483056, 22.433611
Sperone, peperino	*l. Gabinus*	Gabii, Italy	41.887222, 12.715278
Marmo tasio	*m. Thasium*	Island of Thasos (Aliki, Cape Phanari, Cape Vathy, Salaria), Greece	40.743848, 24.768334
Travertino, travertino di Tivoli	*l. Tiburtinus*	Tivoli, Italy	41.980278, 12.798056
Verde antico	*m. Thessalicum, l. Atracius*	Kasambali near Larissa, Greece	39.693889, 22.530556

Italian Name(s)	Ancient Name(s)	Geographic Location	Coordinates
Quarries from where objects identified as modern, or from medieval and Renaissance quarries (as primary use)			
Alabastro di Volterra	—	—	—
Bianco e nero antico	*m. Celticum*	Lez Valley, Aubert, near St. Girons, France	42.966944, 1.100833
Blue Belge, Belgian black	—	Namur, Belgium	—
Botticino	—	Botticino, Lombardy, Italy	45.536389, 10.302500
Brèche Nouvelle	—	—	—
Escalettes	—	Pass of Escalettes, Ariège, France	—
Lumachella nera	—	—	—
Lumachella di San Vitale, lumachella di Verona, astracane di Verona	—	Val Pantena near Verona, Italy	45.32934, 11.04495
Macchia Vecchia di Arzo	—	Arzo, Lugano, Switzerland	45.882285, 8.950296
Pavonazzetto Toscano	—	Carrara, Torano Valley, Italy Seravezza, Apuan Alps, Italy	44.0922, 10.1038 43.994722, 10.227222
Portoro (16th c. quarry)	—	Porto Venere and Palmaria, La Spezia, Liguria, Italy	44.020455, 9.504599
Red Peterhead granite?	—	Peterhead, Aberdeenshire, Scotland	57.463139, -1.795889
Rose de Numidie	—	Kleber (Sidi Benyebka), Oran Province, Algeria	—
Rosso di Levanto	—	Levanto, Liguria, Italy	44.170556, 9.613056
Rosso Veronese, broccato rossa di Verona, Verona mandolato	—	Lessini Mountains near Verona, Italy	—
Verde Alpi	—	—	—
Verde di Prato	—	Monte Ferrato, Prato, Florence, Italy	43.920509, 11.076422

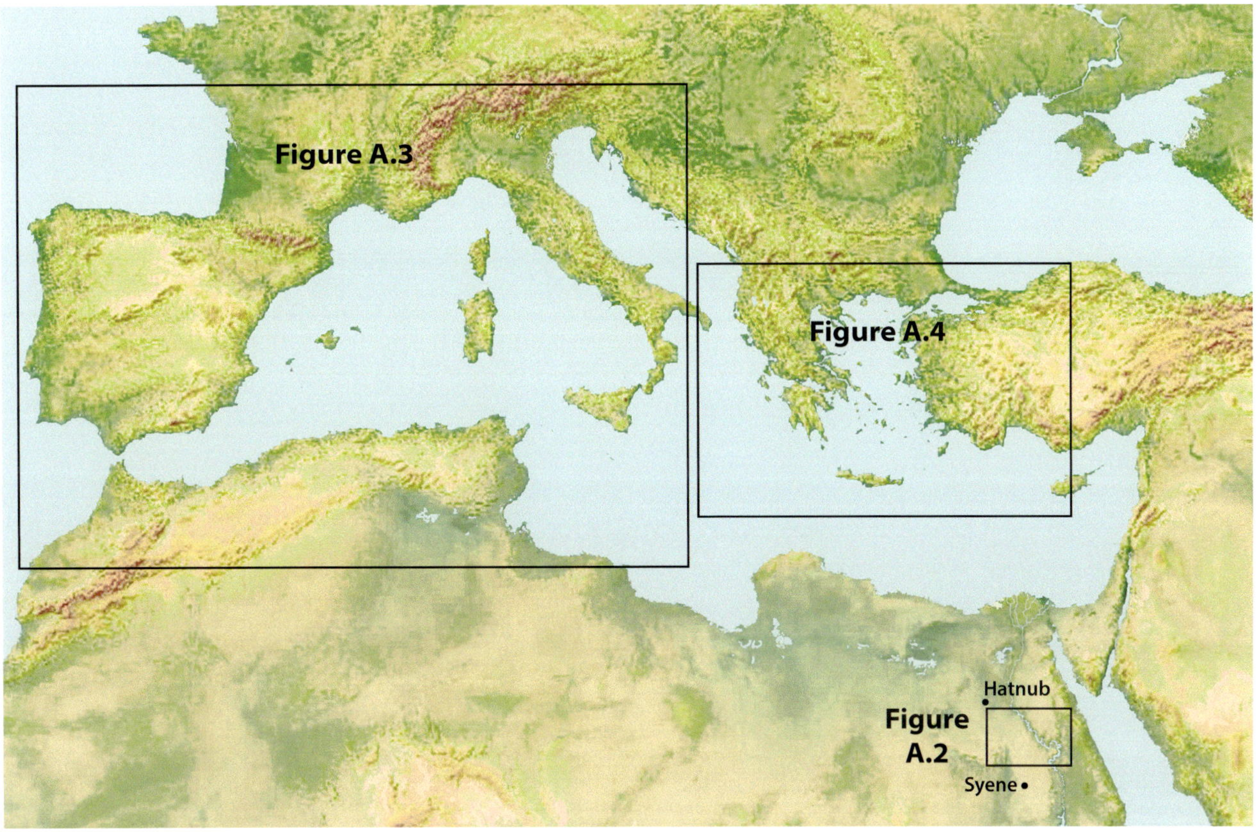

Figure A.1. Key to quarry maps A.2–4. All maps by Lorene Sterner.

Figure A.2. Quarry locations in the Eastern Desert of Egypt.

Figure A.3. Quarry locations in western Europe and North Africa.

Figure A.4. Quarry locations in Greece and Turkey.

Appendix B

CONCORDANCE

KM No.	Cat. No.	Illustrated
592	448	fig. 2.26b; pl. 44
595	701	pl. 67
774	591	fig. 2.34c; pl. 56
775	159	pl. 16
776	65	pl. 7
1089	671	pl. 64
1090	672	pl. 64
1091	673	pl. 64
1845	177	pl. 18
1933	664	pl. 63
1996	559	pl. 53
1997	561	pl. 54
1999	557	pl. 53
2000	558	pl. 53
2001	492	fig. 3.3; pl. 47
2002	495	fig. 3.3; pl. 48
2003	583	fig. 2.8e; pl. 55
2004	600	pl. 57
2005	511	pl. 49
2006	510	fig. 3.2a; pl. 49
2007	494	figs. 2.27a, 3.3; pl. 48
2008	505	pl. 49
2011	607	pl. 58
2012	499	fig. 3.2b; pl. 48
2013	502	pl. 48
2014	503	pl. 48
2015	491	fig. 2.30d; pl. 47
2016	515	fig. 2.38d; pl. 50
2017	497	fig. 3.3; pl. 48
2018	504	fig. 2.35a; pl. 49
2019	493	fig. 3.3; pl. 48
2020	496	fig. 3.3; pl. 48
2021	598	pl. 57
2022	560	fig. 3.2d; pl. 54
2023	562	figs. 2.20e, 3.2e; pl. 54
2025	513	pl. 49
2026	240	pl. 24
2027	239	pl. 24
2028	213	fig. 2.32a; pl. 21
2029	207	pl. 20
2030	218	pl. 22
2031	241	pl. 24
2032	237	pl. 23
2033	238	pl. 23
2034	254	figs. 2.23b, 2.41a; pl. 25
2035	242	pl. 24
2036	243	pl. 24
2037	219	pl. 22
2038	244	fig. 2.38b; pl. 24
2039	245	pl. 24
2040	246	pl. 24
2042	220	pl. 22
2043	101	pl. 10
2044	247	pl. 24
2045	248	pl. 24
2046a	450	pl. 44
2047	223	fig. 2.23a; pl. 22
2048	205	pl. 20
2049	87	pl. 9
2050	206	pl. 20
2051	214	pl. 21
2052	249	pl. 24
2053	208	pl. 21
2054	209	fig. 2.39c; pl. 21
2055	210	pl. 21
2056	250	fig. 2.29e; pl. 25
2057	217	fig. 2.24c; pl. 22
2058	204	pl. 20
2059	251	pl. 25
2060	161	pl. 16
2061	211	pl. 21
2062	221	pl. 22
2063	270	pl. 27
2065	56	pl. 6
2066	57	pl. 6
2067	58	pl. 6
2068	59	pl. 6
2069	62	pl. 6
2070	60	pl. 6
2071	61	pl. 6
2072	13	pl. 2
2073	451	fig. 2.26c; pl. 44
2074	452	fig. 2.26d; pl. 44
2075	453	fig. 2.27b; pl. 44
2076	454	pl. 44
2077	197	pl. 20
2078	102	pl. 10
2079	103	pl. 10
2080	76	pl. 8
2081	104	pl. 10
2082	105	fig. 2.38a; pl. 10
2083	75	pl. 8
2084	79	pl. 8
2085	80	pl. 8
2086	81	pl. 8
2087	77	pl. 8
2088	82	pl. 8
2089	83	pl. 8
2090	273	pl. 27
2091	84	pl. 8
2092	78	pl. 8
2093	337	pl. 33
2094	85	pl. 8
2095	162	pl. 16
2096	484	pl. 47
2097	436	pl. 43
2098	683	pl. 65
2099+2368	354	pl. 35
2100	340	pl. 34
2101	336	fig. 2.29g; pl. 33
2102	339	pl. 34
2103	342	pl. 34
2104	163	pl. 16
2105	355	pl. 35
2106	356	pl. 35
2107	357	pl. 35
2108	298	fig. 2.36c; pl. 30
2109	314	pl. 31
2110	299	pl. 30
2111	281	pl. 28

KM No.	Cat. No.	Illustrated
2112	315	pl. 31
2113	316	pl. 31
2114	303	pl. 30
2115	669	pl. 63
2116	601	pl. 57
2117	319	pl. 32
2118	306	pl. 30
2119	358	pl. 35
2121	320	pl. 32
2122	297	pls. 29–30
2123	296	pl. 29
2124	300	pl. 30
2126	353	pl. 35
2127	186	pl. 18
2128	187	pl. 19
2129	188	pl. 19
2130	189	pl. 19
2131	184	pl. 18
2132	478	fig. 2.5b; pl. 46
2133	190	pl. 19
2134	605	pl. 58
2135	215	pl. 21
2136	410	pl. 40
2137	412	pl. 40
2138	275	pl. 27
2139	422	pl. 41
2140	414	pl. 40
2141	424	pl. 41
2142	413	pl. 40
2143	411	pl. 40
2144	349	pl. 35
2145	426	pl. 42
2146	618	pl. 59
2147	222	pl. 22
2148	695	pl. 66
2149	611	pl. 58
2150	522	pl. 50
2151	175	pl. 17
2152	476	pl. 46
2153	531	pl. 51
2154	533	pl. 51
2155	435	pl. 42
2156	550	fig. 2.29k; pl. 53
2157	554	pl. 53
2158	534	pl. 51
2159	535	fig. 2.24b; pl. 51
2160	431	pl. 42
2161	527	pl. 51
2162	528	pl. 51
2163	663	pl. 63
2164	86	pl. 9
2165	276	pl. 27
2166	406	pl. 40
2167	581	pl. 55
2168	516	pl. 50
2169	523	pl. 50
2170	517	fig. 2.7b; pl. 50
2171	434	fig. 2.41b; pl. 42
2172	432	pl. 42
2173	409	fig. 2.24d; pl. 40
2174	212	pl. 21
2175	118	fig. 2.8a; pl. 11
2176	341	pl. 34
2177	119	fig. 2.30e; pl. 12
2178	216	fig. 2.29c; pl. 21
2179	120	fig. 2.8b; pl. 12
2180	255	pl. 25
2181	121	fig. 2.30a; pl. 12
2182	123	fig. 2.30f; pl. 12
2183	124	figs. 2.31a, 2.37a; pls. 12–13
2184	127	pl. 13
2185	125	pl. 13
2186	126	figs. 2.31b; pl. 13
2187	225	fig. 2.30c; pl. 22
2188	128	pl. 13
2189	129	fig. 2.39b; pl. 13
2190	130	pl. 13
2191	694	pl. 66
2192	131	pl. 13
2193	132	fig. 2.20b; pl. 13
2194	395	pl. 39
2195	479	fig. 2.5a; pl. 46
2196	21	fig. 2.33b; pl. 3
2197	23	pl. 3
2198	14	pl. 2
2199	20	fig. 2.33a; pls. 2, 3
2200	22	pl. 3
2201	641	pl. 61
2202	8	pl. 1
2203	27	pl. 3
2204	19	pl. 2
2205	7	pl. 1
2206	70	fig. 2.29a; pl. 7
2207	69	fig. 2.6b; pl. 7
2208	29	pl. 3
2209	4	pl. 1
2210	30	pl. 4
2211	28	pl. 3
2212	593	pl. 56
2213	594	pl. 56
2214	1	fig. 2.36a; pl. 1
2215	3	fig. 2.37; pl. 1
2216	381	pl. 37
2217	5	pl. 1
2218	388	pl. 38
2219	639	pl. 61
2220	389	pl. 38
2221	390	pl. 38
2222+93466	684	pl. 65
2223	382	pl. 38
2224	271	pl. 27
2225	386	pl. 38
2226	685	pl. 65
2227	686	pl. 65
2228	687	pl. 65
2229	164	pl. 16
2230	475	fig. 2.8c; pl. 46
2231	165	pl. 17
2232	166	pl. 17
2233	167	pl. 17
2234+2436	430	pl. 42
2235	168	pl. 17
2236	169	pl. 17
2237	610	pl. 58
2238	666	pl. 63
2239	651	pl. 62
2240	272	pl. 27
2241	485	pl. 47
2242	415	fig. 2.29j; pl. 40
2243	259	pl. 26
2244	274	pl. 27
2245	269	pl. 27
2246	631	pl. 60
2247	387	pl. 38
2248	423	pl. 41
2249	537	pl. 52
2250	657	pl. 62
2251	536	pl. 51
2252	482	pl. 47
2253	133	pl. 14
2254	226	fig. 2.29d; pl. 22
2255	360	pl. 36
2256	122	fig. 2.30b; pl. 12
2257	361	pl. 36
2258	88	pl. 9
2259	367	fig. 2.29i; pl. 36
2260	350	pl. 35
2261	481	pl. 46
2262	477	pl. 46
2263	363	pl. 36
2264	227	pl. 23
2265	480	pl. 46
2267	596	fig. 2.40; pl. 57
2291	595	pl. 57
2313	228	pl. 23
2314	137	pl. 14
2315	582	fig. 2.8d; pl. 55
2316	71	pl. 7

KM No.	Cat. No.	Illustrated
2317	396	pl. 39
2318	425	pl. 41
2319	193	pl. 19
2320	383	pl. 38
2321	608	pl. 58
2322	609	pl. 58
2323	693	pl. 66
2324	405	pl. 40
2325	365	pl. 36
2326	174	pl. 17
2327	692	pl. 66
2329	307	pl. 31
2330	72	pl. 7
2331	416	pl. 41
2332	17	pl. 2
2333	138	fig. 2.39a; pl. 14
2334	417	pl. 41
2335	26	pl. 3
2336	191	pl. 19
2338	229	pl. 23
2340	501	pl. 48
2341	257	pl. 25
2342	532	pl. 51
2343	15	pl. 2
2344	518	pl. 50
2347	89	pl. 9
2348	552	pl. 53
2349	327	pl. 32
2350	90	pl. 9
2351	397	pl. 39
2352	31	pl. 4
2353	134	pl. 14
2354	16	pl. 2
2355	12	fig. 2.34a; pl. 2
2356	433	pl. 42
2357	538	pl. 52
2358	192	pl. 19
2359	172	pl. 17
2360	230	pl. 23
2361	539	pl. 52
2362	231	pl. 23
2363	553	pl. 53
2364	224	pl. 22
2365	385	fig. 2.21d; pl. 38
2366	169	pl. 17
2367	170	pl. 17
2368	*see under 2099*	
2369	232	pl. 23
2370	551	pl. 53
2371	508	pl. 49
2372	233	pl. 23
2375	234	pl. 23
2382	91	pl. 9
2383	256	pl. 25
2384	92	pl. 9
2385	93	pl. 9
2386	675	pl. 64
2387	455	pl. 44
2388	235	pl. 23
2389	364	pl. 36
2390	418	pl. 41
2391	421	pl. 41
2393	690	pl. 65
2394	94	pl. 9
2395	483	pl. 47
2396	328	pl. 32
2397	95	pl. 9
2398	96	pl. 10
2399	97	fig. 2.29b; pl. 10
2400	98	pl. 10
2402	398	pl. 39
2403	329	pl. 33
2404	236	pl. 23
2405	330	pl. 33
2406	173	pl. 17
2415	99	pl. 10
2421	456	figs. 2.19, 2.39d; pl. 44
2432	338	fig. 2.29h; pls. 33–34
2434	100	pl. 10
2436	*see under 2234*	
2437	366	pl. 36
2439	563	pl. 54
2440	359	pl. 35
2441	519	pl. 50
2442	194	pl. 19
2443	399	pl. 39
2446	196	pl. 20
2506	449	fig. 2.10; pl. 44
2542	599	pl. 57
2543	49	pl. 5
2544	116	pl. 11
2545	326	pl. 32
2546	18	pl. 2
2547	378	pl. 37
2550	490	fig. 2.6h; pl. 47
2552	151	pl. 15
2555	331	pl. 33
2556	152	fig. 2.6f; pl. 15
2558	153	fig. 2.9a; pl. 15
2717	394	pl. 39
2950	702	pl. 68
2951	545	figs. 2.33c, 2.37b; pl. 52
2973	468	fig. 2.28a; pl. 45
3055	704	pl. 70
6729	155	pl. 15
6730	156	fig. 2.30g; pl. 16
6731	375	pl. 37
6732	486	fig. 2.22; pl. 47
6733	661	pl. 63
6734	642	pl. 61
6735	391	pl. 38
6736	392	pl. 38
6737	199	pl. 20
6738	689	pl. 65
6739	640	pl. 61
6741	556	pl. 53
6742	524	fig. 2.20d; pl. 50
6743	548	pl. 52
6744	200	pl. 20
6745	525	pl. 51
6746	589	pl. 56
6747	506	pl. 49
6748	529	fig. 2.6i; pl. 51
6749	696	pl. 66
6750	279	pl. 28
6751	462	fig. 2.26e; pl. 45
6752	674	pl. 64
6753	318	pl. 32
6754	429	pl. 42
6755	420	fig. 2.9d; pl. 41
6756	438	pl. 43
6757	407	pl. 40
6758	408	pl. 40
6759	347	pl. 34
6760	312	pl. 31
6761	285	pl. 28
6762	284	pl. 28
6763	286	fig. 1.4; pl. 28
6764	322	fig. 2.29f; pl. 32
6765	290	pl. 29
6766	310	pl. 31
6767	280	pl. 28
6768	309	pl. 31
6769	282	pl. 28
6770	283	pl. 28
6771	293	pl. 29
6772	157	fig. 2.9c; pl. 16
6773	437	pl. 43
6774+6775	439	pl. 43
6775	*see under 6774*	
6776	659	pl. 62
6777	441	pl. 43
6778	645	pl. 61
6779	440	pl. 43
6780	660	pl. 63
6781	644	pl. 61

KM No.	Cat. No.	Illustrated
6782	402	pl. 39
6783	183	fig. 2.20c; pl. 18
6784	444	pl. 43
6785	637	pl. 61
6786	373	figs. 2.32b; pl. 37
6788	114	pl. 11
6791	115	pl. 11
6794	10	pl. 2
6807	590	fig. 2.13a; pl. 56
6813	158	pl. 16
6819	507	pl. 49
6820	564	pl. 54
6821	565	pl. 54
6822	541	pl. 52
6823	458	fig. 2.27c; pl. 44
6824	459	pla. 44–45
6825	460	pl. 45
6826	461	pl. 45
6828	428	pl. 42
6829	140	pl. 14
6830	178	fig. 2.18b; pl. 18
6831	368	pl. 36
6832	370	pl. 36
6833	108	pl. 11
6834	109	fig. 2.20a; pl. 11
6835	369	pl. 36
6836	400	pl. 39
6837	294	pl. 29
6838	301	pl. 30
6839	308	fig. 2.9b; pl. 31
6840	304	pl. 30
6841	287	pl. 28
6842	317	pl. 32
6843	302	pl. 30
6844	313	pl. 31
6845	278	pl. 27
6846	63	pl. 6
6847	427	pl. 42
6853	32	pl. 4
6854	33	pl. 4
6855	34	pl. 4
6856	35	pl. 4
6858	9	fig. 2.6a; pl. 1
6859	6	pl. 1
6860	2	fig. 2.36b; pl. 1
6861	305	pl. 30
6862	691	fig. 2.41c; pl. 65
6863	371	pl. 37
6864	139	pl. 14
6865	277	pl. 27
6866	321	pl. 32
6867	401	pl. 39
7109	141	pl. 14
23397	586	fig. 2.21f; pl. 56
23398	252	fig. 2.21c; pl. 25
25566	253	pl. 25
25567	106	pl. 10
25571	384	pl. 38
25572	597	pl. 57
25573	136	pl. 14
25574	380	pl. 37
25581	606	pl. 58
25583	362	pl. 36
25584	521	fig. 2.21e; pl. 50
25585	555	pl. 53
25589	198	pl. 20
25590	584	fig. 2.24a; pl. 56
25591	107	pl. 11
25592	195	fig. 2.18c; pl. 19
25593	457	pl. 44
25594	585	pl. 56
25596	509	pl. 49
25598	540	pl. 52
25801	176	fig. 2.18a; pl. 18
29516	374	pl. 37
29517	311	pl. 31
29518	289	pl. 29
29519	25	pl. 3
29574	658	pl. 62
29605	44	pl. 5
29607	260	pl. 26
29608	403	pl. 39
29609	323	pl. 32
29611	442	pl. 43
29612	45	pl. 5
29613	343	pl. 34
29614	67	pl. 7
29615	68	pl. ?
29616	445	pl. 43
29617	291	pl. 29
29618	446	pl. 43
29619	351	pl. 35
29620	603	pl. 57
29623	46	pl. 5
29624	292	pl. 29
29625	263	pl. 26
29626	288	pl. 28
29628	549	pl. 53
29629	419	pl. 41
29635	182	pl. 18
29636	180	pl. 18
29637	526	pl. 51
29639	668	pl. 63
29640	324	pl. 32
29641	372	pl. 37
29642	514	pl. 49
29644	700	pl. 66
29645	379	pl. 37
29646	261	pl. 26
29647	262	pl. 26
29651	571	pl. 54
29652	181	pl. 18
29653	699	fig. 2.7d; pl. 66
29654	463	pl. 45
29661	613	pl. 58
29662	633	pl. 60
29663	667	pl. 63
29664	634	pl. 60
29665	688	pl. 65
29666	670	pl. 63
29667	662	pl. 63
29668	678	pl. 64
29669	615	pl. 59
29670	616	pl. 59
29671	617	pl. 59
29673	620	pl. 59
29674	619	pl. 59
29675	654	pl. 62
29676	655	pl. 62
29677	267	pl. 26
29678	626	pl. 60
29679	630	pl. 60
29680	628	pl. 60
29682	624	pl. 59
29684	629	pl. 60
29686	652	pl. 62
29687	614	pl. 58
29688	587	pl. 56
29689	665	pl. 63
29690	656	pl. 62
29691	632	pl. 60
29692	622	pl. 59
29693	643	pl. 61
29694	646	fig. 2.21g; pl. 61
29695	53	fig. 2.34b; pl. 6
29696	266	pl. 26
29697	676	pl. 64
29698	677	pl. 64
29699	604	pl. 57
29700	635	pl. 60
29701	636	pl. 60
29702	623	pl. 59
29703	621	pl. 59
29704	650	pl. 62
29705	647	pl. 61
29706	335	pl. 33
29707	648	pl. 61
29708	649	pl. 62
29709	627	pl. 60

KM No.	Cat. No.	Illustrated
29710	679	pl. 64
29711	680	pl. 64
29712	653	pl. 62
29713	681	pl. 64
29714	588	pl. 56
29715	638	pl. 61
29716	625	pl. 59
29717	612	pl. 58
29718	682	pl. 65
29719	36	pl. 4
29720	74	pls. 7–8
29725	37	pl. 4
29726	38	pl. 4
29727	39	pl. 4
29729	154	fig. 2.28c; pl. 15
29730	469	fig. 2.28d; pl. 45
29731	40	pl. 4
29732	41	pl. 4
29733	42	pl. 5
29734	43	pl. 5
29737	470	fig. 2.28b; pls. 45–46
29875	54	pl. 6
29876	55	pl. 6
91752	52	fig. 2.21b; pl. 5
93466	*see under 2222*	
93472	498	fig. 3.2c; pl. 48
93520	24	pl. 3
93532	567	fig. 2.20f; pl. 54
93539	568	fig. 2.13b; pl. 54
93540	573	fig. 2.13e; pl. 55
93541	569	fig. 2.13c; pl. 54
93542	570	fig. 2.13d; pl. 54
93543	142	fig. 2.12a; pl. 14
93544	143	fig. 2.12b; pl. 14
93545	574	fig. 2.13f; pl. 55
93546	572	fig. 2.13g; pl. 55
93547	575	fig. 2.13h; pl. 55
93548	576	fig. 2.13i; pl. 55
93549	577	fig. 2.13j; pl. 55
93550	578	fig. 2.13k; pl. 55
93551	579	fig. 2.13l; pl. 55
93552	148	fig. 2.12f; pl. 15
93553	144	fig. 2.12c; pl. 14
93554	145	fig. 2.12g; pl. 15
93555	146	fig. 2.12d; pl. 15
93556	147	fig. 2.12e; pl. 15
93557	149	pl. 15
93558	150	fig. 2.12h; pl. 15
93559	464	fig. 2.11a; pl. 45
93560	465	fig. 2.11b; pl. 45
93561	344	fig. 2.6g; pl. 34
93562	466	fig. 2.7a; pl. 45
93563	467	pl. 45
93564	345	pl. 34
93565	201	pl. 20
93566	47	pl. 5
93567	48	fig. 2.21a; pl. 5
93568	11	pl. 2
93569	580	pl. 55
93780	471	fig. 2.28e; pl. 46
93781	472	fig. 2.28f; pl. 46
93782	473	fig. 2.28g; pl. 46
93783	447	fig. 2.26a; pl. 43
93785	268	pls. 26–27
93786	73	pl. 7
93787	203	pl. 20
93788	332	pl. 33
93790	112	fig. 2.6d; pl. 11
93791	113	fig. 2.6e; pl. 11
93793	325	pl. 32
93794	512	pl. 49
93795	602	pl. 57
93837	544	pl. 52
93838	697	pl. 66
93840	543	pl. 52
93843	443	pl. 43
93844	542	pl. 52
93845	698	pl. 66
93847	202	pl. 20
93848	64	pl. 7
93849	547	pl. 52
93850	546	fig. 2.7c; pl. 52
93851	346	pl. 34
93852	376	pl. 37
93853	566	pl. 54
93854	377	pl. 37
93855	393	pl. 39
93856	264	pl. 26
93857	265	fig. 2.38c; pl. 26
93858	110	fig. 2.6c; pl. 11
93859	474	pl. 46
93865	488	fig. 2.25b; pl. 47
93866	489	fig. 2.25c; pl. 47
93867	487	fig. 2.25a; pl. 47
93884	179	pl. 18
93885	258	pl. 26
93901	50	pl. 5
93912	703	pl. 69
93913	111	pl. 11
1963.5.12	66	pl. 7
1969.5.1	51	pl. 5
1969.5.2	348	pl. 35
1969.5.3a–c	160	pl. 16
1969.5.3d	592	pl. 56
1969.5.4	333	pl. 33
1969.5.5	334	pl. 33
1969.5.6	117	pl. 11
1969.5.7	404	pl. 39
T2008.58	295	pl. 29
T2008.59	185	pl. 18
T2008.60	530	pl. 51
T2008.61	352	pl. 35
T2008.62	135	pl. 14
T2008.63	520	pl. 50
T2018.1	500	pl. 48

Bibliography

Abbreviations

ASMOSIA I — Herz, Norman, and Marc Waelkens, eds. *Classical Marble: Geochemistry, Technology, Trade.* NATO ASI Series, Series E, Applied Sciences 153. Dordrecht and Boston: Kluwer Academic Publishers, 1988. [Technically, this is not an ASMOSIA volume, but for the sake of convenience we are referring to it as such.]

ASMOSIA II — Waelkens, Marc, Norman Herz, and Luc Moens, eds. *Ancient Stones: Quarrying, Trade and Provenance.* ASMOSIA II; Acta Archaeologica Lovaniensia, Monographiae 4. Leuven: Leuven University Press, 1992.

ASMOSIA III — Maniatis, Yannis, Norman Herz, and Yannis Basiakos, eds. *The Study of Marble and Other Stones Used in Antiquity.* ASMOSIA III. London: Archetype, 1995.

ASMOSIA IV — Schvoerer, Max, ed. *Archéomatériaux: marbres et autres roches. Actes de la IVe conférence internationale de l'Association pour l'étude des marbres et autres roches utilisés dans le passé, Bordeaux-Talence, 9-13 Octobre 1995.* ASMOSIA IV. Talence: Centre de Recherche en Physique Appliquée à l'Archéologie, 1999.

ASMOSIA V — Herrmann, John J., Jr., Norman Herz, and Richard Newman, eds. *Interdisciplinary Studies on Ancient Stone. Proceedings of the 5th International Conference of the Association for the Study of Marble and Other Stones in Antiquity, Museum of Fine Arts, Boston, 1998.* ASMOSIA V. London: Archetype, 2002.

ASMOSIA VI — Lazzarini, Lorenzo, ed. *Interdisciplinary Studies on Ancient Stone. Proceedings of the 6th International Conference, Venice, June 15-18 2000.* ASMOSIA VI. Padua: Bottega d'Erasmo, 2002.

ASMOSIA VII — Maniatis, Yannis, ed. *Proceedings of the 7th International Conference of Association for the Study of Marble and Other Stones in Antiquity, Thassos, 15–20 September 2003.* ASMOSIA VII; Bulletin de Correspondance Hellénique, Supplément 51. Athens: Ecole française d'Athènes, 2009.

ASMOSIA VIII — —

ASMOSIA IX — Gutiérrez Garcia-Moreno, Anna, Pilar Lapuente Mercadal, and Isabel Rodà de Llanza, eds. *Interdisciplinary Studies on Ancient Stone: Proceedings of the 9th Association for the Study of Marbles and Other Stones in Antiquity (ASMOSIA) Conference (Tarragona 2009).* ASMOSIA IX; Documenta 23. Tarragona: Institut Català d'Arqueologia Clàssica, 2012.

ASMOSIA X — Pensabene, Patrizio, and Eleonora Gasparini, eds. *Interdisciplinary Studies on Ancient Stone: Proceedings of the 10th International Conference of ASMOSIA, Association for the Study of Marble and Other Stones in Antiquity, Rome, 21–26 May 2012.* 2 vols. ASMOSIA X. Rome: L'Erma di Bretschneider, 2015.

ASMOSIA XI — Matetić Poljak, Daniela, and Katja Marasović, eds. *Interdisciplinary Studies on Ancient Stone: Proceedings of the 11th International Conference of ASMOSIA, Split, 18–22 May 2015.* ASMOSIA XI. Split: Arts Academy in Split and University of Split, 2018.

Adak, Mustafa, and Musa Kadıoğlu

2017 "Die Steinbrüche von Teos und 'Marmor Luculleum.'" *Philia* 3: 1–43.

Alumni Association of the University of Michigan

1944 "News … by Classes." In *The Michigan Alumnus*, Vol. 51, 222. Ann Arbor: University of Michigan.

Ambrogi, Annarena

1995 *Vasche di età romana in marmi bianchi e colorati.* Studia archaeologica 79. Rome: L'Erma di Bretschneider.

2005 *Labra di età romana in marmi bianchi e colorati.* Studia archaeologica 136. Rome: L'Erma di Bretschneider.

Andreoli, A., Fede Berti, Lorenzo Lazzarini, and Raffaella Pierobon Benoit

2002 "New Contributions on Marmor Iassense." In ASMOSIA VI, 13–18.

Angelelli, Claudia

2000 "VRC00 Preliminary Report — Wall and Floor Lithic Revetments (1992–1999 Findings)." *Annual Reports of Institute of Cultural Exchange-Faculty of Letters, The University of Tokyo* 16: 88–101.

Antonelli, Fabrizio

2002 "I marmi della Gallia e dell'Iberia importati a Roma." In *I marmi colorati della Roma imperiale (Roma, Mercati di Traiano, 28 settembre 2002–19 gennaio 2003)*, edited by Marilda De Nuccio and Lucrezia Ungaro, 267–76. I fori imperiali. Venice: Marsilio.

Antonelli, Fabrizio, G. Gentili, Alberto Renzulli, and Maria Letizia Amadori

2003 "Provenance of the Ornamental Stones Used in the Baroque Church of S. Pietro in Valle (Fano, Central Italy) and Commentary on Their State of Conservation." *Journal of Cultural Heritage* 4/4: 299–312.

Antonelli, Fabrizio, and Lorenzo Lazzarini

2015 "An Updated Petrographic and Isotopic Reference Database for White Marbles Used in Antiquity." *Rendiconti Lincei* 26/4: 399–413.

Antonelli, Fabrizio, Lorenzo Lazzarini, Stefano Cancelliere, and D. Dessandier

2009 "Minero-Petrographic and Geochemical Characterization of 'Greco Scritto' Marble from Cap de Garde, near Hippo Regius (Annaba, Algeria)." *Archaeometry* 51/3: 351–65.

2010 "On the White and Coloured Marbles of the Roman Town of Cuicul (Djemila, Algeria)." *Archaeometry* 52/4: 575–96.

Antonelli, Fabrizio, Lorenzo Lazzarini, Léopold Rasplus, and Bruno Turi

2002 "Petrographic and Geochemical Characterization of Cipollino Mandolato Marble from the French Central Pyrenees." In ASMOSIA V, 77–90.

Asgari, Nuşin

1977 "Die halbfabrikate kleinasiatischer Girlandensarkophage und ihre Herkunft." *Archäologischer Anzeiger* 1997: 329–80.

1978 "Roman and Early Byzantine Marble Quarries of Proconnesus." In *The Proceedings of the Xth International Congress of Classical Archaeology, Ankara-İzmir, 23-30/IX/1973*, edited by Ekrem Akurgal, 1:467–80. Ankara: Türk Tarih Kurumu.

1988 "The Stages of Workmanship of the Corinthian Capital in Proconnesus and Its Export Form." In ASMOSIA I, 115–26.

1990 "Objects de marbre finis, semi-finis et inachevés proconnèse." In *Pierre éternelle du Nil au Rhin: carrières et préfabrication*, edited by Marc Waelkens, 106–26. [Brussels]: Crédit communal.

1992 "Observations on Two Types of Quarry-Items from Proconnesus: Column-Shafts and Column-Bases." In ASMOSIA II, 73–80.

Asgari, Nuşin, and Thomas Drew-Bear

2002 "The Quarry Inscriptions of Prokonnesos." In ASMOSIA V, 1–19.

Asgari, Nuşin, and Keith Matthews

1995 "The Stable Isotope Analysis of Marble from Proconnesus." In ASMOSIA III, 123–29.

Asimakopoulou-Atzaka, Panagiôta

1980 *E technike opus sectile sten entoíchia diakósmese: symbole ste meléte tes technikes apó ton 10 méchri ton 70 m.ch. ai. me báse ta mnemeía kai ta keímena.* Byzantiná Mnemeía 4. Thessalonike: Kéntro Byzantinon Ereunon - APTH.

Aston, Barbara G.

1994 *Ancient Egyptian Stone Vessels: Materials and Forms.* Studien zur Archäologie und Geschichte Altägyptens 5. Heidelberg: Heidelberger Orientverlag.

Aston, Barbara G., James A. Harrell, and Ian Shaw

2000 "Stone." In *Ancient Egyptian Materials and Technology*, edited by Paul T. Nicholson and Ian Shaw, 5–77. Cambridge: Cambridge University Press.

Attanasio, Donato

2003 *Ancient White Marbles: Analysis and Identification by Paramagnetic Resonance Spectroscopy.* Studia Archaeologica 122. Rome: L'Erma di Bretschneider.

Attanasio, Donato, Mauro Brilli, and Matthias Bruno

2008 "The Properties and Identification of Marble from Proconnesos (Marmara Island, Turkey): A New Database Including Isotopic, EPR and Petrographic Data." *Archaeometry* 50/5: 747–74.

Attanasio, Donato, Mauro Brilli, Matthias Bruno, Lucrezia Ungaro, and M. Vitti

2012 "Rome: White Marbles in the Fora of Caesar, Augustus, and Trajan." In ASMOSIA IX, 331–43.

Attanasio, Donato, Mauro Brilli, and Neil Ogle

2006 *The Isotopic Signature of Classical Marbles.* Studia Archaeologica 145. Rome: L'Erma di Bretschneider.

Attanasio, Donato, Matthias Bruno, Walter Prochaska, and Ali Bahadir Yavuz

2017 "Ancient 'Black' Decorative Stones and the Ephesian Origin of Sculptural Bigio Antico." *Archaeometry* 59/5: 794–814. https://doi.org/10.1111/arcm.12278.

Attanasio, Donato, Matthias Bruno, and Ali Bahadir Yavuz

2009 "Quarries in the Region of Aphrodisias: The Black and White Marbles of Göktepe (Muğla)." *Journal of Roman Archaeology* 22: 312–48. https://doi.org/10.1017/S1047759400020729.

Attanasio, Donato, John J. Herrmann Jr., Robert H. Tykot, and Annewies van den Hoek

2012 "Characterization and Distribution of Marble from Cap de Garde and Mt. Filfila, Algeria." In ASMOSIA IX, 300–09.

Attanasio, Donato, Ali Bahadir Yavuz, Matthias Bruno, John J. Herrmann Jr., Robert H. Tykot, and Annewies van den Hoek

2012 "On the Ephesian Origin of Greco Scritto Marble." In ASMOSIA IX, 245–54.

Baccini Leotardi, Paola

1979 *Marmi di cava rinvenuti ad Ostia e considerazioni sul commercio dei marmi in età romana.* Scavi di Ostia 10. Rome: Istituto Poligrafico e Zecca dello Stato, Libreria.

1989 *Nuove testimonianze sul commercio dei marmi in età imperiale.* Studi pubblicati dall'Istituto Italiano per la Storia Antica 44. Rome: Istituto Italiano per la Storia Antica.

Baker-Penoyre, John

1902–03 "Annual Meeting of Subscribers." *Annual of the British School at Athens* 9: 391–97.

Bald Romano, I., Elaine K. Gazda, and S. Pike

2009 "The Hartwig-Kelsey Fragments: A Pentelic Marble Case Study." In ASMOSIA VII, 273–89.

Ball, Larry F.

2002 "How Did the Romans Install Revetment?" *American Journal of Archaeology* 106/4: 551–73.

Ballance, M.H.

1966 "The Origin of Africano." *Papers of the British School at Rome* 34: 79–81.

Banti, Luisa

1931 "Antiche lavorazione nelle cave lunensi." *Studi Etruschi* 5: 475–97.

Barker, Simon J.

2010 "Roman Builders — Pillagers or Salvagers? The Economics of Deconstruction and Reuse." In *Arqueología de la Construcción II: Los Procesos Constructivos en el Mundo Romano; Italia y Provincias Orientales,* edited by Stefano Camporeale, Hélène Dessales, and Antonio Pizzo, 127–42. Anejos de Archivo Español de Arqueología 57. Mérida: Consejo Superior de Investigaciones Científicas, Instituto de Arqueología.

2012 "Roman Marble Salvaging." In ASMOSIA IX, 22–30.

Barker, Simon J., and J. Clayton Fant

2018 "Recent Work on the Stone at the Villa Arianna and the Villa San Marco (Castellamare di Stabia) and Their Context within the Vesuvian Area." In ASMOSIA XI, 65–78.

2019a "Lithic Decoration: Sources, Styles, Repairs, and Spoliation." In *Oplontis: Villa A ("of Poppaea") at Torre Annunziata, Italy,* Vol. 2: *The Decorations: Painting, Stucco, Pavements, Sculptures,* edited by John R. Clarke and Nayla K. Muntasser, ch. 17. New York: The Humanities E-Book Series of the American Council of Learned Societies. https://hdl.handle.net/2027/heb.90048.

2019b "Stone Sources and Types: Report on Provenance Analyses for the Sculpture and Architectural Marbles from Oplontis." In *Oplontis: Villa A ("of Poppaea") at Torre Annunziata, Italy,* Vol. 2: *The Decorations: Painting, Stucco, Pavements, Sculptures,* edited by John R. Clarke and Nayla K. Muntasser, ch. 18. New York: The Humanities E-Book Series of the American Council of Learned Societies. https://hdl.handle.net/2027/heb.90048.

Barker, Simon J., J. Clayton Fant, and Courtney A. Ward

2015 "Relazione Preliminare sulla Decorazione Marmorea a Villa Arianna e San Marco, Stabiae." *Rivista Di Studi Pompeiani* 24: 124–34.

Barker, Simon J., J. Clayton Fant, Courtney A. Ward, and Brittany Amiet

2014 "The Marmo Al Mare Project 2013: Marble Décor at the Roman Villas of Stabiae." *Fold&r Fastionline Documents & Research* 298: 1–13.

Barker, Simon J., and Simona Perna

2018 "Alabaster: Quarrying and Trade in the Roman World; Evidence from Pompeii and Herculaneum." ASMOSIA XI, 45–64.

2024 "Alabaster: An Interdisciplinary Study of the Sources and Uses of Calcite-Alabaster across Archaeological Contexts." Proceedings of an International Workshop Held at the Norwegian Institute in Rome, 9th–10th May 2019. Tournhout, Belgium: Brepols.

Barker, Simon J., Simona Perna, J. Clayton Fant, Lorenzo Lazzarini, and Igor M. Villa

2018 "Archaeometric Analysis of the Alabaster Thresholds of Villa A, Oplontis (Torre Annunziata, Italy) and New Sr and Pb Isotopic Data for *Alabastro Ghiaccione del Circeo.*" In ASMOSIA XI, 215–30.

Bartelletti, Antonio, and Alessia Amorfini

2003 "Le brecce policrome apuane nell'Antichità." In *Ante et post Luna: splendore e ricchezza dei marmi apuani, I: l'evo antico*, 63–77. Acta Apuana 2. Seravezza: Parco regionale delle Alpi Apuane.

Becatti, Giovanni
1969 *Edificio con opus sectile fuori Porta Marina.* Scavi di Ostia 6. Rome: Istituto poligrafico dello Stato, Libreria dello Stato.

Belli, Francesco
1842 Catalogo della collezione di pietre usate dagli antichi per costruire ed adornare le loro fabbriche, ora posseduta dal Conte Stefano Karolyi. Rome: Tipografia Mugnoz.

Belli Pasqua, Roberta
1995 *Sculture di età romana in "basalto."* Xenia antiqua. Monografie 2. Rome: L'Erma di Bretschneider.

Berger, Ernst, ed.
1990 *Antike Kunstwerke aus der Sammlung Ludwig,* Vol. 3: *Skulpturen.* Veröffentlichungen des Antikenmuseums Basel, 4/3. Mainz am Rhein: Von Zabern.

Bernard, Hélène, Jean-Claude Bessac, Paul Mardikian, and Michel Feugère
1998 "L'épave romaine de marbre de Porto Novo." *Journal of Roman Archaeology* 11: 53–81.

Bernard, Seth G.
2010 "Pentelic Marble in Architecture at Rome and the Republican Marble Trade." *Journal of Roman Archaeology* 23: 35–54.

Bernard, Seth G., and Scott Pike
2015 "Isotopic Analysis of Marble from the Stoa of Attalos in the Athenian Agora and the Hellenistic Quarries of Mount Pentelikon." In ASMOSIA X, 1:451–60.

Beste, Heinz-Jürgen, Stepfan Arnold, and Markus Wolf
2015 "La decorazione architettonica e parietale della natatio." In *Le terme di Diocleziano: la certosa di Santa Maria degli Angeli,* edited by Anna Friggeri and Marina Magnani Cianetti, 186–211. Milan: Electa.

Betti, Marco, and Carlotta Brovadan
2012 "Per il Gran Principe Ferdinando: Tre opere del Museo dell'Opificio delle Pietre Dure e alcuni cenni sul suo gusto per le 'Arti Minori.'" *OPD Restauro* 24: 263–72.

Blake, Marion Elizabeth
1947 *Ancient Roman Construction in Italy from the Prehistoric Period to Augustus: A Chronological Study Based in Part upon the Material Accumulated by the Late Dr. Esther Boise Van Deman.* Carnegie Institution of Washington Publication 570. Washington, D.C.: Carnegie Institution of Washington.
1958 "Esther Boise Van Deman." In *Dictionary of American Biography,* Supplement 2, edited by Edward T. James and Robert Livingston Schuyler, 676–77. New York: Scribner.

Blanc, Philippe, Philippe Bromblet, and Lise Leroux
2012 "White Marbles Provenance Study of Caesar's Head and Other Major Sculptures Found in the Rhône River at Arles (France)." In ASMOSIA IX, 401–06.

Bodel, John P.
1983 *Roman Brick Stamps in the Kelsey Museum.* Kelsey Museum Studies 6. Ann Arbor: University of Michigan Press.

Boldrighini, Francesca, ed.
2007 *Elementi architettonici e di rivestimento: voce "definizione dell'oggetto": proposte di terminologia per la catalogazione dei reperti archeologici mobili del Lazio.* Arcata: archeologia e catalogazione 1. Rome: Regione Lazio, Ufficio Musei.

Bonanni, Alessandro
1998 "Interraso marmore (Plin., *N.H.*, 35, 2): Esempi della tecnica decorativa a intarsio in età Romana." In *Marmi antichi II: cave e tecnica di lavorazione, provenienze e distribuzione,* edited by Patrizio Pensabene, 259–92. Studi Miscellanei 31. Rome: L'Erma di Bretschneider.

Borghini, Gabriele, ed.
1989 *Marmi antichi.* Materiali della cultura artistica 1. Rome: De Luca.
1997 *Marmi antichi.* 2nd ed. Materiali della cultura artistica 1. Rome: De Luca.

Bragantini, Irene, Mariette de Vos, Franca Parise Badoni, and Valeria Sampaolo, eds.
1986 *Pitture e pavimenti di Pompei 3. Regioni VII, VIII, IX, Indici delle Regioni I–IX.* Rome: Istituto Centrale per il Catalogo e la Documentazione.

Breuer, Christine, and Pierre-Louise Gatier
2001 *Antike Skulpturen: Bestandskatalog des Badischen Landesmuseums Karlsruhe.* Karlsruhe: Badisches Landesmuseum.

Brilli, Mauro, Fabrizio Antonelli, Francesca Giustini, Lorenzo Lazzarini, and Patrizio Pensabene
2010 "Black Limestones Used in Antiquity: The Petrographic, Isotopic and EPR Database for Provenance Determination." *Journal of Archaeological Science* 37/5: 994–1005. https://doi.org/10.1016/j.jas.2009.11.032.

Brilli, Mauro, Francesca Giustini, and Musa Kadıoğlu
2019 "Black Limestone Used in Antiquity: Recognizing the Limestone of Teos." *Archaeometry* 61/2: 282–95. https://doi.org/10.1111/arcm.12426.

Brilli, Mauro, Francesca Giustini, Pier Matteo Barone, Ali Elmoughazi Fayek, and Giuseppe Scardozzi
2018 "Characterizing the *Alabastro listato* or *fiorito* of Hierapolis in Phrygia: A Simple Method to Identify Its Provenance Using Carbon Stable Isotopes." *Archaeometry* 60/3: 403–18.

Brindley, William
1897 "Verde Antico and the Old Quarries." *Transactions of the Royal Institute of British Architects* 3 (Third Series): 267.

Broughton, T. Robert S.
1959 "Roman Asia." In *An Economic Survey of Ancient Rome,* edited by Tenney Frank, 4:499–950. Paterson: Pageant Books. First published in 1938.

Brown, V. Max, and James A. Harrell
1995 "Topographical and Petrological Survey of Ancient Roman Quarries in the Eastern Desert of Egypt." In ASMOSIA III, 221–34.
1998 "Aswan Granite and Granodiorite." *Göttinger Miszellen* 164: 33–39.

Brun, Jean-Pierre
2003 "Chronologie de l'équipement de la route à époque gréco-romaine." In *La route de Myos Hormos: l'armée romaine dans le désert oriental d'Egypte,* edited by Héléne Cuvigny, 187–205. Fouilles de l'Institut français d'archéologie orientale du Caire 48. Cairo: Institut Français d'Archéologie Orientale.

Bruno, Matthias
1998 "Su una cava d'alabastro del Circeo in località 'La Batteria.'" In *Marmi Antichi II: cave e tecnica di lavorazione, provenienze e distribuzione*, by Patrizio Pensabene, 213–22. Studi Miscellanei 31. Rome: L'Erma di Bretschneider.
2002a "Alabastro Quarries near Hierapolis (Turkey)." In ASMOSIA VI, 19–24.
2002b "Considerazioni sulle cave, sui metodi di estrazione, di lavorazione e sui trasporti." In *I marmi colorati della Roma imperiale: catalogo della mostra* (*Roma, Mercati di Traiano, 28 settembre 2002–19 gennaio 2003*), 179–94. Venice: Marsilio.
2002c "Il mondo delle cave in Italia: considerazioni su alcuni marmi e pietre usati nell'antichità." In *I marmi colorati della Roma imperiale: catalogo della mostra* (*Roma, Mercati di Traiano, 28 settembre 2002–19 gennaio 2003*), 277–90. Venice: Marsilio.
2017 "*Tituli picti* su due fronti di cava nel distretto di Bacakale a *Docimium* (Iscehisar, Afyonkarahisar)." *Journal of Roman Archaeology* 30: 469–89. https://doi.org/10.1017/S1047759400074213.

Bruno, Matthias, Stefano Cancelliere, Carlo Gorgoni, Lorenzo Lazzarini, Paolo Pallante, and Patrizio Pensabene
2002 "Provenance and Distribution of White Marbles in Temples and Public Buildings of Imperial Rome." In ASMOSIA V, 289–300.

Bruno, Matthias, Lucia Conti, Patrizio Pensabene, and Bruno Turi
2002 "Pompeii after the AD 62 Earthquake: Historical, Isotopic, and Petrographic Studies of Quarry Blocks in the Temple of Venus." In ASMOSIA V, 282–88.

Bruno, Matthias, H. Elçi, Ali Bahadir Yavuz, and Donato Attanasio
2012 "Unknown Ancient Marble Quarries of Western Asia Minor." In ASMOSIA IX, 562–72.

Bruno, Matthias, and Lorenzo Lazzarini
1999 "Discovery of the Sienese Provenance of *breccia dorata*, and *breccia gialla fibrosa*, and the Origin of *breccia rossa appenninica*." In ASMOSIA IV, 77–82.

Bruno, Matthias, and Paolo Pallante
2002 "The 'Lapis Taenarius' Quarries of Cape Tainaron." In ASMOSIA VI, 163–76.

Bruto, M. Luisa, and Cinzia Vannicola
1990 "Ricostruzione e tipologia delle crustae parietali in età imperiale." *Archeologia classica* 42: 325–76.

Bruzza, Luigi
1870 "Iscrizioni dei marmi Grezzi." *Annali dell'Instituto di corrispondenza archeologica* 42: 106–204.

Bugini, Roberto, and Luisa Folli
2013 "Three Important Stones of Italian Baroque and Rococo Architecture: *Macchiavecchia*, *Broccatello*, and *Rosso di Arzo* (Ticino, Switzerland)." *Marmora* 8: 127–46.

Bülow-Jacobsen, Adam
2008 "An Ostracon from the Quarry at Umm Balad." In *Sixty Five Papyrological Texts: Presented to Klaas A. Worp on the Occasion of His 65th Birthday*, edited by F.A.J. Hoogendijk and Brian P. Muhs, 311–16. Papyrologica Lugduno-Batava 33. Leiden and Boston: Brill.
2009 *Mons Claudianus: Ostraca Graeca et Latina IV: The Quarry-Texts, O.Claud. 632–896.* Institut Français d'Archéologie Orientale, Documents de Fouilles 47. Cairo: Institut Français d'Archéologie Orientale.

Bull-Simonsen Einaudi, Karin
1983 "Esther Boise Van Deman: un'archeologa Americana." In *L'Archeologia in Roma capitale tra sterro e scavo*, 41–47. Venice: Marsilio Editori.

Burrell, Barbara
2012 "Phrygian for Phrygians: Semiotics of 'Exotic' Local Marble." In ASMOSIA IX, 780–86.

Butler, Orma F.
1930 "Report of the Museum of Classical Archaeology, 1928–1929." University of Michigan Press.

Candler, Henry Ernest
1933 *A Century and One: Life Story of William Robert Candler.* New York: Putnam.

Carlson, Deborah N., and William Aylward
2010 "The Kızılburun Shipwreck and the Temple of Apollo at Claros." *American Journal of Archaeology* 114/1: 145–59.

Chidiroglou, M.
2009 "New Data on the Ancient Quarries in Southern Euboea, Greece." In ASMOSIA VIII, 73–91.

Christol, Michel, and Thomas Drew-Bear
1986 "Documents latins de Phrygie." *Tyche* 1: 41–87.
1988 "Inscriptions de Dokimeion." In *Anatolia Antiqua - Eski Anadolu 1987*, 83–137. Varia Anatolica 1. Istanbul: Institut Français d'Études Anatoliennes.
1991 "Les carrières de Dokimeion à l'époque sévérienne." *Epigraphica* 53: 113–74.
2005 "De Lepcis Magna à Aezani: Hesperus Procurateur de Phrygie et l'administration des carrières de marbre." In *Le monde romain à travers l'épigraphie: méthodes et pratiques. Actes du XXIV*e *Colloque international de Lille, 8–10 novembre 2001*, edited by Janine Desmulliez and Christine Hoët-Van Cauwenberghe, 189–216. Université de Lille 3, Travaux et recherches. Villeneuve-d'Ascq: Université Charles-de-Gaulle - Lille 3.

Ciampini, Emanuele Marcello
2004 *Gli obelischi iscritti di Roma*. Rome: Istituto poligrafico e Zecca dello Stato, Libreria dello Stato.

Claridge, Amanda, ed.
2010 *Rome: An Oxford Archaeological Guide*. 2nd ed., revised and expanded. Oxford: Oxford University Press.

Clarke, John R.
2014 "Domus/Single Family House." In *A Companion to Roman Architecture*, edited by Roger B. Ulrich and Caroline K. Quenemoen, 342–62. Blackwell Companions to the Ancient World. Malden, MA: Wiley Blackwell.

Clarke, John R., and Nayla K. Muntasser, eds.
2019 *Oplontis: Villa A ("of Poppaea") at Torre Annunziata, Italy*, Vol. 2: *The Decorations: Painting, Stucco, Pavements, Sculptures*. New York: The Humanities E-Book Series of the American Council of Learned Societies. https://hdl.handle.net/2027/heb.90048.

Coarelli, Filippo
1982 *Lazio*. Guide archeologiche Laterza 5. Rome: Laterza.

Cohen, Getzel M., and Martha Sharp Joukowsky, eds.
2004 *Breaking Ground: Pioneering Women Archaeologists*. Ann Arbor: University of Michigan Press.

Çolak, Mumtaz, and Lorenzo Lazzarini
2002 "Quarries and Characterisation of a Hitherto Unknown Alabaster and Marble from Thyatira (Akhisar, Turkey)." In ASMOSIA VI, 35–40.

Cooper, Frederick A.
1988 "The Quarries of Mount Taygetos in the Peloponnesos, Greece." In ASMOSIA I, 65–76.

Corremans, M., Patrick Degryse, Dagmara Wielgosz, and Marc Waelkens
2012 "The Import and the Use of White Marble and Coloured Stone for Wall and Floor Revetment at Sagalassos." In ASMOSIA IX, 38–51.

Corsi, Faustino
1828 *Delle pietre antiche, libri quattro di Faustino Corsi romano*. Rome: Giuseppe Salviucci, 1828.
1845 *Delle pietre antiche: trattato*. 3rd ed. Rome: Gaetano Puccinelli.

Couyat, J., and P. Montet
1912 *Les inscriptions hiéroglyphiques et hiératiques du Ouâdi Hammâmât*. Mémoires publiés par les membres de l'Institut français d'archéologie orientale du Caire 34. Cairo: Institut Français d'Archéologie Orientale.

Cramer, Thomas
2004 "Multivariate Herkunftanalyse von Marmor auf petrographischer und geochemischer Basis: Das Beispiel kleinasiatischer archaischer, hellenistischer und römischer Marmorobjecte der Berliner Antikensammlung und ihre Zuordnung zu mediterranen und anatolischen Marmorlagerstätten." Doctoral dissertation, Berlin: Technische Universität Berlin.

Cramer, Thomas, Klaus Germann, and Wolf-Dieter Heilmeyer
2002 "Petrographic and Geochemical Characterization of the Pergamon Altar Marble in the Pergamon Museum, Berlin." In ASMOSIA VI, 285–92.

Cuvigny, Hélène
2021 *Rome in Egypt's Eastern Desert*. 2 vols. ISAW Monographs. New York: New York University Press.

Dalla Rosa, Alberto
2016 "From Exploitation to Integration: Imperial Quarries, Estates and Freedmen, and the Integration of Rural Phrygia." *Studi Ellenistici* 30: 305–30.

Davies, Paul, David Hemsoll, and Mark Wilson Jones
1987 "The Pantheon: Triumph of Rome or Triumph of Compromise?" *Art History* 10/2: 133–53.

Davies, Penelope J. E.
2017 *Architecture and Politics in Republican Rome*. Cambridge: Cambridge University Press.

De Fine Licht, Kjeld
1974 Untersuchungen an den Trajansthermen zu Rom. Analecta Romana Instituti Danici 7. Copenhagen: Munksgaard.

De Nuccio, Marilda, and Lucrezia Ungaro, eds.
2002 *I marmi colorati della Roma imperiale* (*Roma, Mercati di Traiano, 28 settembre 2002–19 gennaio 2003*). I fori imperiali. Venice: Marsilio.

DeLaine, Janet
1997 *The Baths of Caracalla: A Study in the Design, Construction, and Economics of Large-Scale Building Projects in Imperial Rome.* Journal of Roman Archaeology, Supplementary Series 25. Portsmouth: Journal of Roman Archaeology.

Delbrueck, Richard
1932 *Antike Porphyrwerke.* Studien zur Spätantiken Kunstgeschichte im Auftrage des Deutschen Archäologischen Instituts 6. Berlin, Leipzig: Walter de Gruyter.

Dell'Acqua, Antonio
2014 "Materiali lapidei e marmorei dall'area capitolina di Brescia: studio e catalogazione dei frammenti di rivestimento parietale." In *Atti del XIX colloquio dell'associazione italiana per lo studio e la conservazione del mosaico,* edited by Claudia Angelelli, 455–66. Tivoli: Scripta manent.

Dessandier, David, Fabrizio Antonelli, Lorenzo Lazzarini, M. Varti-Matarangas, L. Leroux, M. Hamiane, C. Riache, and C. Khalfallah
2012 "An Introductory Study to the Ornamental and Building Stones of the Djemila (Algeria) Archaeological Site." In ASMOSIA IX, 68–74.

Digelmann, Patrick
2003 "Le dépôt de marbre du parking Pasteur. Éléments de décoration trouvés dans une domus à Aix-en-Provence (Bouches-du-Rhône)." *Revue archéologique de Narbonnaise* 36: 231–47.

Dodge, Hazel, and Bryan Ward-Perkins, eds.
1992 *Marble in Antiquity: Collected Papers of J.B. Ward-Perkins.* Archaeological Monographs of the British School at Rome 6. London: British School at Rome.

Dolci, Enrico
1988 "Marmora Lunensia: Quarrying Technology and Archaeological Use." In ASMOSIA I, 77–84.
1998 "Una cava lunense scoperta di recente a Carrara: il sito della Scalocchiella." *Quaderni di Studi Lunensi,* n.s., 4: 115–38.

Dolci, Enrico, and Leila Nista, eds.
1992 *Marmi antichi da collezione: la raccolta Grassi del Museo Nazionale Romano.* Carrara: Museo civico del marmo.

Dolci, Enrico, and Mario Nistri
1980 *Carrara, cave antiche: materiali archeologici, relazione delle campagne di rilevamento dei beni culturali del territorio promosse dal Comune di Carrara, anni 1977, 1978, 1979.* Carrara: Comune, Assessorato alla cultura, Assessorato al marmo.

Drew-Bear, Thomas
1994 "Nouvelles inscriptions de Dokimeion." *Mélanges de l'École française de Rome* 106/2: 747–844.

Dubois, Charles
1908 *Etude sur l'administration et l'exploitation des carrières, marbres, porphyre, granit, etc., dans le monde romain.* Paris: Fontemoing.

Dubois-Pélerin, Éva
2007 "Les décors de parois précieux en Italie au Ier siècle ap. J.-C.: sources littéraires et données archéologiques." *Mélanges de l'École française de Rome* 119/1: 103–24.

Ducrot, Enrico
1998 "Approccio metodologico per il riconoscimento della pietra antica: l'esempio della 'Puddinga di Hereke.'" In *Marmi antichi II: cave e tecnica di lavorazione, provenienze e distribuzione,* edited by Patrizio Pensabene, 79–86. Studi Miscellanei 31. Rome: L'Erma di Bretschneider.

Dunbabin, Katherine M.D.
1999 *Mosaics of the Greek and Roman World.* Cambridge: Cambridge University Press.

Dworakowska, Angelina
1972 "Carrières antiques à Scyros et dans les Îles de Carie: matériaux pour l'inventaire." *Archeologia Warszawa* 23: 7–21.
1975 *Quarries in Ancient Greece.* Bibliotheca Antiqua 14. Wrocław: Zakład Narodowy im. Ossolińskich.
1983 *Quarries in Roman Provinces.* Bibliotheca Antiqua 16. Wrocław: Zakład Narodowy im. Ossolińskich.
1990 "Once Again on Marmor Luculleum." In *Marble: Art Historical and Scientific Perspectives on Ancient Sculpture,* edited by Marion True and Jerry Podany, 253–62. Malibu: The J. Paul Getty Museum.

Einaudi, Karin
1988 "Esther Van Deman and the Roman Forum." *Places* 5/1: 62–70.

Ellis, S. E., R. A. Higgins, and Richard Hope Simpson
1968 "The Façade of the Treasury of Atreus at Mycenae." *The Annual of the British School at Athens* 63: 331–36.

Erim, Kenan T., and Joyce Reynolds
1970 "The Copy of Diocletian's Edict on Maximum Prices from Aphrodisias in Caria." *Journal of Roman Studies* 60/1: 120–41. https://doi.org/10.1017/S0075435800043306.

Evangelista, Patrizia, and Lorenzo Lazzarini
1998 "La collezione ex Kircheriana di diaspri Siciliani del Museo di Mineralogia alla 'Sapienza.'" In *Marmi antichi II: cave e tecnica di lavorazione, provenienze e distribuzione,* edited by Patrizio Pensabene, 391–409. Studi Miscellanei 31. Rome: L'Erma di Bretschneider.

Falcone, Roberto, and Lorenzo Lazzarini
1998 "Note storico-scientifiche sul broccatello di Spagna." In *Marmi antichi II: cave e tecnica di lavorazione, provenienze e distribuzione,* edited by

Patrizio Pensabene, 87–97. Studi Miscellanei 31. Rome: L'Erma di Bretschneider.

Fant, J. Clayton
1985 "Four Unfinished Sarcophagus Lids at Docimium and the Roman Imperial Quarry System in Phrygia." *American Journal of Archaeology* 89/4: 655–62. https://doi.org/10.2307/504206.
1988 "The Roman Emperors in the Marble Business: Capitalists, Middelmen or Philanthropists?" In ASMOSIA I, 147–58.
1989 *Cavum Antrum Phrygiae: The Organization and Operations of the Roman Imperial Marble Quarries in Phrygia.* BAR International Series 482. Oxford, England: BAR.
1993 "Ideology, Gift, and Trade: A Distribution Model for the Roman Imperial Marbles." In *The Inscribed Economy: Production and Distribution in the Roman Empire in the Light of Instrumentum Domesticum. The Proceedings of a Conference Held at the American Academy in Rome on 10–11 January 1992*, edited by William V. Harris, 145–70. Journal of Roman Archaeology, Supplementary Series 6. Ann Arbor: University of Michigan Press.
1999 "Augustus and the City of Marble." In ASMOSIA IV, 277–80.
2001 "Rome's Marble Yards." *Journal of Roman Archaeology* 14: 167–98.
2009 "White Marbles in the Summer Triclinium of the Casa del Bracciale d'Oro, Pompeii." In ASMOSIA VII, 53–57.
2016 "A Milestone in the History of the Roman Trade in Stone. Review of 'The Economics of the Roman Stone Trade,' by Ben Russell (Oxford Studies on the Roman Economy; Oxford 2013)." *Journal of Roman Archaeology* 29: 701–11.
2022 "The Early History of Roman Quarrying in Egypt and *Granito della colonna* from Wadi Umm Shegilat in the Eastern Desert." *Marmora* 18: 39–56.
forth. "Pliny's *Marmor Augusteum*." *Journal of Egyptian Archaeology* 110.

Fant, J. Clayton, Stefano Cancelliere, Lorenzo Lazzarini, M. Preite Martinez, and Bruno Turi
2002 "White Marbles at Pompeii: Sampling the Casa dei Vettii." In ASMOSIA VI, 309–16.

Fant, J. Clayton, B. Russell, and Simon J. Barker
2013 "Marble Use and Reuse at Pompeii and Herculaneum: The Evidence from the Bars." *Papers of the British School at Rome* 81: 181–209.

Farr, Jason, Fabrizio Marra, and Nicola Terrenato
2015 "Geochemical Identification Criteria for 'Peperino' Stones Employed in Ancient Roman Buildings: A *Lapis Gabinus* Case Study." *Journal of Archaeological Science: Reports* 3: 41–51.

Ficoroni, Francesco
1744 *Le vestigia e rarità di Roma antica ricercate, e spiegate da Francesco de' Ficoroni aggregato alla Reale accademia di Francia.* 2 vols. Rome: Girolamo Mainardi.

Filippi, Fedora
2005a "Le cornici." In *Palazzo Altemps: i colori del fasto; la Domus del Gianicolo e i suoi marmi*, edited by Fedora Filippi, 72–77. Milan: Electa.
2005b "Gli intarsi." In *Palazzo Altemps: i colori del fasto; la Domus del Gianicolo e i suoi marmi*, edited by Fedora Filippi, 78–79. Milan: Electa.

Fogagnolo, Stefania
2002 "Scoperta di frammenti di colonne colossali dal foro della pace." In *I marmi colorati della roma imperiale (Roma, Mercati di Traiano, 28 Settembre 2002–19 Gennaio 2003)*, edited by Marilda De Nuccio and Lucrezia Ungaro, 136–37. I Fori Imperiali. Venice: Marsilio.

Fratini, Fabio, Elena Pecchioni, Emma Cantisani, Fabrizio Antonelli, Marco Giamello, Marco Lezzerini, and Roberta Canova
2015 "Portoro, the Black and Gold Italian 'Marble.'" *Rendiconti Lincei* 26/4: 415–23.

Gans, Ulrich-Walter
1992 *Korinthisierende Kapitelle der römischen Kaiserzeit: Schmuckkapitelle in Italien und den nordwestlichen Provinzen.* Arbeiten zur Archäologie. Cologne: Böhlau.

Gazda, Elaine K., and John R. Clarke, eds.
2016 *Leisure and Luxury in the Age of Nero: The Villas of Oplontis near Pompeii.* Kelsey Museum Publications 14. Ann Arbor: Kelsey Museum of Archaeology.

Gazda, Elaine K., and Anne E. Haeckl
1996 *Images of Empire: Flavian Fragments in Rome and Ann Arbor Rejoined.* Ann Arbor: Kelsey Museum of Archaeology.

Germann, Klaus, Gottfried Gruben, H. Knoll, V. Valis, and F. J. Winkler
1988 "Provenance Characteristics of Cycladic (Paros and Naxos) Marbles: A Multivariate Geological Approach." In ASMOSIA I, 251–62.

Giacchero, Marta
1974 *Edictum Diocletiani et collegarum de pretiis rerum venalium: in integrum fere restitutum e latinis graecisque fragmentis.* Pubblicazioni dell'Istituto di storia antica e scienze ausiliarie dell'Università di Genova 8. Genoa: Istituto di storia antica e scienze ausiliarie.

Gnoli, Raniero
1971 *Marmora romana.* Rome: Edizioni dell'Elefante.
1988 *Marmora romana.* 2nd, revised and expanded ed. Rome: Edizioni dell'Elefante.

Gorgoni, Carlo, Lorenzo Lazzarini, Paolo Pallante, and Bruno Turi

2002 "An Updated and Detailed Mineropetrographic and C-O Stable Isotopic Reference Database for the Main Mediterranean Marbles Used in Antiquity." In ASMOSIA V, 115–31.

Goyon, Georges

1957 *Nouvelles inscriptions rupestres du Wadi Hammamat.* Paris: Imprimerie Nationale, Librarie d'Amerique et d'Orient Adrien-Maisonneuve.

Granino Cecere, Maria Grazia, and Vincenza Morizio

2007 "Nuove testimonianze sul'amministrazione dei marmi nella Roma imperiale." In *Supplying Rome and the Empire: The Proceedings of an International Seminar Held at Siena-Certosa di Pontignano on May 2–4, 2004, on Rome, the Provinces, Production and Distribution,* edited by Emanuele Papi, 127–38. Journal of Roman Archaeology, Supplementary Series 69. Portsmouth: Journal of Roman Archaeology.

Grieco, Roberto

2008 *Micromosaici Romani / Roman Micromosaic.* Rome: Gangemi Editore SpA.

Gros, Pierre

1973 "Hermodorus et Vitruve." *Mélanges de l'École française de Rome, Antiquité* 85: 137–61.

1976 "Les premières générations d'architectes hellénistiques à Rome." In *L'Italie préromaine et la Rome républicaine: mélanges offerts à Jacques Heurgon,* 387–410. Collection de l'École française de Rome 27. Rome: École française de Rome.

Guéraud, Octave

1942 "Ostraca Grecs et Latins de l'Wâdi Fawâkhir." *Bulletin de l'Institut Français d'Archéologie Orientale* 41: 141–96.

Guidobaldi, Federico

1990 "L'intarsio marmoreo nella decorazione parietale e pavimentale di età romana." In *Il marmo nella civiltà romana. La produzione e il commercio,* edited by Enrico Dolci, 55–81. Lucca: Cassa di risparmio.

2000a "La lussuosa aula presso Porta Marina a Ostia: la decorazione in opus sectile dell'aula." In *Aurea Roma: dalla città pagana alla città cristiana,* edited by Serena Ensoli and Eugenio La Rocca, 251–62. Rome: L'Erma di Bretschneider.

2000b "La produzione di mosaici e sectilia pavimentali e parietali." In *Christiana loca: lo spazio cristiano nella Roma del primo millennio,* edited by Letizia Pani Ermini, 275–82. Rome: Fratelli Palombi.

2009 "Sectilia pavimenta tardoantichi e paleocristiani a piccolo modulo dell'Italia Settentrionale." *Rivista di Archeologia Cristiana* 85: 355–420.

Guidobaldi, Federico, and Claudia Angelelli

2002 "Frammenti di lastre da *opus sectile* come materiali di scavo: criteri di individuazione, classificazione ed edizione." In *I mosaici: cultura, tecnologia, conservazione. Atti del XVIII convegno internazionale scienza e beni culturali bressanone, 2–5 luglio 2002,* edited by Guido Biscontin and Guido Driussi, 155–63. Venice: Arcadia richerche.

Guidobaldi, Federico, and Antonio Salvatori

1988 "The Introduction of Polychrome Marbles in Late Republican Rome: The Evidence from Mosaic Pavements with Marble Insertions." In ASMOSIA I, 171–75.

1994 "La decorazione pavimentale di Villa Adriana: tipologie, distribuzione e restauri." In *Fifth International Colloquium on Ancient Mosaics: Held at Bath, England, on September 5–12, 1987,* edited by Peter Johnson, Roger Ling, and D. J. Smith, 1:103–22. Journal of Roman Archaeology Supplementary Series 9. Ann Arbor: Journal of Roman Archaeology.

Guidobaldi, Maria Paola, and Domenico Esposito

2009 "Le nuove ricerche archeologiche nella villa dei papiri di Ercolano." *Cronache Ercolanesi* 39: 331–70.

Gutiérrez Garcia-Moreno, Anna

2009 *Roman Quarries in the Northeast of Hispania (Modern Catalonia).* Documenta 10. Tarragona: Institut Català d'Arqueologia Clàssica.

Gutiérrez Garcia-Moreno, Anna, Pilar Lapuente Mercadal, and Isabel Rodà de Llanza, eds.

2012 *Interdisciplinary Studies on Ancient Stone: Proceedings of the 9th Association for the Study of Marbles and Other Stones in Antiquity (ASMOSIA) Conference (Tarragona 2009).* ASMOSIA IX; Documenta 23. Tarragona: Institut Català d'Arqueologia Clàssica.

Hall, Alan, and Marc Waelkens

1982 "Two Dokimeian Sculptors in Iconium." *Anatolian Studies* 32: 151–55. https://doi.org/10.2307/3642678.

Harrell, James A.

1990 "Misuse of the Term 'Alabaster' in Egyptology." *Göttinger Miszellen* 119: 37–42.

2002 "Pharaonic Stone Quarries in the Egyptian Deserts." In *Egypt and Nubia: Gifts of the Desert,* edited by Renée F. Friedman, 232–43. London: British Museum Press.

Harrell, James A., V. Max Brown, and Lorenzo Lazzarini

1999 "Two Newly Discovered Roman Quarries in the Eastern Desert of Egypt." In ASMOSIA IV, 285–92.

2002 "Breccia Verde Antica: Source, Petrology and Ancient Uses." In ASMOSIA VI, 207–18.

Harrell, James A., and Lorenzo Lazzarini
2002 "A New Variety of *Granito Bianco e Nero* from Wadi Barud, Egypt." In ASMOSIA V, 47–51.

Harrell, James A., Lorenzo Lazzarini, and Matthias Bruno
2002 "Reuse of Ancient Ornamental Stones in Medieval Cairo, Egypt." In ASMOSIA VI, 89–96.

Harrell, James A., and Steven E. Sidebotham
2018 "Wadi Barud North: A New Roman Quarry and Stone from Egypt's Eastern Desert." *Marmora* 14: 53–75.

Heberdey, Rudolf
1904 "Daitis: Ein Beitrag zum Ephesischen Artemiscult." *Jahreshefte des Österreichischen Archäologischen Instituts in Wien* 7: 210–15.
1905 "Die Proconsules Asia unter Traian." *Jahreshefte des Österreichischen Archäologischen Instituts in Wien* 8: 231–37.

Herrmann, John J., Vincent Barbin, and Aristotelis Mentzos
1999 "The Exportation of Marble from Thasos in Late Antiquity: The Quarries of Aliki and Cape Fanari." In *Thasos: matiéres premiéres et technologie de la préhistoire á nos jours,* edited by Arthur Muller, 75–90. Atina: Ecole Francaise d'Athenes.

Herrmann, John J., Norman Herz, and Richard Newman, eds.
2002 *Interdisciplinary Studies on Ancient Stone. Proceedings of the 5th International Conference of the Association for the Study of Marble and Other Stones in Antiquity, Museum of Fine Arts, Boston, 1998.* ASMOSIA V. London: Archetype.

Herrmann, John J., and Annewies van den Hoek
2014 "Elite Marble from Mauritania to Rome: Alabastro a Pecorella from Bouhanifia, Algeria." In *Centro y periferia en el mundo clásico: 18 Congreso Internacional Arqueología Clásica / Centre and Periphery in the Ancient World: 18th International Congress of Classical Archaeology,* edited by José María Álvarez Martínez, Trinidad Nogales Basarrate, and Isabel Rodà de Llanza, 2:1319–23. Mérida: Museo Nacional de Arte Romano.

Herrmann, John J., Annewies van den Hoek, and Richard Newman
2002 "New Sculptures in Thasian Dolomite: The Black Sea, Tunisia, and Questions of Style." In ASMOSIA VI, 357–62.

Herrmann, John J., Jr., Annewies van den Hoek, and Robert H. Tykot
2012 "Alabastro a Pecorella, Aïn Tekbalet, and Bou Hanifia, Algeria: Preliminary Report." In ASMOSIA IX, 463–70.

Herrmann, John J., and Richard Newman
1995 "The Exploration of Dolomitic Sculptural Marble from Thasos: Evidence from Mediterranean and Other Collections." In ASMOSIA III, 73–86.

Herz, Norman
1987 "Carbon and Oxygen Isotopic Ratios: A Data Base for Classical Greek and Roman Marble." *Archaeometry* 29/1: 35–43.
1988 "The Carbon and Oxygen Isotopic Data Base for Classical Marble." In ASMOSIA I, 305–14.

Herz, Norman, and Marc Waelkens, eds.
1988 *Classical Marble: Geochemistry, Technology, Trade.* NATO ASI Series, Series E, Applied Sciences 153. [ASMOSIA I.] Dordrecht and Boston: Kluwer Academic Publishers, 1988.

Higgins, Michael Denis, and Reynold Alleyne Higgins
1996 *A Geological Companion to Greece and the Aegean.* Ithaca: Cornell University Press.

Hinsdale, Burke A., and Isaac N. Demmon
1906 *History of the University of Michigan.* Ann Arbor: University of Michigan Press.

Hirschfeld, Otto
1905 *Die kaiserlichen Verwaltungsbeamten bis auf Diocletian.* 2nd ed. Berlin: Weidmann.

Hirt, Alfred Michael
2010 *Imperial Mines and Quarries in the Roman World: Organizational Aspects, 27 BC–AD 235.* Oxford Classical Monographs. Oxford and New York: Oxford University Press.

İlhan, Sinan
2014 "Restaurierung im Marmorsaal 31." In *Hanghaus 2 in Ephesos: Die Wohneinheit 6; Baubefund, Ausstattung, Funde,* edited by Elisabeth Rathmayr and Hilke Thür, 803–29. Forschungen in Ephesos, VIII/9. Vienna: Austrian Academy of Sciences Press.

Istituto commercio estero
1982 *Marmi italiani: guida tecnica.* 2 vols. Milan: Vallardi.

Iversen, Erik
1968 *Obelisks in Exile,* Vol. 1: *The Obelisks of Rome.* Copenhagen: Gad.

Jackson, Marie, and Fabrizio Marra
2006 "Roman Stone Masonry: Volcanic Foundations of the Ancient City." *American Journal of Archaeology* 110/3: 403–36.

Jongste, Peter F. B., J. B. Jansen, Luc Moens, Paul De Paepe, and Marc Waelkens
1992 "The Use of Marble in Latium between 70 and 150 A.D. Icpaes for Determination of the Provenance of White Marbles." In ASMOSIA II, 263–67.

Kayser, François
1993 "Nouveaux Textes Grecs du Ouadi Hammamat." *Zeitschrift für Papyrologie und Epigraphik* 98: 111–56.

Keil, Josef
1951 *Die Johanneskirche.* Forschungen in Ephesos, IV/3. Vienna: Österreichisches Archäologisches Institut.

Khanoussi, Mustapha
1991 "Nouveaux documents sur la présence militaire dans la colonie julienne augustéenne de Simitthus (Chemtou, Tunisie)." *Comptes rendus des séances de l'Académie des Inscriptions et Belles-Lettres* 135/4: 825–39. https://doi.org/10.3406/crai.1991.15051.

Kiilerich, Bente
2014 "The Opus Sectile from Porta Marina at Ostia and the Aesthetics of Interior Decoration." In *Production and Prosperity in the Theodosian Period*, edited by Ine Jacobs, 169–214. Interdisciplinary Studies in Ancient Culture and Religion 14. Leuven: Peeters.

Klein, Michael J.
1988 *Untersuchungen zu den kaiserlichen Steinbrüchen an Mons Porphyrites und Mons Claudianus in der östlichen Wüste Ägyptens.* Habelts Dissertationsdrucke. Reihe Alte Geschichte 26. Bonn: R. Habelt.

Klemm, Rosemarie, and Dietrich D. Klemm
1993 *Steine und Steinbrüche im alten Ägypten.* Berlin and New York: Springer.
2008 *Stones and Quarries in Ancient Egypt.* London: British Museum Press.

Koch, Guntram, and Hellmut Sichtermann
1982 *Römische Sarkophage.* Handbuch der Archäologie. Munich: Beck.

Kokkorou-Alevras, Georgia, Achilleas Chatziconstantinou, Alexis Efstathopoulos, Eleni Zavvou, N. Themos, Konstantinos Kopanias, and Eirene A. Poupaki
2009 "Ancient Quarries in Laconia." *British School at Athens Studies* 16, special issue, "Sparta and Laconia: From Prehistory to Pre-Modern": 169–79.

Korres, Manolis
1995 *From Pentelicon to the Parthenon: The Ancient Quarries and the Story of a Half-Worked Column Capital of the First Marble Parthenon.* Athens: Melissa.
2000 "Upogeia latomeia tes Parou." In *Paria lithos: latomeia, marmaro kai ergasteria glyptikes tes Paros; praktika 1. Diethnous Synedriou Archaiologias Parou kai Kykladon, Paroikia, Paros 2–5 oktovriou 1997*, edited by Demetrio Schilardi and Dora Katsonopoulou, 61–82. Archaiologikes kai historikes meletes 1. Athens: D.U. Schilardi.

Lambraki, Anna
1980 "Le cipolin de la karystie: contribution a l'étude des marbres de la grèce exploités aux époques romaine et paléochrétienne." *Revue Archéologique*, n.s., 1: 31–62.

Lanciani, Rodolfo Amedeo
1897 *The Ruins and Excavations of Ancient Rome: A Companion Book for Students and Travelers.* Boston: Houghton, Mifflin & Company.

Lassale, Jacques Dubarry de
2000 *Identifying Marble.* France: H. Vial.

Lazzarini, Lorenzo
2002a "La determinazione della provenienza delle pietre decorative usate dai romani." In *I marmi colorati della Roma imperiale (Roma, Mercati di Traiano, 28 Settembre 2002-19 Gennaio 2003)*, edited by Marilda De Nuccio and Lucrezia Ungaro, 223–65. I Fori Imperiali. Venice: Marsilio.
2002b "The Origin and Characterization of Breccia Nuvolata, Marmor Sagarium, and Marmor Triponticum." In ASMOSIA V, 58–67.
2007 *Poikiloi lithoi, versiculores maculae: i marmi colorati della Grecia antica; storia, uso, diffusione, cave, geologia, caratterizzazione scientifica, archeometria, deterioramento.* Marmora. Supplemento 1. Pisa: F. Serra.
2009 "The Distribution and Re-Use of the Most Important Coloured Marbles in the Provinces of the Roman Empire." In ASMOSIA VII, 459–84.
2012 "I marmi e le pietre del pavimento Marciano." In *Il manto di pietra della Basilica di San Marco a Venezia: storia, restauri, geometrie del pavimento*, edited by Ettore Vio, 51–107. Venice: Cicero.
2014 "Le lumachelle veronesi: storia d'uso e diffusione dal XV al XX secolo, geologia, cave antiche e moderne." *La Rivista di Engramma* 117: 1–11.
2019 "I diaspri siciliani: genesi, composizione e storia di alcune tra le pietre più usate dall'Opificio fiorentino." In *Pietre colorate molto vaghe e belle; arte senza tempo dal Museo dell'Opificio delle Pietre Dure*, edited by Sandra Rossi, Peter Assmann, and Anna Patera, 85–107. Mantova: Tre Lune.

Lazzarini, Lorenzo, ed.
2002 *Interdisciplinary Studies on Ancient Stone. Proceedings of the 6th International Conference, Venice, June 15–18 2000.* ASMOSIA VI. Padua: Bottega d'Erasmo.

Lazzarini, Lorenzo, M. Agus, and Stefano Cara
2006 "The Ancient Quarries of the *Neri Antichi* (Black Limestones) from Zeugitania (Tunisia)." Marmora 2: 59–70.

Lazzarini, Lorenzo, Fabrizio Antonelli, Stefano Cancelliere, Bruno Turi, and M. Varti-Matarangas
2002 "Marmor Chalcidicum (Fior di Pesco): Source, History of Use and Scientific Characterization." In ASMOSIA VI, 233–40.

Lazzarini, Lorenzo, and Stefano Cancelliere
2000 "Characterization of the White Marble of Two Unpublished Ancient Roman Quarries on the Island of Fourni Ad Skyros (Greece)." *Periodico di Mineralogia* 69/1: 49–62.

Lazzarini, Lorenzo, and M. Mariottini
2012 "A First Study of Some Lumachelle (Fossiliferous Stones) Used in Roman Antiquity." In ASMOSIA IX, 445–51.

Lazzarini, Lorenzo, and Bruno Turi
1999 "Characterisation and Differentiation of the Skyros Marbles (Greece) and the Medici's Breccias (Italy)." In ASMOSIA IV, 117–24.

Lazzarini, Lorenzo, Igor M. Villa, and D. Visonà
2006 "Caratterizzazione e identificazione di alabastri usati in antico." In *Archeometria del costruito: l'edificato storico; materiali, strutture e rischio sismico. Atti del convegno nazionale di archeometria, Ravello, 6–7 febbraio 2003*, edited by Gino Mirocle Crisci and Caterina Gattuso, 273–81. Scienze e materiali del patrimonio culturale 9. Bari: Edipuglia.

Lazzarini, Lorenzo, D. Visonà, M. Giamello, and Igor M. Villa
2012 "Archaeometric Characterisation of One Tunisian and Two Italian Calcareous Alabasters Used in Antiquity." In ASMOSIA IX, 436–44.

Lindsay, Wallace M., ed.
1913 *Sexti Pompei Festi, De verborum significatu quae supersunt cum Pauli epitome.* Leipzig: B.G. Teubneri.

Long, Leah E.
2012 "Urbanism, Art, and Economy: The Marble Quarrying Industries of Aphrodisias and Roman Asia Minor." Doctoral dissertation, University of Michigan.

Lucci, Maria Luisa
1964 "Il porfido nell'antichità." *Archeologia Classica* 16: 226–71.

Mackensen, Michael
2005 *Simitthus III. Militärlager oder Marmorwerkstätten: neue Untersuchungen im Ostbereich des Arbeits- und Steinbruchlagers von Simitthus/Chemtou.* Deutsches Archäologisches Institut; Institut National du Patrimoine, Tunis. Mainz am Rhein: Philipp von Zabern.

Maischberger, Martin
1997 *Marmor in Rom: Anlieferung, Lager- und Werkplätze in der Kaiserzeit.* Palilia 1. Wiesbaden: Ludwig Reichert.

Mango, Cyril A.
1972 *The Art of the Byzantine Empire, 312–1453: Sources and Documents.* Englewood Cliffs: Prentice-Hall.

Maniatis, Yannis, ed.
2009 *Proceedings of the 7th International Conference of Association for the Study of Marble and Other Stones in Antiquity, Thassos, 15–20 September 2003.* ASMOSIA VII; Bulletin de Correspondance Hellénique, Supplément 51. Athens: Ecole française d'Athènes.

Maniatis, Yannis, Norman Herz, and Yannis Basiakos, eds.
1995 *The Study of Marble and Other Stones Used in Antiquity.* ASMOSIA III. London: Archetype.

Mannoni, Tiziano
1987 "I primi probabili impieghi del marmo lunense e il Portus Lunae." In *Studi lunensi e prospettive sull'occidente romano: Proceedings of the Convegno italo-francese di studi lunensi e prospettive sull'occidente romano held at Villa Marigola, Lerici, September 1985*, 395–403. Quaderni del Centro Studi Lunensi 10–12. Luni: Il Centro.

Mari, Zaccaria
2004 "La villa di Traiano ad Arcinazzo Romano." *The Journal of Fasti Online* 2004.

Matetić Poljak, Daniela, and Katja Marasović, eds.
2018 *Interdisciplinary Studies on Ancient Stone: Proceedings of the 11th International Conference of ASMOSIA, Split, 18–22 May 2015.* ASMOSIA XI. Split: Arts Academy in Split and University of Split.

Matthews, K. J., Luc Moens, S. Walker, Marc Waelkens, and Paul De Paepe
1992 "The Re-Evaluation of Stable Isotope Data for Pentelic Marble." In ASMOSIA II, 203–12.

Maxfield, Valerie A.
2001 "Stone Quarrying in the Eastern Desert with Particular Reference to Mons Claudianus and Mons Porphyritie." In *Economies beyond Agriculture in the Classical World,* edited by David J. Mattingly and John Salmon, 143–70. Leicester-Nottingham Studies in Ancient Society 9. London: Routledge.

Maxfield, Valerie A., and David P. S. Peacock
2001 *The Roman Imperial Quarries: Survey and Excavation at Mons Porphyrites, 1994–1998,* Vol. 1: *Topography and Quarries.* Excavation Memoir 67. London: Egypt Exploration Society.
2007 *The Roman Imperial Quarries: Survey and Excavation at Mons Porphyrites, 1994–1998,* Vol. 2: *The Excavations.* Excavation Memoir 82. London: Egypt Exploration Society.

Mayer, Marc, and Isabel Rodà
1999 "El Broccatello de Tortosa: Testimonios Arqueológicos." *Pallas* 50: 43–52.

Mazeran, René
1998 "Le transport maritime des marbres sous l'empire romain: les données des tranches de sciage de l'épave de Porto Nuovo près de Porto-Vecchio (Corse du Sud)." In *Méditerranée Antique.* Paris: Editions du Comité des travaux historiques et scientifiques.

McAlpine, Lynley J.
2016 "Luxury in Fantasy and Reality: Exotic Marble in Villa A." In *Leisure and Luxury in the Age of Nero: The Villas of Oplontis near Pompeii,* edited by Elaine K. Gazda and John R. Clarke, 111–18.

Kelsey Museum Publications 14. Ann Arbor: Kelsey Museum of Archaeology.

Menella, G.
1990 "L'imprenditoria private nelle cave di Carrara alla luce di *CIL* XI 6946." In *Miscellanea di studi archeologici e di antichita* 3, edited by Fernando Rebecchi, 133–40. Modena: Aedes Muratoriana.

Mielsch, Harald
1985 *Buntmarmore aus Rom im Antikenmuseum Berlin*. Berlin: Staatliche Museen Preussischer Kulturbesitz.

Mitchell, Stephen
1976 "Requisitioned Transport in the Roman Empire: A New Inscription from Pisidia." *Journal of Roman Studies* 66: 106–31. https://doi.org/10.2307/299783.

Moens, Luc, Paul De Paepe, and Marc Waelkens
1992 "Multidisciplinary Research and Cooperation: Keys to a Successful Provenance Determination of White Marble." In ASMOSIA II, 247–54.

Monna, Dario, and Patrizio Pensabene
1977 *Marmi dell'Asia Minore*. Rome: Consiglio nazionale delle ricerche.

Moss, Christopher Frederick
1988 "Roman Marble Tables." Doctoral dissertation, Princeton University.

Nylander, Carl
1968 "Bemerkungen zur Steinbrüchgeschichte von Assuan." *Archäologischer Anzeiger* 1968/1: 6–10.

Osborne, Robin
2015 "De-Contextualising and Re-Contextualising: Why Mediterranean Archaeology Needs to Get Out of the Trench and Back into the Museum." *Journal of Mediterranean Archaeology* 28/2: 241–61.

Østby, Erik, ed.
2014a *Tegea I: Investigations in the Temple of Athena Alea 1991–94*. Papers and Monographs from the Norwegian Institute at Athens 3. Athens: Norwegian Institute at Athens.
2014b *Tegea II: Investigations in the Sanctuary of Athena Alea 1990–94 and 2004*. Papers and Monographs from the Norwegian Institute at Athens 3. Athens: Norwegian Institute at Athens.

Öztürk, Arzu
2005 "Zur Scaenae Frons des Theaters in Ephesus." *Architectura* 35: 4–14.

Palagia, O.
2002 "A New Metope from Bassai." In ASMOSIA VI, 375–82.

Papageorghakis, I.
1964 "Ta archaia latomeia tou karystiou marmarou [The ancient quarries of the Karystos marble]." *Praktika tes Akademias Athenon* 39: 262–78.

Paribeni, Roberto
1920 *Le Terme de Diocleziano e il Museo Nazionale Romano*. Guide dei Musei Italiani. Rome: E. Cuggiani.

Paribeni, Emanuela, and Simonetta Segenni, eds.
2015 *Notae lapicidinarum dalle cave di Carrara*. Pisa: Pisa University Press.

Peacock, David P. S., Olwen Williams-Thorpe, R. S. Thorpe, and A. G. Tindle
1994 "Mons Claudianus and the Problem of the 'Granito del Foro': A Geological and Geochemical Approach." *Antiquity* 68/259: 209–30.

Peacock, David P. S., and Valerie A. Maxfield
1997 *Mons Claudianus: Survey and Excavation, 1987–1993*, Vol. 1: *Topography and Quarries*. Fouilles de l'Institut Français d'archéologie Orientale du Caire 37. Cairo: Institut Français d'Archéologie Orientale.

Pedley, John Griffiths
2012 *The Life and Work of Francis Willey Kelsey: Archaeology, Antiquity, and the Arts*. Ann Arbor: University of Michigan Press.

Pensabene, Patrizio
1976 "Sull'impiego del marmo di Cap de Garde. Condizioni giuridiche e significato economico delle cave in età imperiale." In *In memoria di Giovanni Becatti*, edited by Lucia Guerrini, 179 90. Studi miscellanei 22. Rome: De Luca.
1994 *Le vie del marmo: i blocchi di cava di Roma e di Ostia, il fenomeno del marmo nella Roma antica*. Itinerari ostiensi 7. Rome: Ministero per i beni culturali e ambientali, Soprintendenza archeologica di Ostia.
1998 "Il fenomeno del marmo nella Roma Tardo-Repubblicana e Imperiale." In *Marmi antichi II: cave e tecnica di lavorazione, provenienze e distribuzione*, edited by Patrizio Pensabene, 333–90. Studi Miscellanei 31. Rome: L'Erma di Bretschneider.
2002 "Il fenomeno del marmo del mondo romano." In *I marmi colorati della Roma imperiale* (Roma, Mercati di Traiano, 28 settembre 2002–19 gennaio 2003), edited by Marilda De Nuccio and Lucrezia Ungaro, 3–67. I fori imperiali. Venice: Marsilio.
2010 "Cave di marmo bianco e pavonazzetto in Frigia: sulla produzione e sui dati epigrafici." *Marmora* 6: 71–134.
2013 *I marmi nella Roma antica*. Biblioteca di testi e studi 890. Rome: Carocci editore.

Pensabene, Patrizio, ed.
1998 *Marmi Antichi II: cave e tecnica di lavorazione, provenienze e distribuzione*. Studi Miscellanei 31. Rome: L'Erma di Bretschneider.

Pensabene, Patrizio, and Matthias Bruno
1998a "Aggiornamenti, nuove acquisizioni e riordino dei marmi di cava dal canale di Fiumicino." In

Marmi antichi II: cave e tecnica di lavorazione, provenienze e distribuzione, edited by Patrizio Pensabene, 1–22. Studi miscellanei 31. Rome: L'Erma di Bretschneider.

1998b *Il marmo e il colore: guida fotografica, i marmi della Collezione Podesti.* Rome: L'Erma di Bretschneider.

Pensabene, Patrizio, and Eleonora Gasparini, eds.

2015 *Interdisciplinary Studies on Ancient Stone: Proceedings of the 10th International Conference of ASMOSIA, Association for the Study of Marble and Other Stones in Antiquity, Rome, 21–26 May 2012.* 2 vols. ASMOSIA X. Rome: L'Erma di Bretschneider.

Pensabene, Patrizio, and Lorenzo Lazzarini

1998 "Il problema del bigio antico e del bigio morato: contributo allo studio delle cave di Teos e di Chios." In *Marmi antichi II: cave e tecnica di lavorazione, provenienze e distribuzione*, edited by Patrizio Pensabene, 141–74. Studi miscellanei 31. Rome: L'Erma di Bretschneider.

Pensabene, Patrizio, T. Semeraro, Lorenzo Lazzarini, Bruno Turi, and M. Soligo

1999 "The Provenance of the Marbles from the Depository of the Temple of the *Fabri Navales* at Ostia." In ASMOSIA IV, 147–56.

Perna, Simona

2015 "Fabri Luxuriae: Production and Consumption of Coloured Stone Vases in the Roman Period." In ASMOSIA X, 2:1021–30.

Perna, Simona, and Simon J. Barker

2017 "I rivestimenti pavimentali di Pompei ed Ercolano e l'uso dell'alabastro in epoca Romana (II sec. a.C.–I sec. d.C.): nuove indagini." In *Atti del XXII Colloquio dell'Associazione Italiana per lo Studio e la Conservazione del Mosaico, Matera, 16–19 marzo 2016*, edited by Claudia Angelelli, Daniela Massara, and Andrea Paribeni, 73–82. Tivoli (Rome): Scripta manent.

Pike, Scott

1999 "Preliminary Results of a Systematic Characterization Study of Mount Pentelikon, Attica, Greece." In ASMOSIA IV, 165–70.

2009 "A Stable Isotope Database for the Ancient White Marble Quarries of Mount Pentelikon, Greece." In ASMOSIA VII, 699–708.

Pliny the Elder

1991 *Natural History, A Selection.* Translated by John F. Healy. Penguin Classics. London and New York: Penguin Books.

Pollini, John, Norman Herz, Kyriaki Polikreti, and Yannis Maniatis

1998 "Parian *Lychnites* and the Prima Porta Statue: New Scientific Tests and the Symbolic Value of the Marble." *Journal of Roman Archaeology* 11: 275–84. https://doi.org/10.1017/S1047759400017311.

Price, Monica T.

2007 *The Sourcebook of Decorative Stone: An Illustrated Identification Guide.* Toronto: Firefly Books.

Prochaska, Walter

2010 "Tracing the Origin of Marbles by the Chemical Analysis of Inclusion Fluids from the Ancient Metropolis of Ephesos / Asia Minor." Kiel.

Rakob, Friedrich, ed.

1993 *Simitthus I. Die Steinbrüche und die antike Stadt.* Deutsches Archäologisches Institut; Institut National du Patrimoine, Tunis 1. Mainz am Rhein: Philipp von Zabern.

1994 *Simitthus II. Der Tempelberg und das römische Lager.* Deutsches Archäologisches Institut; Institut National du Patrimoine, Tunis 2. Mainz am Rhein: Philipp von Zabern.

Reeves, Nicholas

2015 "Tutankhamun's Mask Reconsidered." In *The Art and Culture of Ancient Egypt: Studies in Honor of Dorothea Arnold*, edited by Adela Oppenheim and Ogden Goelet Jr., 511–26. Bulletin of the Egyptological Seminar 19. New York: The Egyptological Seminar of New York.

Ritti, Tullia, Klaus Grewe, and Paul Kessener

2007 "A Relief of a Water-Powered Stone Saw Mill on a Sarcophagus at Hierapolis and Its Implications." *Journal of Roman Archaeology* 20: 139–63.

Robinson, David M.

1926 "Roman Sculptures from Colonia Caesarea (Pisidian Antioch)." *The Art Bulletin* 9/1: 5–69.

Rockwell, Peter

1993 *The Art of Stoneworking: A Reference Guide.* Cambridge: Cambridge University Press.

Röder, Josef

1965 "Zur Steinbruchsgeschichte des Rosengranits von Assuan." *Archäologischer Anzeiger* 3: 467–552.

1971 "Marmor Phrygium: Die Antiken Marmorbrüche von İscehisar in Westanatolien." *Jahrbuch des Deutschen Archäologischen Instituts* 86: 253–312.

Romano, I. B., Scott Pike, and Elaine K. Gazda

2012 "The Use and Symbolism of Pentelic Marble in Domitianic Rome." In ASMOSIA IX, 772–79.

Root, Margaret Cool

1982 *Wondrous Glass: Reflections on the World of Rome, c. 50 B.C.–A.D. 650.* Ann Arbor: Kelsey Museum of Archaeology.

Russell, Ben

2013 *The Economics of the Roman Stone Trade.* Oxford Studies on the Roman Economy. Oxford: Oxford University Press.

Sacchi, Furio

2015 "La decorazione dell'edificio." In *La villa romana dei Nonii Arrii a Toscolano Maderno*, edited by Elisabetta Roffia, 127–48. Milan: Edizioni ET.

Saguì, Lucia

1998 *Storie al caleidoscopio: i vetri della collezione Gorga; un patrimonio ritrovato.* Rome: Museo dell'Arte Classica Gipsoteca.

Saguì, Lucia, Paola Santopadre, and Marco Verità

2012 "Glass Technology Colours, Forms and Shaping in the 2nd Century Opus Sectile Glass Materials from the Villa of Lucius Verus in Rome." In *Annales du 18e congrès de l'association internationale pour l'histoire du verre,* edited by Despoina Ignatiadou and Anastasios Antonaras, 133–38. Thessaloniki: AIHV.

Sapelli Ragni, Marina

2009 *Anzio e Nerone: tesori dal British Museum e dai Musei Capitolini.* Rome: Gangemi.

Savalli, A., P. Pesaresi, and Lorenzo Lazzarini

2015 "Casa del Rilievo di Telefo and *Opus Sectile* at Herculaneum." In ASMOSIA X, 1:349–62.

Scardozzi, Giuseppe

2012 "Ancient Marble and Alabaster Quarries near Hierapolis in Phrygia (Turkey): New Data from Archaeological Surveys." In ASMOSIA IX, 573–83.

2016 "The Alabaster Quarries of Hierapolis." In *Ancient Quarries and Building Sites in Asia Minor: Research on Hierapolis in Phrygia and Other Cities in South-Western Anatolia; Archaeology, Archaeometry, Conservation,* edited by Tommaso Ismaelli and Giuseppe Scardozzi, 141–66. Bibliotheca Archaeologica 45. Bari: Edipuglia.

2019 "The Provenance of Marbles and Alabasters Used in the Monuments of Hierapolis in Phrygia (Turkey): New Information from a Systematic Review and Integration of Archaeological and Archaeometric Data." *Heritage* 2/1: 519–52. https://doi.org/10.3390/heritage2010035.

Schmid, Stephan G.

2009 "Marmor chalcidicum vel Eretrianum: réflexions sur l'économie et le commerce d'Érétrie à l'époque impériale." In ASMOSIA VII, 527–42.

Schvoerer, Max, ed.

1999 *Archéomatériaux: marbres et autres roches. Actes de la IVe conférence internationale de l'Association pour l'étude des marbres et autres roches utilisés dans le passé, Bordeaux-Talence, 9-13 Octobre 1995.* ASMOSIA IV. Talence: Centre de Recherche en Physique Appliquée à l'Archéologie.

Shoe Meritt, L.

1971 "Esther Boise Van Deman." In *Notable American Women, 1607–1950: A Biographical Dictionary,* edited by Edward T. James, Janet Wilson James, and Paul S. Boyer, 3:507–08. Cambridge, MA: Belknap Press.

Sidebotham, Steven E.

1996 "Newly Discovered Sites in the Eastern Desert." *The Journal of Egyptian Archaeology* 82: 181–92.

Sidebotham, Steven E., Hans Barnard, James A. Harrell, and R. S. Tomber

2001 "The Roman Quarry and Installations in Wadi Umm Wikala and Wadi Semna." *The Journal of Egyptian Archaeology* 87: 135–70.

Sodini, Jean-Pierre, Anna Lambraki, and Tony Koželj

1980 *Les carrières de marbre d'Aliki à l'époque paléochrétienne.* Études Thasiennes 9. Athens: École française d'Athènes.

Spencer, Edward B. T.

1938 "The Stones Used in the Construction and Decoration of Ancient Rome." *The Classical Journal* 33/5: 271–79.

Sutherland, Jeanne

2013 *Karystian Cipollino Marble: Its Export from Euboea and Distribution.* BAR International Series 2578. Oxford: Archaeopress.

Talalay, Lauren E., and Margaret Cool Root

2015 *Passionate Curiosities: Tales of Collectors and Collections from the Kelsey Museum.* Kelsey Museum Publications 13. Ann Arbor: Kelsey Museum of Archaeology.

Talamo, E.

2000 "Statua di Agrippina Minore come orante." In *Aurea Roma: dalla città pagana alla città cristiana,* edited by Serena Ensoli and Eugenio La Rocca, 599–600. Rome: L'Erma di Bretschneider.

Tedeschi Grisanti, Giovanna

2004 "Marmi antichi colorati Toscani: problemi di identificazione e di reimpiego." In *Ante et post Lunam: splendore e ricchezza dei marmi apuani,* edited by Antonio Bartelletti and Emanuela Paribeni, 79–83. Seravezza: Parco regionale delle Alpi Apuane.

Tomber, R. S.

1996 "Provisioning the Desert: Pottery Supply to Mons Claudianus." In *Archaeological Research in Roman Egypt. Proceedings of the Seventeenth Classical Colloquium of the Department of Greek and Roman Antiquities, British Museum, Held on 1–4 December 1993,* edited by Donald M. Bailey. Journal of Roman Archaeology, Supplementary Series 19. Ann Arbor: Journal of Roman Archaeology.

Tregenza, Leo Arthur

1951 "The Curator Inscription and Other Recently Found Fragments from Wadi Semna." *Bulletin of the Faculty of Arts, Fouad I University* 13/2: 39–52.

Tuck, Steven L.

2005 *Latin Inscriptions in the Kelsey Museum: The Dennison and De Criscio Collections.* Kelsey Museum Studies 9. Ann Arbor: University of Michigan Press.

Turchiano, Maria
2008 "I pannelli in *opus sectile* di Faragola (Ascoli Satriano) tra archeologia e archeometria." In *Atti del XIII Colloquio dell'Associazione italiana per lo studio e la conservazione del mosaico,* edited by Claudia Angelelli and Federica Rinaldi, 59–70. Tivoli: Scripta manent.

Türk, N., S. Çakici, D. M. Uz, S. Akça, and K. Geyik
1988 "The Geology, Quarrying Technology and Use of Beylerköy Marbles in Western Turkey." In ASMOSIA I, 85–89.

Van Deman, Esther Boise
1909 *The Atrium Vestae.* Carnegie Institution of Washington, Publication 108. Washington, D.C.: Carnegie Institution.
1925 "Untitled Report." In Carnegie Institution of Washington Year Book: July 1, 1924, to June 30, 1925, 271–73. Washington, D.C.: Carnegie Institution of Washington.

Van der Veen, Marijke
1997 "High Living in Rome's Distant Quarries." *British Archaeology* 28: 6–7.

Van Rengen, Wilfried
1995 "A New Paneion at Mons Porphyrites." *Chronique d'Egypte* 70/139–140: 240–45.

Vanhove, Doris
1996 *Roman Marble Quarries in Southern Euboea and the Associated Road Systems.* Monumenta Graeca et Romana 8. Leiden: E.J. Brill.

Vermeule, Cornelius C., Walter Cahn, and Rollin van N. Hadley
1977 *Sculpture in the Isabella Stewart Gardner Museum.* Boston: The Trustees.

Violante, Sabrina
2013 "Rivestimenti parietali e pavimentali, elementi architettonici e di arredo minore." In *Evan Gorga: la collezione di archeologia,* edited by Alessandra Capodiferro, 151–70. Milan: Electa.

Viscogliosi, Alessandro
1993 "Apollo, Aedes in Circo." In *Lexicon Topographicum Urbis Romae,* edited by E.M. Steinby, 2nd ed., 1:51–53. Rome: Edizioni Quasar.
1996 *Il tempio di Apollo in Circo e la formazione del linguaggio architettonico augusteo.* Bullettino della Commissione archeologica comunale di Roma, Supplementi 3. Rome: L'Erma di Bretschneider.

Volpe, Giuliano, Giuliano De Felice, and M. Turchiano
2005 "I rivestimenti marmorei, i mosaici e i pannelli in opus sectile della villa tardoantica di Faragola (Ascoli Satriano, Foggia)." In *Atti del X Colloquio dell'Associazione italiana per lo studio e la conservazione del mosaico (AISCOM),* edited by Claudia Angelelli, 61–78. Tivoli: Scripta manent.

von Bissing, F.W.
1940 "Die Alabastra der hellenistischen und römischen Zeit." *Studi Etruschi* 14: 99–146.

Waelkens, Marc
1982 *Dokimeion: die Werkstatt der repräsentativen kleinasiatischen Sarkophage; Chronologie und Typologie ihrer Produktion.* Archäologische Forschungen 11. Berlin: Mann.
1985 "From a Phrygian Quarry: The Provenance of the Statues of the Dacian Prisoners in Trajan's Forum at Rome." *American Journal of Archaeology* 89/4: 641–53. https://doi.org/10.2307/504205.
1986 *Die kleinasiatischen Türsteine: typologische und epigraphische Untersuchungen der kleinasiatischen Grabreliefs mit Scheintür.* Mainz am Rhein: Philipp von Zabern.

Waelkens, Marc, Norman Herz, and Luc Moens, eds.
1992 *Ancient Stones: Quarrying, Trade and Provenance.* ASMOSIA II; Acta Archaeologica Lovaniensia, Monographiae 4. Leuven: Leuven University Press.

Waelkens, Marc, Philippe Muchez, Lieven Loots, Patrick Degryse, Lutgarde Vandeput, Semih Ercan, Luc Moens, and Paul De Paepe
2002 "Marble and the Marble Trade at Sagalassos (Turkey)." In ASMOSIA V, 370–80.

Walker, Susan
1988 "From West to East: Evidence for a Shift in the Balance of Trade in White Marbles." In ASMOSIA I, 187–95.

Waywell, G.B., and J.J. Wilkes
1995 "Excavations at the Ancient Theatre of Sparta 1992–4: Preliminary Report." *The Annual of the British School at Athens* 90: 435–60.
1999 "Excavations at the Ancient Theatre of Sparta 1995–1998: Preliminary Report." *The Annual of the British School at Athens* 94: 437–55.

Waywell, G. B., J. J. Wilkes, and S.E.C. Walker
1998 "The Ancient Theatre at Sparta." In *Sparta in Laconia: Proceedings of the 19th British Museum Classical Colloquium, Held with the British School at Athens and King's and University Colleges, London, 6-8 December 1995,* edited by William G. Cavanagh and S.E.C. Walker, 97–111. British School at Athens Studies 4. London: British School at Athens.

Welch, Katherine
2004 "Esther B. Van Deman, 1867–1937." In *Breaking Ground: Pioneering Women Archaeologists,* edited by Getzel M. Cohen and Martha Sharp Joukowsky, 68–108. Ann Arbor: University of Michigan Press.

West, Andrew F., and Richard Norton
1903 "Eighth Annual Report of the Managing Committee of the American School of Classical Studies in Rome." *American Journal*

of Archaeology 7, Supplement Annual Reports 1902–1903: 21–32.

Wielgosz-Rondolino, Dagmara
forth. *Marmora Asiatica. Polish Studies on Marbles from Asia Minor. Proceedings of the 19th International Congress of Classical Archaeology. Bonn, 22–26 May 2018.*

Wielgosz-Rondolino, Dagmara, M. Bojanowski, M. Gładki, A. Jaremek, and M. Ruszkowski
2015 "Dokimeion AD 2014: The First Season of the Project 'Marmora Asiatica towards Archaeopetrology in Poland'." Paper presented at the 11th International Conference of ASMOSIA, Association for the Study of Marble and Other Stones in Antiquity, Split, 18–22 May 2015.

Williams Thorpe, Olwen
2008 "A Thousand and One Columns: Observations on the Roman Granite Trade in the Mediterranean Area." *Oxford Journal of Archaeology* 27/1: 73–89.

Williams-Thorpe, Olwen, M. C. Jones, Philip J. Potts, and I. J. Rigby
2001 "Geology, Mineralogy and Characterization Studies of Imperial Porphyry." In *The Roman Imperial Quarries: Survey and Excavation at Mons Porphyrites, 1994–1998*, Vol. 1: *Topography and Quarries*, edited by Valerie A. Maxfield and David P. S. Peacock, 305–18. Excavation Memoir 67. London: Egypt Exploration Society.

Wilson Jones, Mark
2000 *Principles of Roman Architecture.* New Haven: Yale University Press.

Yavuz, Ali Bahadir, Matthias Bruno, and Donato Attanasio
2011 "An Updated, Multi-Method Database of Ephesos Marbles, Including White, Greco Scritto and Bigio Varieties." *Archaeometry* 53/2: 215–40.

Zöldföldi, Judit, and Zs. Kasztovszky
2009 "Provenance Study of Lapis Lazuli by Non-Destructive Prompt Gamma Activation Analysis (PGAA)." In ASMOSIA VII, 677–91.

Index A

STONE TYPES

The numbers listed in this index refer to catalogue entries.

Index B

OBJECT TYPES

The numbers listed in this index refer to catalogue entries.

Index C

PLACES WHERE ACQUIRED

The numbers listed in this index refer to catalogue entries. For maps, see figures 2.1 and 2.2, pp. 17–19.

North America

Palestine

Spain

Tunisia

Turkey

Unknown

Index D

COLLECTORS

The numbers listed in this index refer to catalogue entries.

Index E

STONE SOURCES

The numbers listed in this index refer to catalogue entries. For maps, see figures A.1–4, pp. 163–64.

North America

Scotland

Spain

Switzerland

Tunisia

Turkey

PLATES

cat. no. 1, face

cat. no. 1, profile

cat. no. 2, front

cat. no. 2, edge

cat. no. 3, curved face

cat. no. 3, profile

cat. no. 4

cat. no. 5

cat. no. 6

cat. no. 7

cat. no. 8

cat. no. 9

5 cm

cat. no. 10

5 cm

cat. no. 11

5 cm

cat. no. 12, face

5 cm

cat. no. 12, profile

5 cm

cat. no. 13

5 cm

cat. no. 14

5 cm

cat. no. 15

5 cm

cat. no. 16

5 cm

cat. no. 17

5 cm

cat. no. 18

5 cm

cat. no. 19

5 cm

cat. no. 20, face

cat. no. 20, interior

cat. no. 20, profile

cat. no. 21, front

cat. no. 21, top

cat. no. 22

cat. no. 23

cat. no. 24

cat. no. 25

cat. no. 26

cat. no. 27

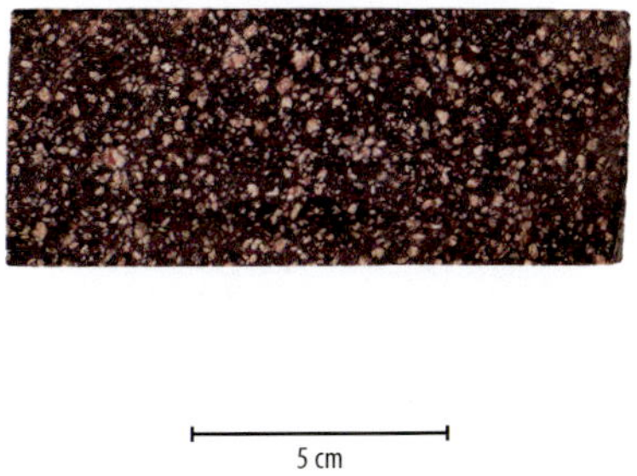

cat. no. 28

cat. no. 29

cat. no. 30

cat. no. 31

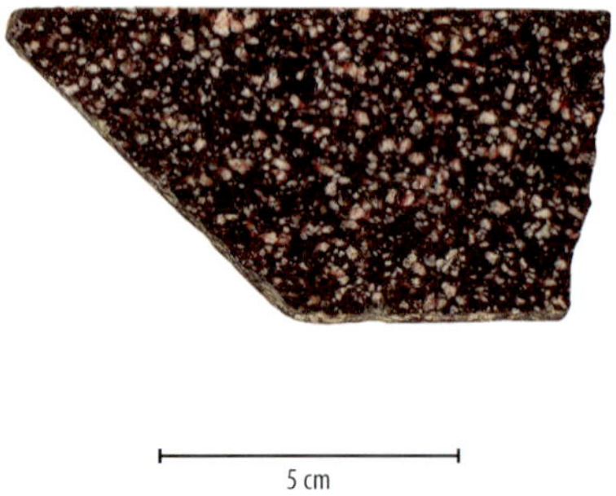

cat. no. 32

cat. no. 33

cat. no. 34

cat. no. 35

cat. no. 36

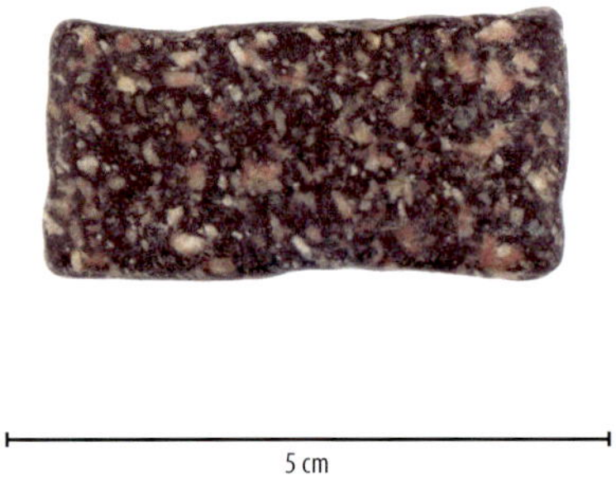

cat. no. 37

cat. no. 38

cat. no. 39

cat. no. 40

cat. no. 41

cat. no. 42

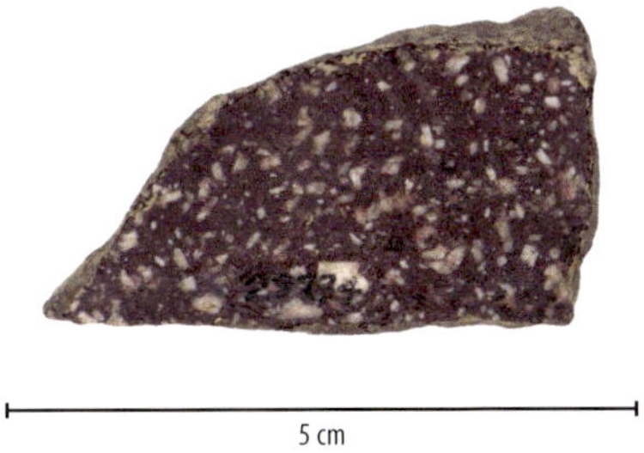

cat. no. 43

cat. no. 44

cat. no. 45

cat. no. 46

cat. no. 47

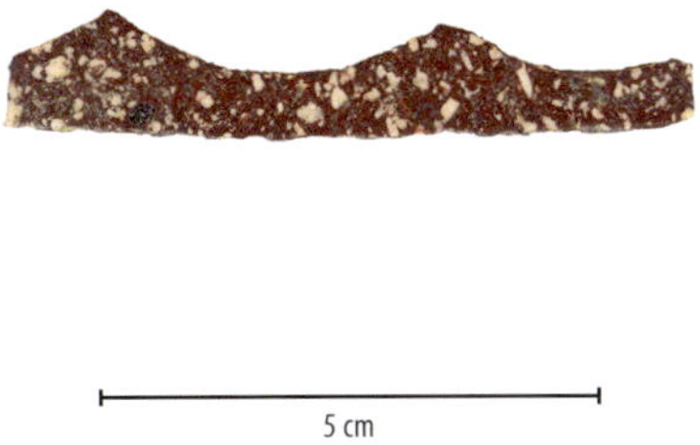

cat. no. 48

cat. no. 49

cat. no. 50

cat. no. 51

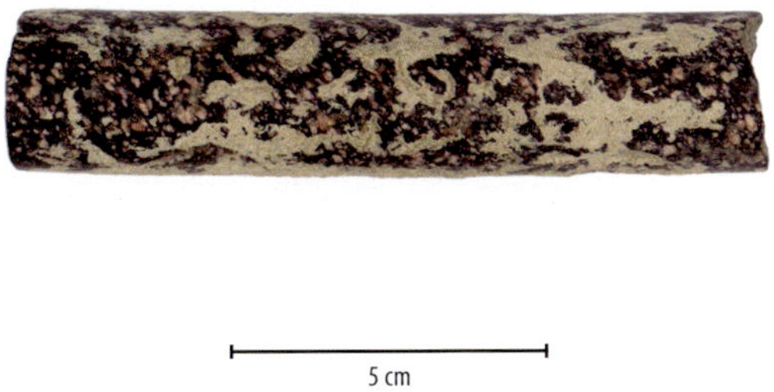

cat. no. 52, front

cat. no. 52, profile

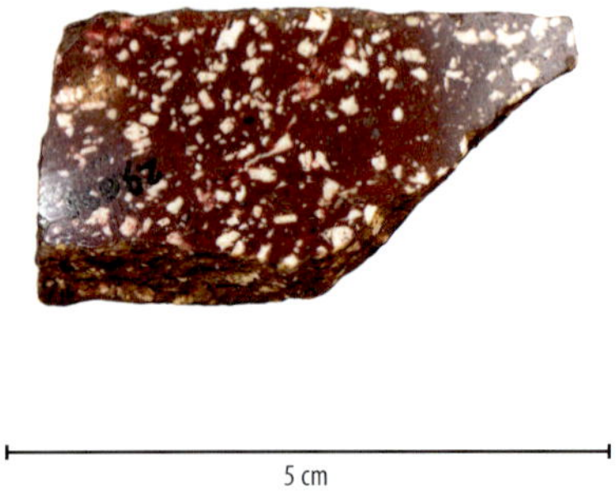

cat. no. 53, face

cat. no. 53, profile

cat. no. 54

cat. no. 55

cat. no. 56

cat. no. 57

cat. no. 58

cat. no. 59

cat. no. 60

cat. no. 61

cat. no. 62

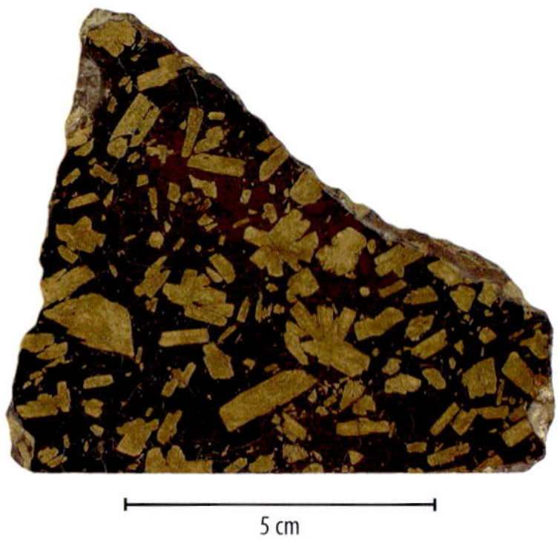

cat. no. 63

cat. no. 64

cat. no. 65

cat. no. 66

cat. no. 67

cat. no. 68

cat. no. 69

cat. no. 70

cat. no. 71

cat. no. 72

cat. no. 72, profile

cat. no. 73

cat. no. 74, face

cat. no. 74, back

cat. no. 75

cat. no. 76

cat. no. 77

cat. no. 78

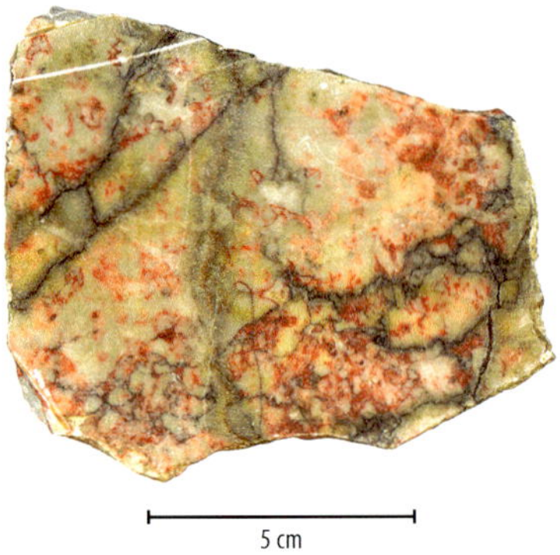

cat. no. 79

cat. no. 80

cat. no. 81

cat. no. 82

cat. no. 83

cat. no. 84

cat. no. 85

cat. no. 86

cat. no. 87

cat. no. 88

cat. no. 89

cat. no. 90

cat. no. 91

cat. no. 92, front

cat. no. 92, back

cat. no. 93

cat. no. 94

cat. no. 95, front

cat. no. 95, back

cat. no. 96

cat. no. 97

cat. no. 98

cat. no. 99

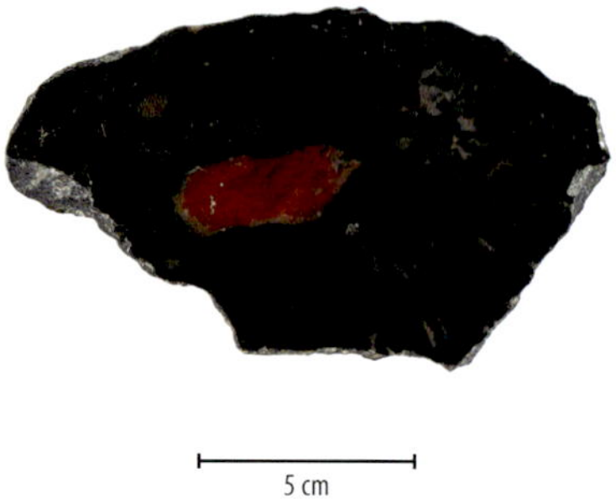

cat. no. 100

cat. no. 101

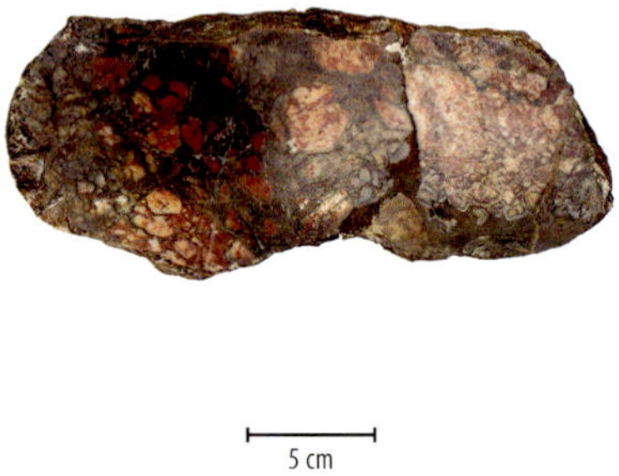

cat. no. 102

cat. no. 103

cat. no. 104

cat. no. 105, front

cat. no. 105, edge

cat. no. 106

cat. no. 107

cat. no. 108

cat. no. 109

cat. no. 110

cat. no. 111

cat. no. 112

cat. no. 113

cat. no. 114

cat. no. 115

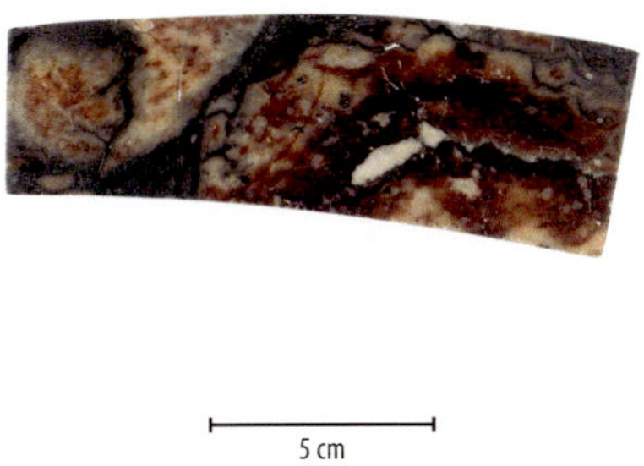

cat. no. 116

cat. no. 117

cat. no. 118

5 cm

cat. no. 119

5 cm

cat. no. 119, polished side

5 cm

cat. no. 119, back

5 cm

cat. no. 120

5 cm

cat. no. 121, face

5 cm

cat. no. 121, carved face

5 cm

cat. no. 122, face

5 cm

cat. no. 122, profile

5 cm

cat. no. 123

5 cm

cat. no. 123, side profile

5 cm

cat. no. 123, top end

5 cm

cat. no. 124

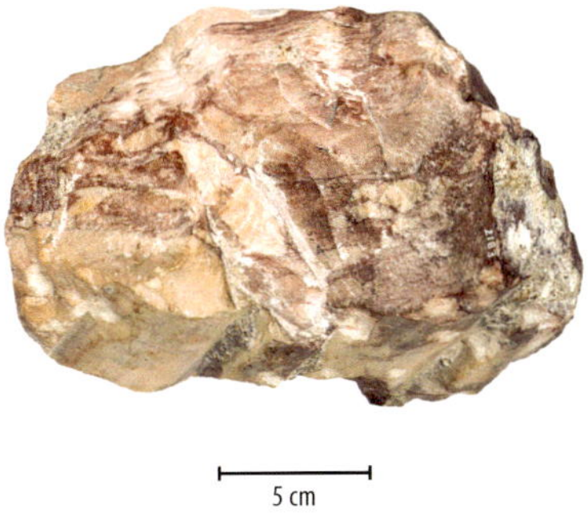

cat. no. 124, angle view

cat. no. 124, side

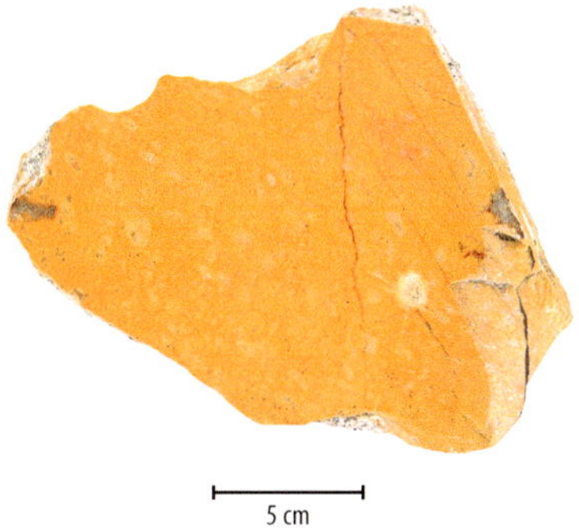

cat. no. 125

cat. no. 126, faces

cat. no. 126, side profile

cat. no. 126, top profile

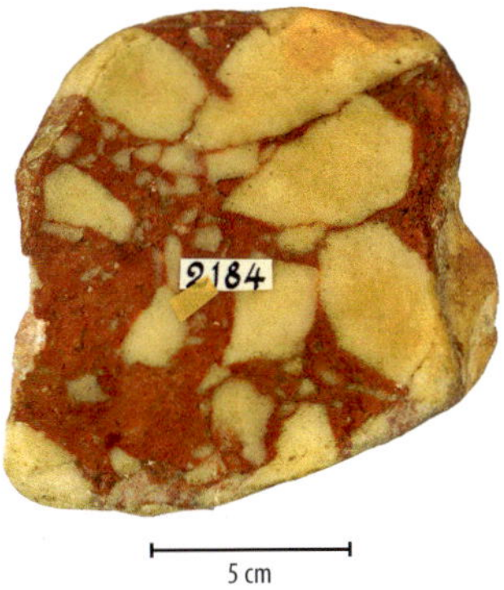

cat. no. 127

cat. no. 128

cat. no. 129

cat. no. 130

cat. no. 131

cat. no. 132

5 cm

cat. no. 133

5 cm

cat. no. 134

5 cm

cat. no. 135

5 cm

cat. no. 136

5 cm

cat. no. 137

5 cm

cat. no. 138

5 cm

cat. no. 139

5 cm

cat. no. 140

5 cm

cat. no. 141

5 cm

cat. no. 142

5 cm

cat. no. 143

5 cm

cat. no. 144

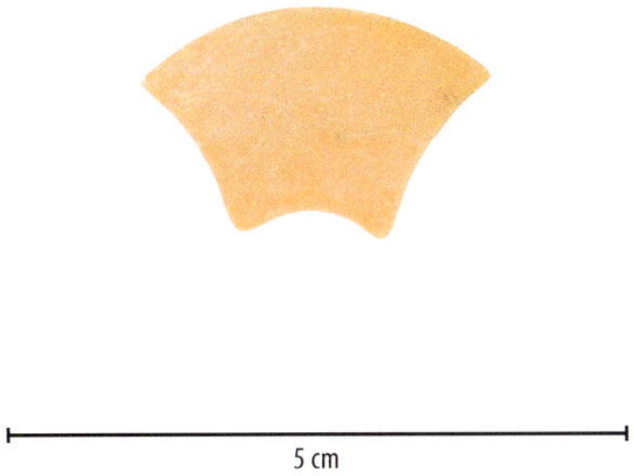

cat. no. 145

cat. no. 146

cat. no. 147

cat. no. 148

cat. no. 149

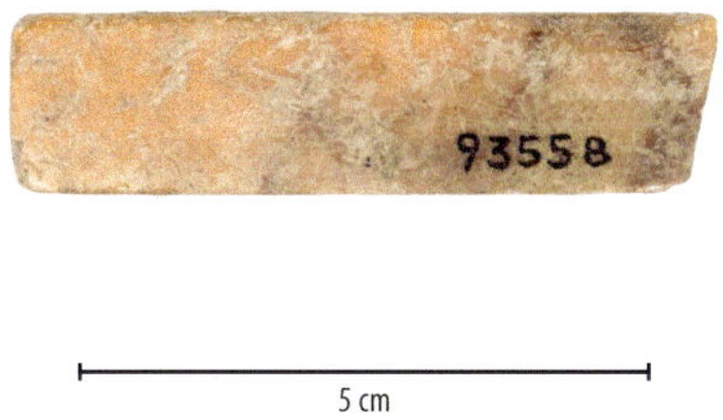

cat. no. 150

cat. no. 151, face

cat. no. 151, back

cat. no. 152

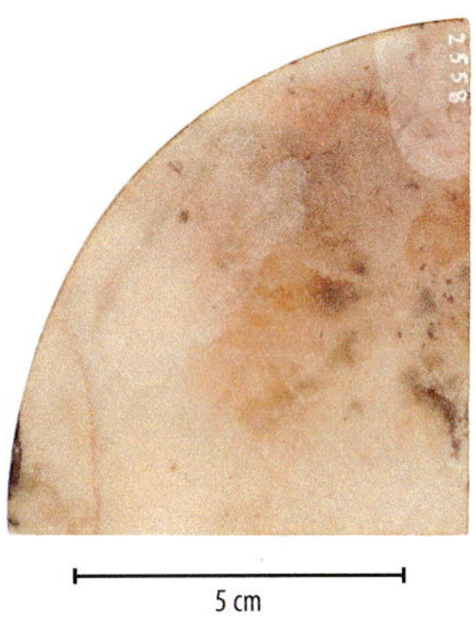

cat. no. 153

cat. no. 154

cat. no. 155

cat. no. 156

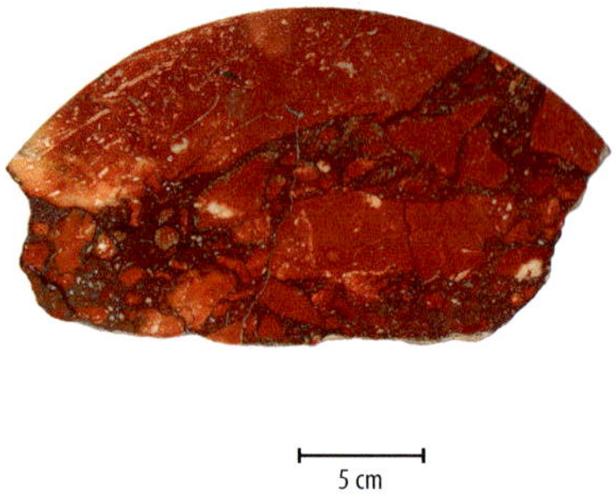

cat. no. 157

cat. no. 158

cat. no. 159

cat. no. 160, top

cat. no. 160, bottom

cat. no. 160, side

cat. no. 161

cat. no. 162, face

cat. no. 162, profile

cat. no. 163

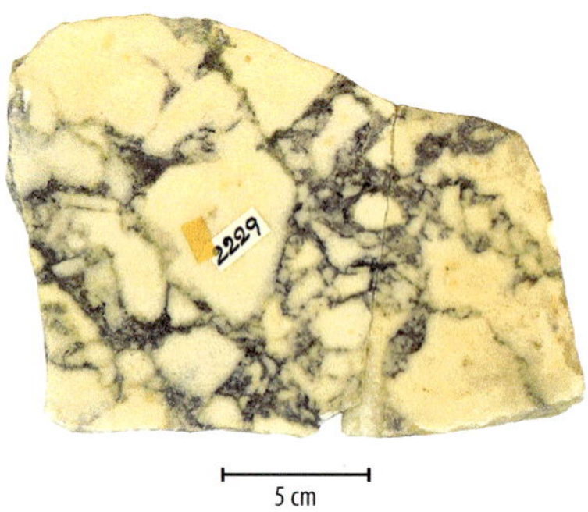

cat. no. 164

5 cm

cat. no. 165

5 cm

cat. no. 166

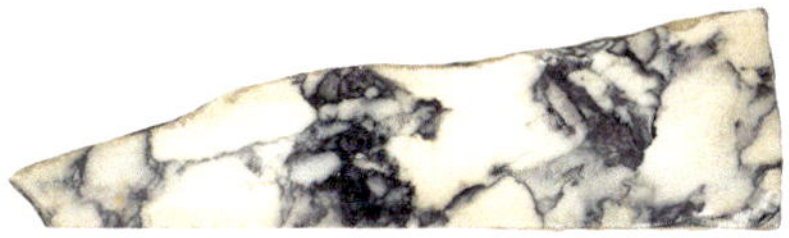

5 cm

cat. no. 167

5 cm

cat. no. 168

5 cm

cat. no. 169

5 cm

cat. no. 170

5 cm

cat. no. 171

5 cm

cat. no. 172

5 cm

cat. no. 173, face

5 cm

cat. no. 173, back

5 cm

cat. no. 174

5 cm

cat. no. 175

5 cm

cat. no. 176, face

5 cm

cat. no. 176, back

5 cm

cat. no. 177

5 cm

cat. no. 178

5 cm

cat. no. 179

5 cm

cat. no. 180

5 cm

cat. no. 181

5 cm

cat. no. 182

5 cm

cat. no. 183

5 cm

cat. no. 184

5 cm

cat. no. 185

5 cm

cat. no. 186

cat. no. 187

cat. no. 188

cat. no. 189

cat. no. 190, face

cat. no. 190, profile

cat. no. 190, edge with traces of cement

cat. no. 191

cat. no. 192

cat. no. 193

cat. no. 194

cat. no. 195, face

cat. no. 195, back

cat. no. 196

cat. no. 197

cat. no. 198

cat. no. 199

cat. no. 200

cat. no. 201

cat. no. 202

cat. no. 203

cat. no. 204

cat. no. 205

cat. no. 206

cat. no. 207

cat. no. 208

cat. no. 209, face

cat. no. 209, back

cat. no. 210, face

cat. no. 210, back

cat. no. 211

cat. no. 212

cat. no. 213, face

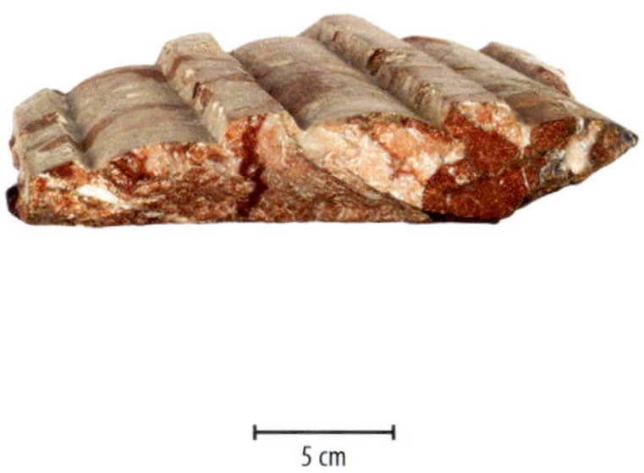

cat. no. 213, profile

cat. no. 214

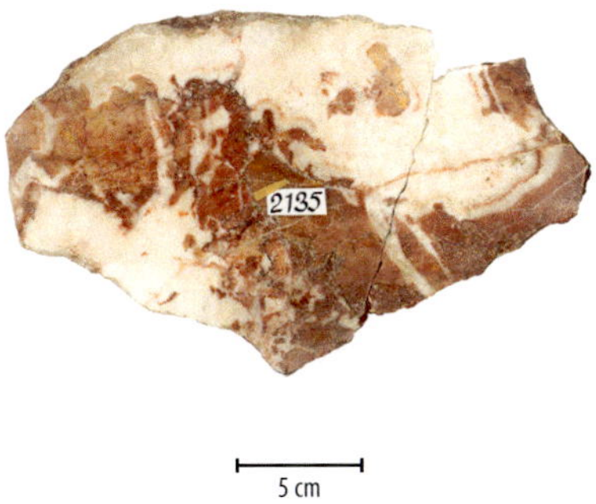

cat. no. 215

cat. no. 216

5 cm

cat. no. 217

5 cm

cat. no. 218

5 cm

cat. no. 219

5 cm

cat. no. 220

5 cm

cat. no. 221

5 cm

cat. no. 222

5 cm

cat. no. 223

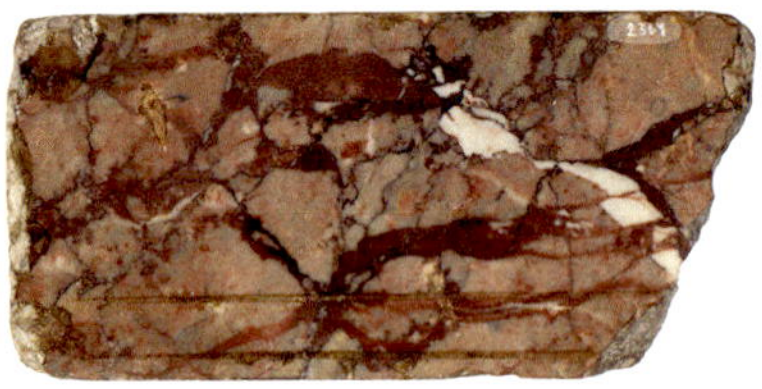

5 cm

cat. no. 224

5 cm

cat. no. 225, face

5 cm

cat. no. 225, back

5 cm

cat. no. 226, face

5 cm

cat. no. 226, profile

5 cm

cat. no. 227

5 cm

cat. no. 228

5 cm

cat. no. 229

5 cm

cat. no. 230

5 cm

cat. no. 231

5 cm

cat. no. 232

5 cm

cat. no. 233

5 cm

cat. no. 234

5 cm

cat. no. 235

5 cm

cat. no. 236

5 cm

cat. no. 237

5 cm

cat. no. 238

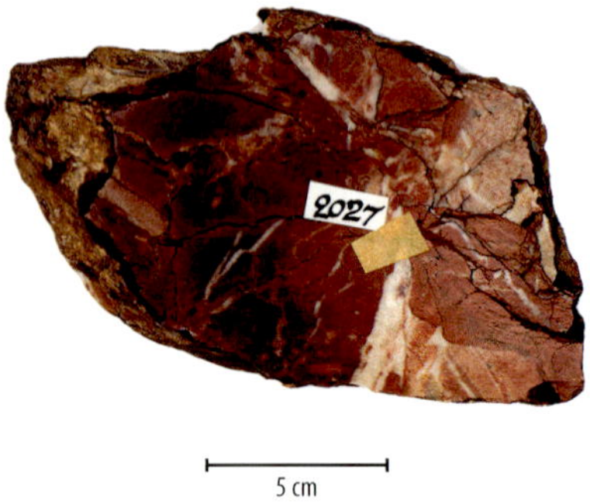

cat. no. 239

cat. no. 240

cat. no. 241

cat. no. 242

cat. no. 243

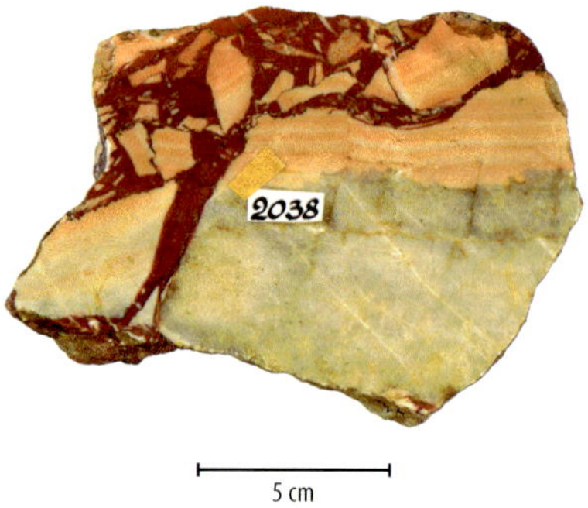

cat. no. 244, face

cat. no. 244, back

cat. no. 245

cat. no. 246

cat. no. 247

cat. no. 248

cat. no. 249

cat. no. 250, face

cat. no. 250, profile

cat. no. 251, face

cat. no. 251, back

cat. no. 252, top and bottom

cat. no. 252, edge

cat. no. 252, profile

cat. no. 253

cat. no. 254, face and edge

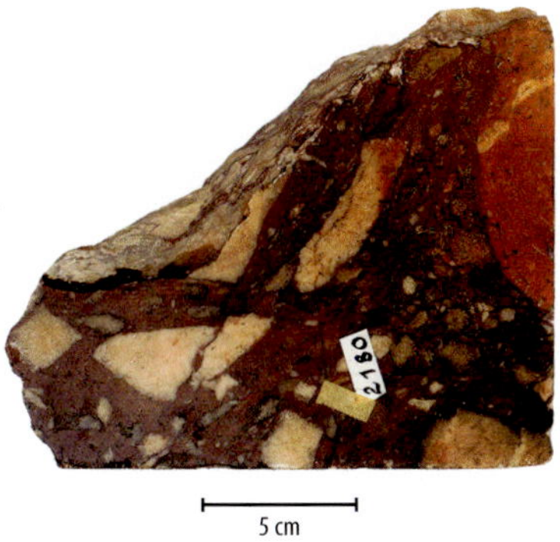

cat. no. 255

cat. no. 256

cat. no. 257

cat. no. 258

cat. no. 259

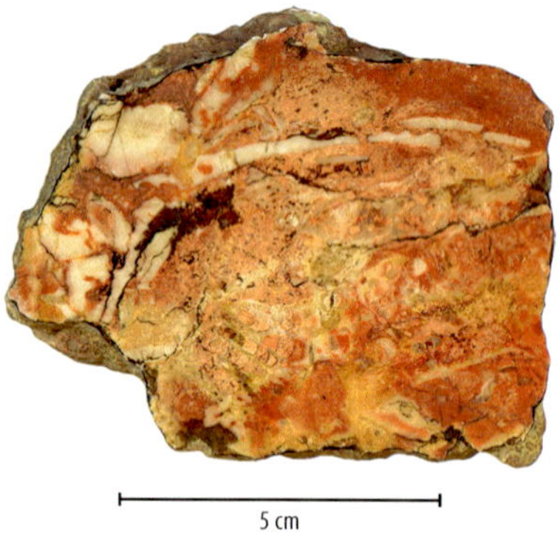

cat. no. 260

cat. no. 261

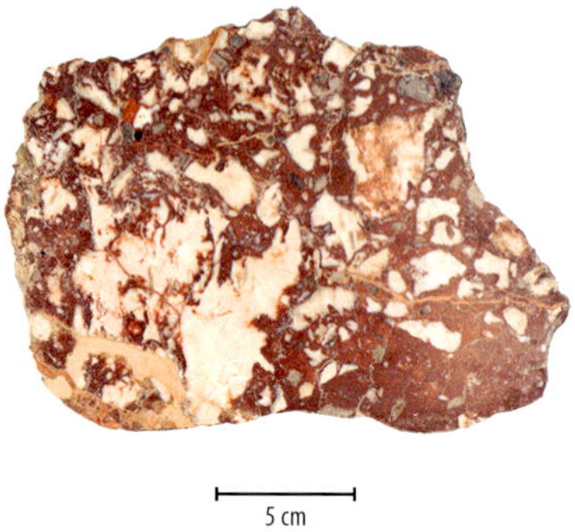

cat. no. 262

cat. no. 263

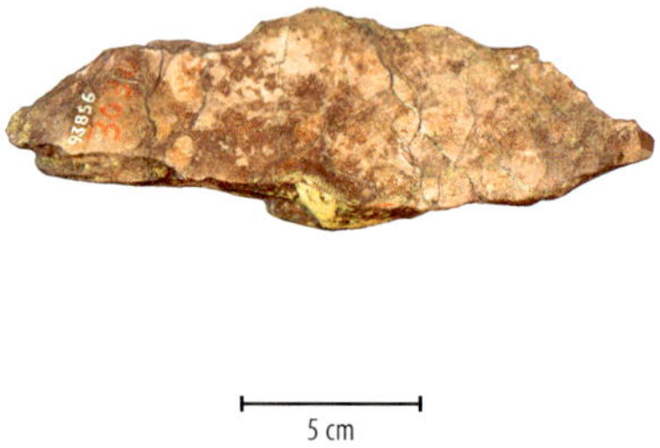

cat. no. 264

cat. no. 265, face

cat. no. 265, back

cat. no. 266

cat. no. 267

cat. no. 268, face

cat. no. 268, edge

cat. no. 269

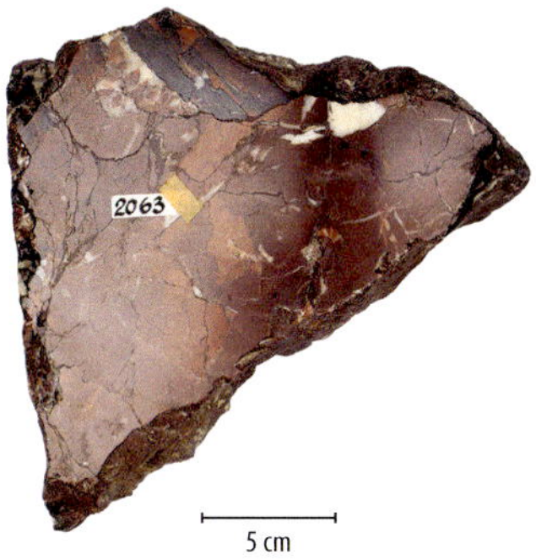

cat. no. 270, face

cat. no. 270, back

cat. no. 271

cat. no. 272

cat. no. 273

cat. no. 274

cat. no. 275

cat. no. 276

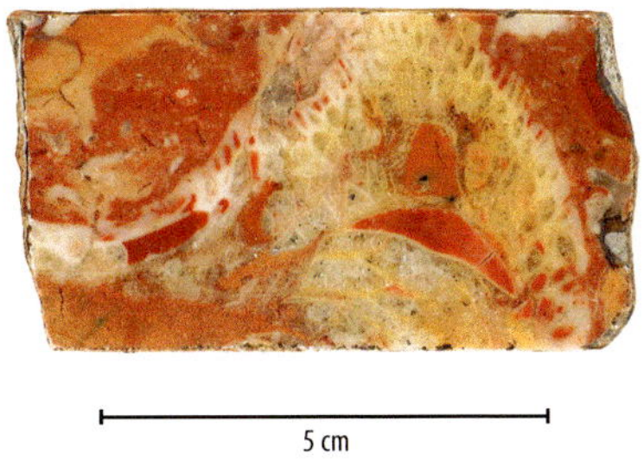

cat. no. 277

cat. no. 278

Plate 28

5 cm

cat. no. 279

5 cm

cat. no. 280

5 cm

cat. no. 281, front

5 cm

cat. no. 281, back and profile

5 cm

cat. no. 282

5 cm

cat. no. 283

5 cm

cat. no. 284

5 cm

cat. no. 285

5 cm

cat. no. 286

5 cm

cat. no. 287, face

5 cm

cat. no. 287, corner

5 cm

cat. no. 288

5 cm

cat. no. 289

5 cm

cat. no. 290

5 cm

cat. no. 291

5 cm

cat. no. 292

5 cm

cat. no. 293

5 cm

cat. no. 294, face

5 cm

cat. no. 294, profile

5 cm

cat. no. 295

5 cm

cat. no. 296, face

5 cm

cat. no. 296, back

5 cm

cat. no. 296, edge

5 cm

cat. no. 297, face

cat. no. 297, back

cat. no. 298, front

cat. no. 298, back

cat. no. 298, bottom

cat. no. 299

cat. no. 300

cat. no. 301

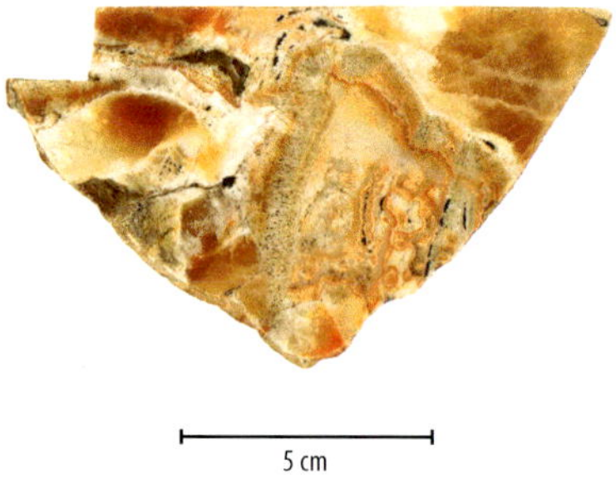

cat. no. 302

cat. no. 303

cat. no. 304

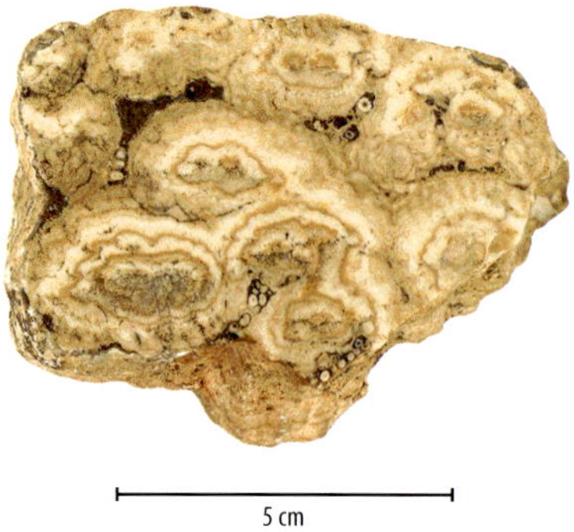

cat. no. 305

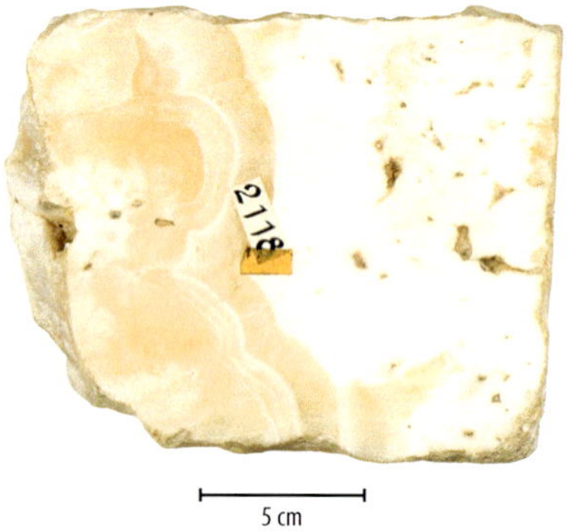

cat. no. 306

cat. no. 307, face

cat. no. 307, edge 1

cat. no. 307, edge 2

cat. no. 308

cat. no. 309

cat. no. 310

cat. no. 311

cat. no. 312

cat. no. 313

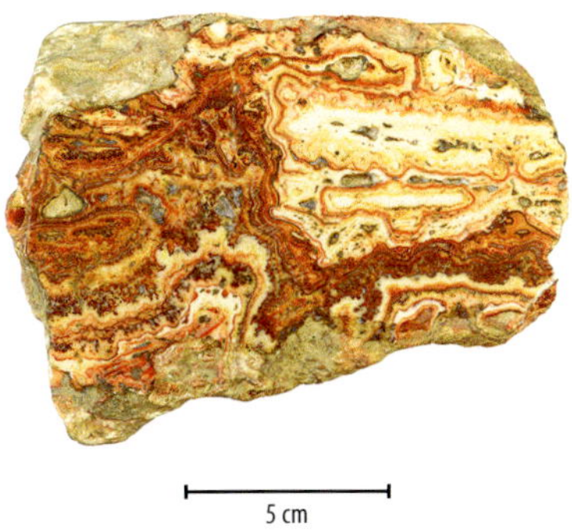

cat. no. 314

cat. no. 315

cat. no. 316

5 cm

cat. no. 317

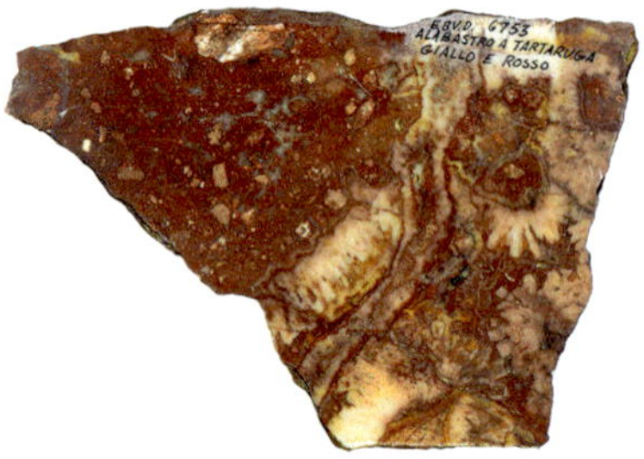

5 cm

cat. no. 318

5 cm

cat. no. 319

5 cm

cat. no. 320

5 cm

cat. no. 321

5 cm

cat. no. 322

5 cm

cat. no. 323

5 cm

cat. no. 324

5 cm

cat. no. 325

5 cm

cat. no. 326

5 cm

cat. no. 327

5 cm

cat. no. 328

cat. no. 329

cat. no. 330

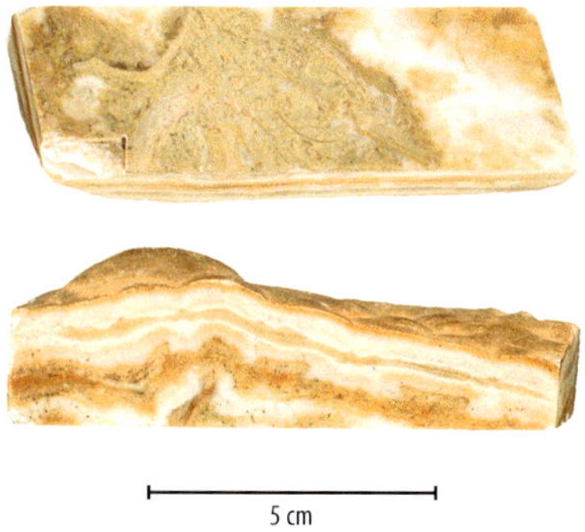

cat. no. 331, face and profile

cat. no. 332, face

cat. no. 332, back

cat. no. 333

cat. no. 334

cat. no. 335

cat. no. 336, face

cat. no. 336, profile

cat. no. 337

cat. no. 338, top

cat. no. 338, profile

cat. no. 338, back

cat. no. 339

cat. no. 340

cat. no. 341

cat. no. 342

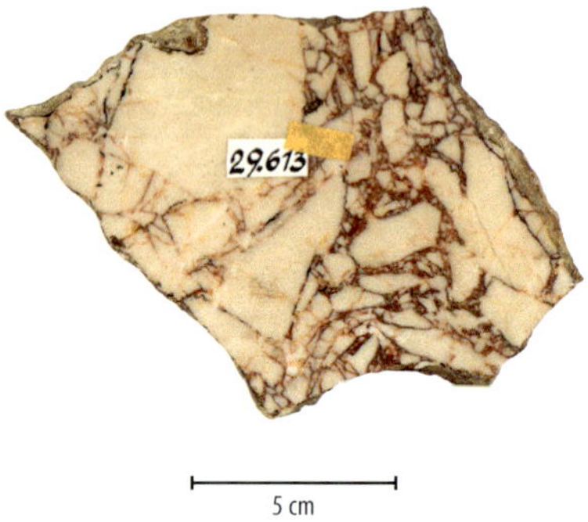

cat. no. 343

cat. no. 344

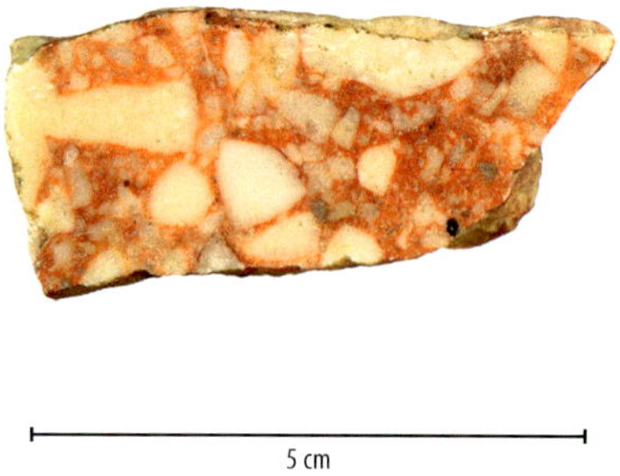

cat. no. 345, face

cat. no. 345, worked face

cat. no. 346

cat. no. 347

cat. no. 348

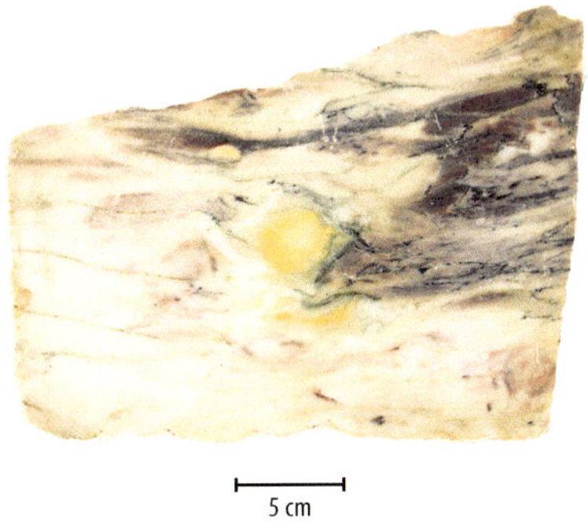

cat. no. 349

cat. no. 350

cat. no. 351

cat. no. 352

cat. no. 353

cat. no. 354

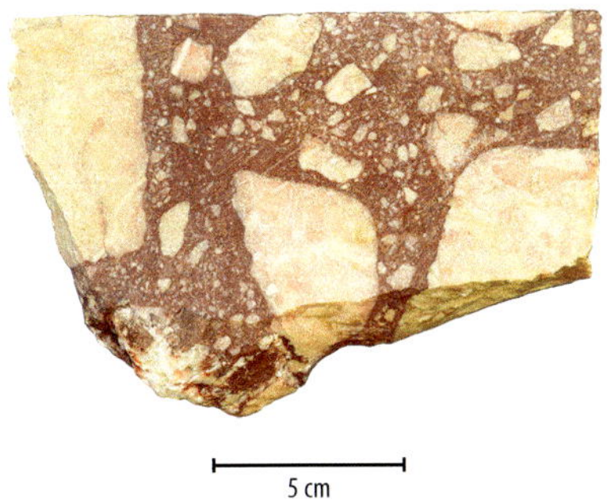

cat. no. 355

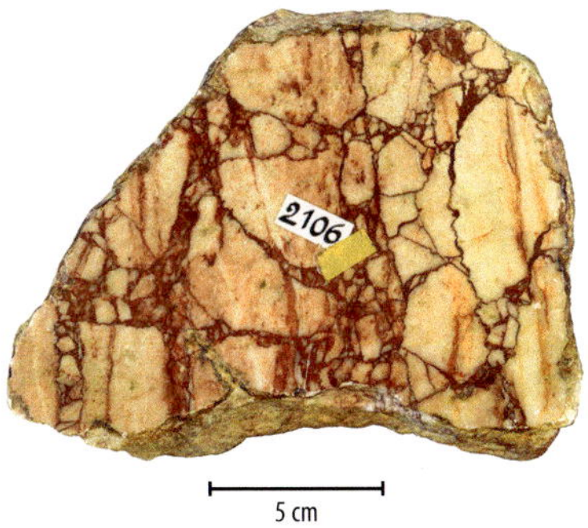

cat. no. 356

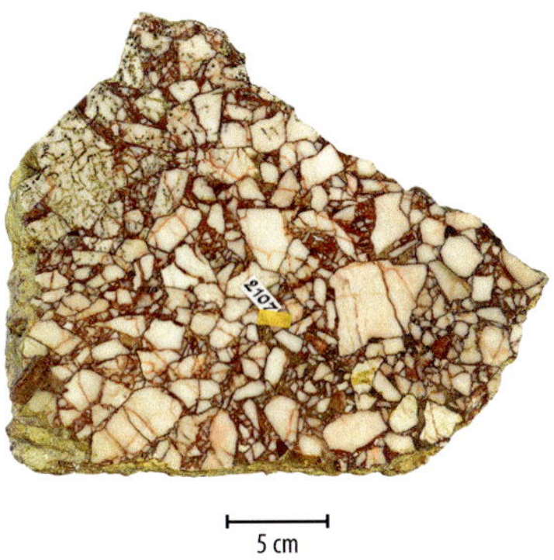

cat. no. 357

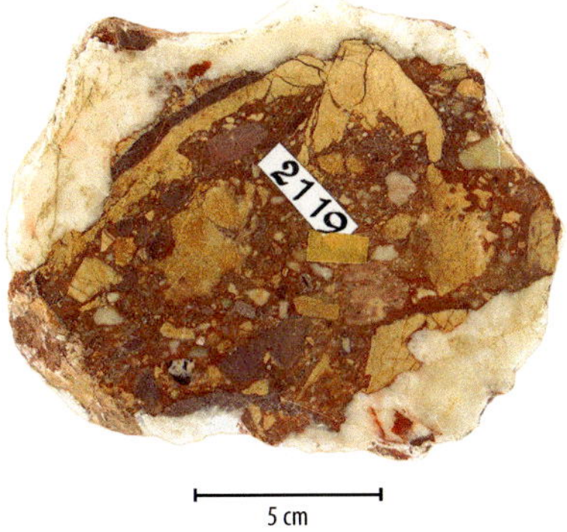

cat. no. 358

cat. no. 359

cat. no. 360

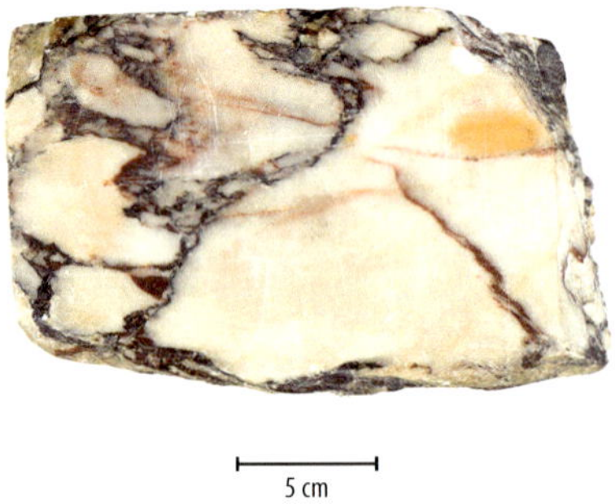

cat. no. 361

cat. no. 362

cat. no. 363

cat. no. 364

cat. no. 365

cat. no. 366

cat. no. 367, face

cat. no. 367, profile

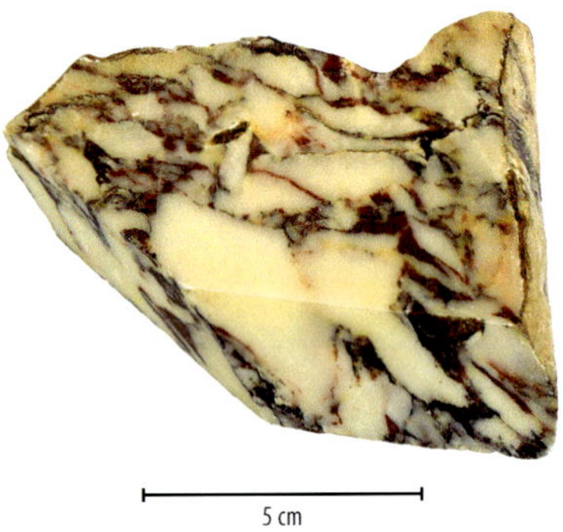

cat. no. 368

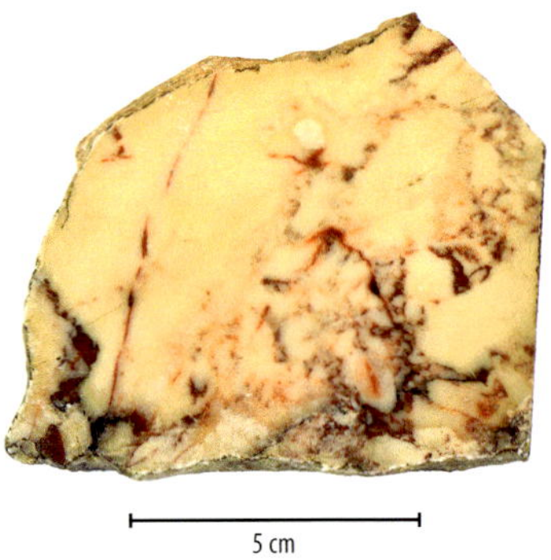

cat. no. 369

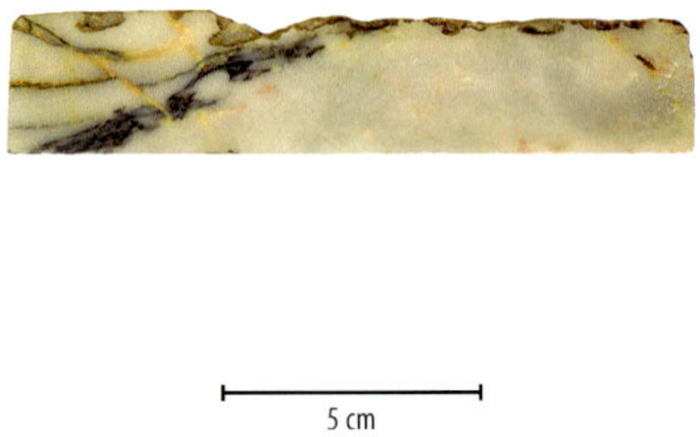

cat. no. 370

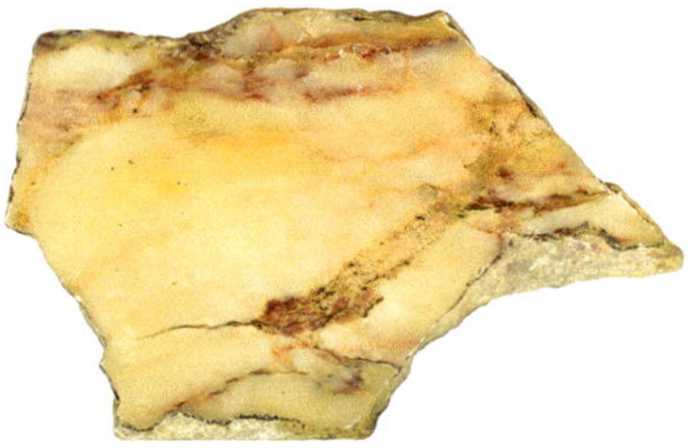

5 cm

cat. no. 371

5 cm

cat. no. 372

5 cm

cat. no. 373, face

5 cm

cat. no. 373, back

5 cm

cat. no. 374

5 cm

cat. no. 375

5 cm

cat. no. 376

5 cm

cat. no. 377

5 cm

cat. no. 378

5 cm

cat. no. 379

5 cm

cat. no. 380

5 cm

cat. no. 381

cat. no. 382

cat. no. 383

cat. no. 384

cat. no. 385

cat. no. 386

cat. no. 387

cat. no. 388, face 1

cat. no. 388, face 2

cat. no. 389

cat. no. 390

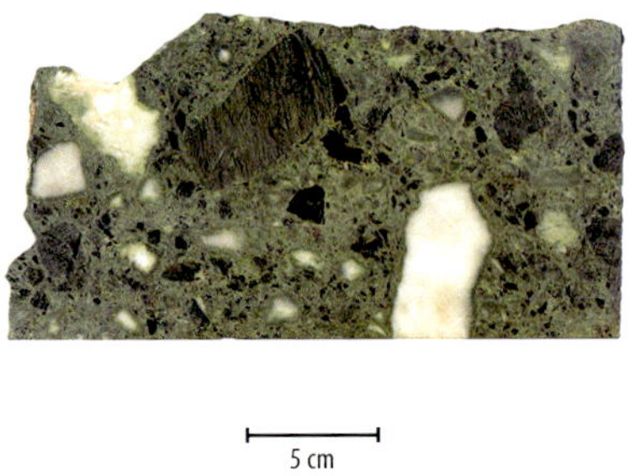

cat. no. 391

cat. no. 392

cat. no. 393

cat. no. 394

cat. no. 395

cat. no. 396

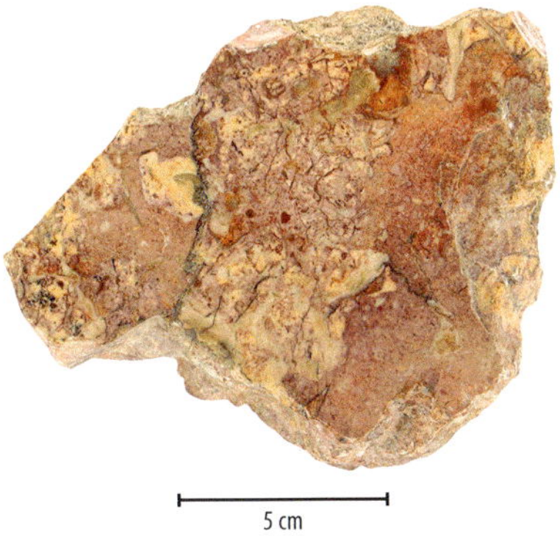

cat. no. 397

cat. no. 398

cat. no. 399

cat. no. 400

cat. no. 401

cat. no. 402

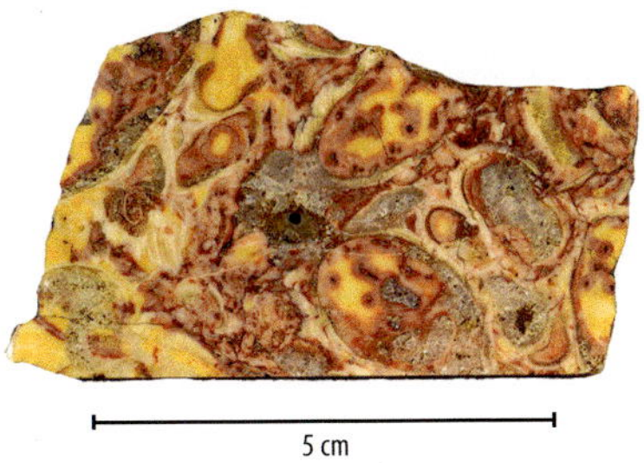

cat. no. 403

cat. no. 404

cat. no. 405

cat. no. 406

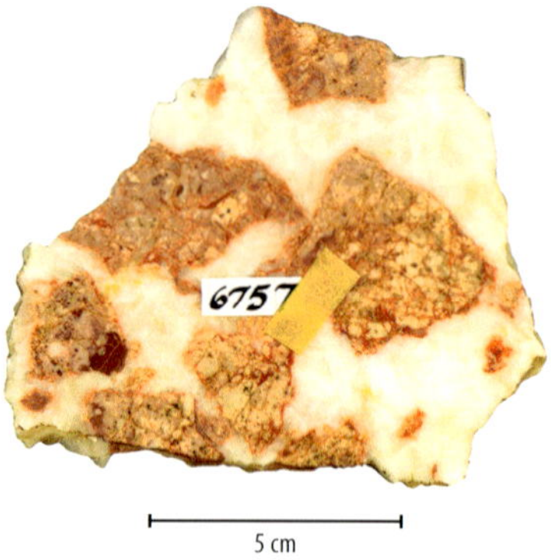

cat. no. 407

cat. no. 408

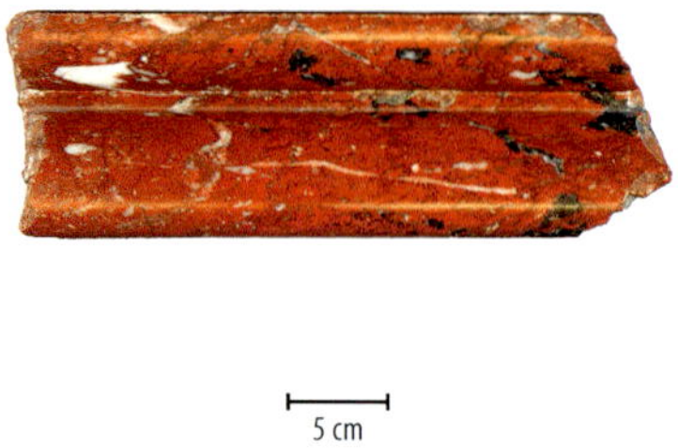

cat. no. 409

cat. no. 410

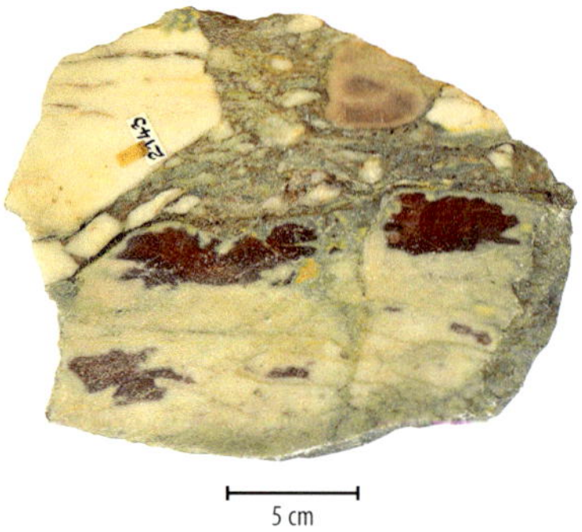

cat. no. 411

cat. no. 412

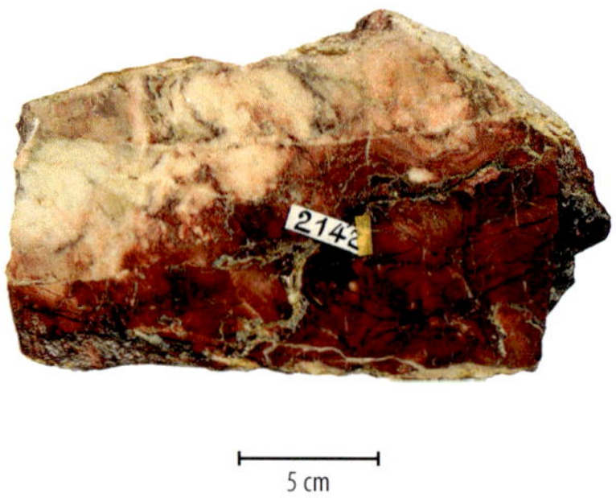

cat. no. 413

cat. no. 414

cat. no. 415, edge

cat. no. 415, profile

5 cm

cat. no. 416, face

5 cm

cat. no. 416, profile

5 cm

cat. no. 417, face

5 cm

cat. no. 417, front

5 cm

cat. no. 418

5 cm

cat. no. 419

5 cm

cat. no. 420

5 cm

cat. no. 421

5 cm

cat. no. 422

5 cm

cat. no. 423

5 cm

cat. no. 424

5 cm

cat. no. 425

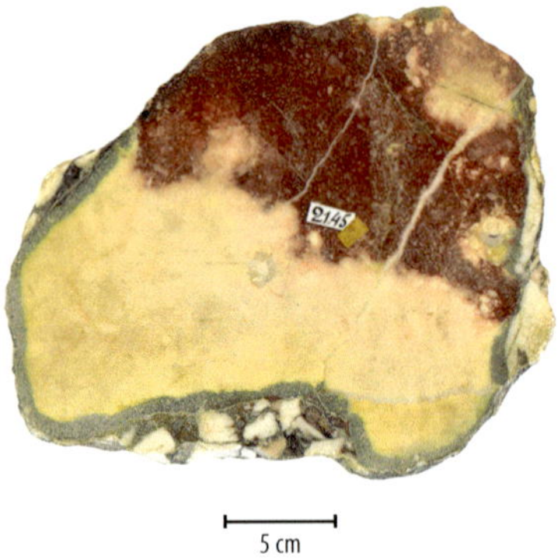

cat. no. 426

cat. no. 427

cat. no. 428

cat. no. 429

cat. no. 430

cat. no. 431

cat. no. 432

cat. no. 433

cat. no. 434, face

cat. no. 434, edge

cat. no. 434, profile

cat. no. 435

5 cm

cat. no. 436

5 cm

cat. no. 437

5 cm

cat. no. 438

5 cm

cat. no. 439

5 cm

cat. no. 440

5 cm

cat. no. 441

5 cm

cat. no. 442

5 cm

cat. no. 443

5 cm

cat. no. 444

5 cm

cat. no. 445

5 cm

cat. no. 446

5 cm

cat. no. 447

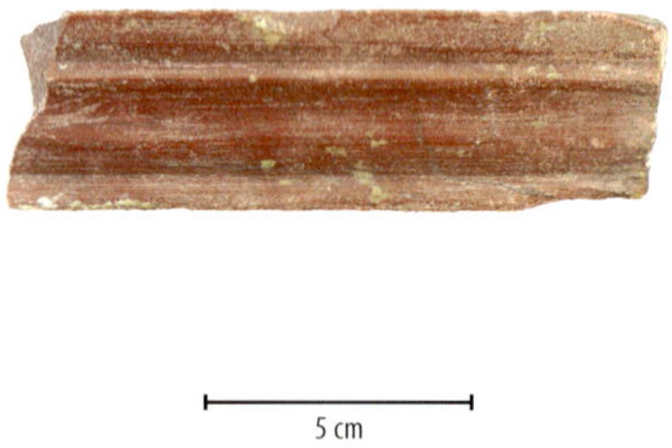

cat. no. 448

cat. no. 449

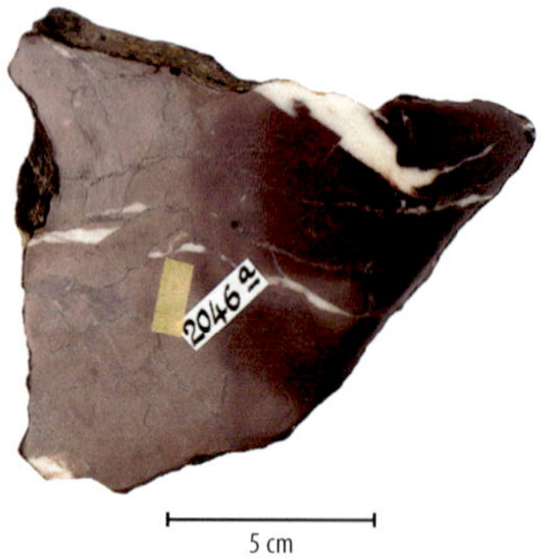

cat. no. 450

cat. no. 451

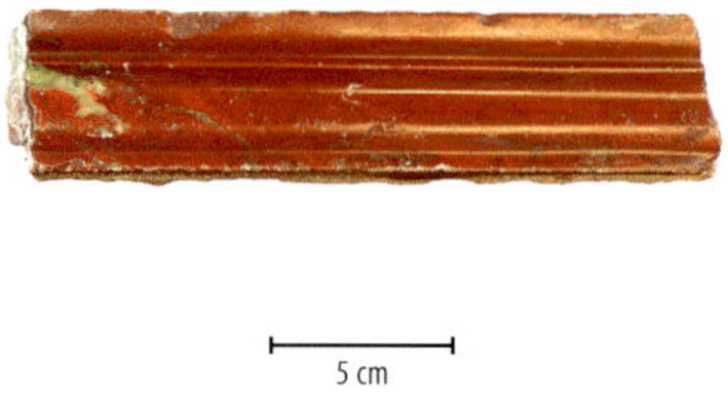

cat. no. 452

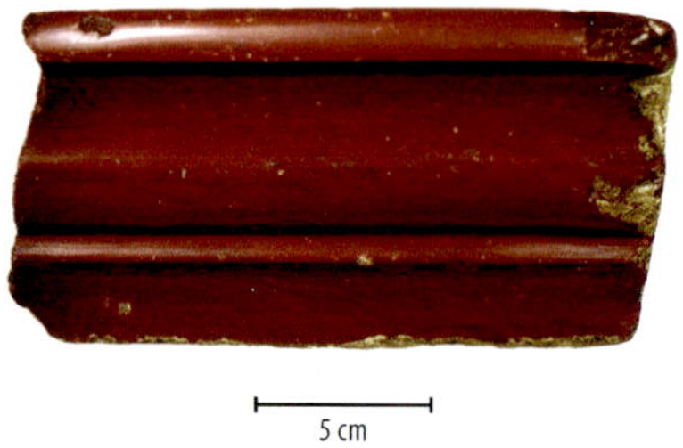

cat. no. 453

cat. no. 454

cat. no. 455

cat. no. 456

cat. no. 457

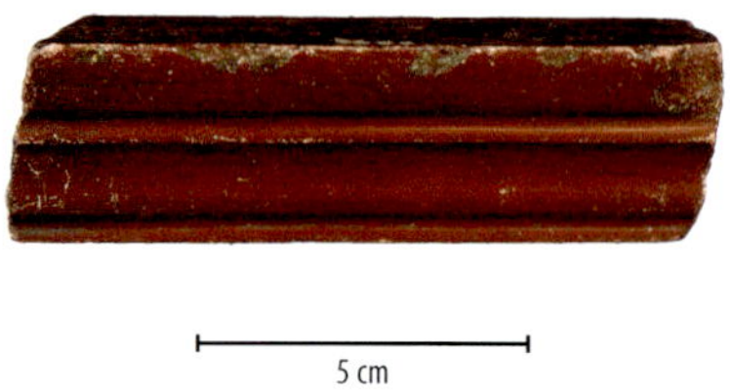

cat. no. 458

cat. no. 459, front

cat. no. 459, back

cat. no. 460

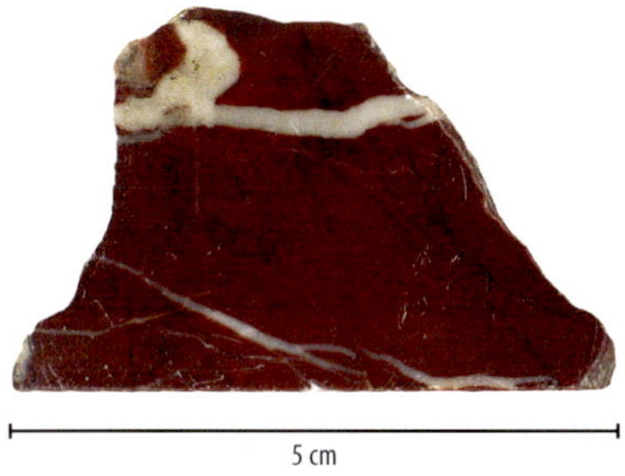

cat. no. 461

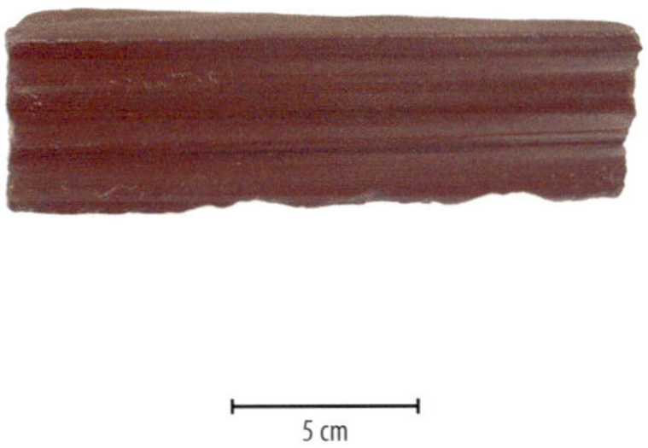

cat. no. 462

cat. no. 463

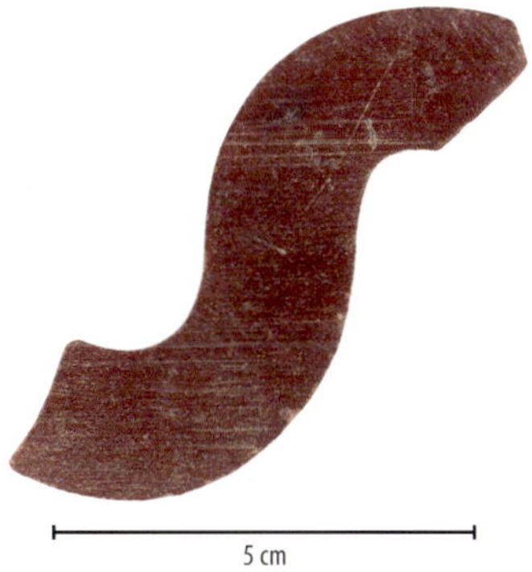

cat. no. 464

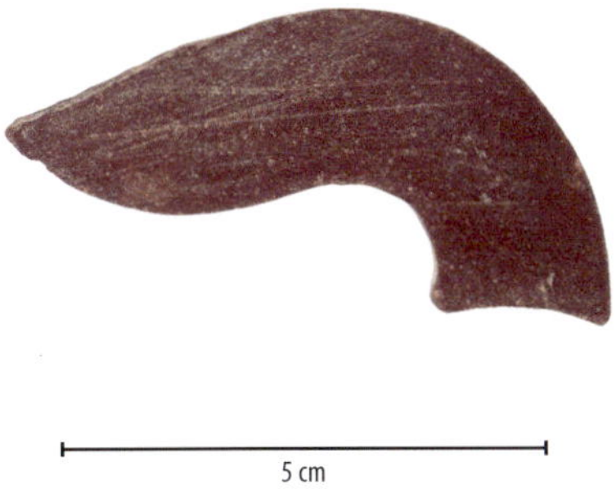

cat. no. 465

cat. no. 466

cat. no. 467

cat. no. 468

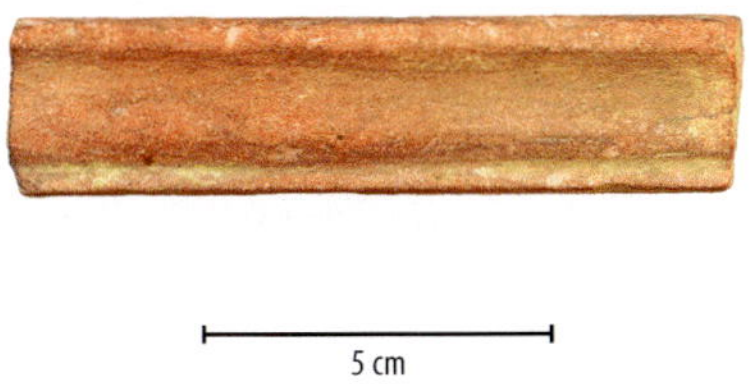

cat. no. 469

cat. no. 470,
front and back edge

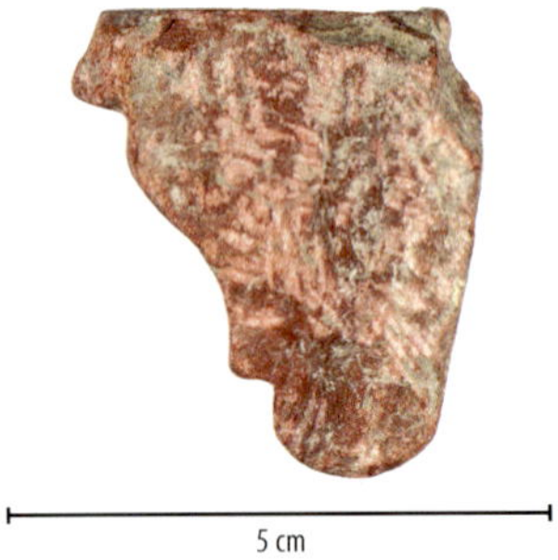

cat. no. 470, profile

cat. no. 471

cat. no. 472

cat. no. 473

cat. no. 474

cat. no. 475

cat. no. 476

cat. no. 477

cat. no. 478

cat. no. 479

cat. no. 480

cat. no. 481

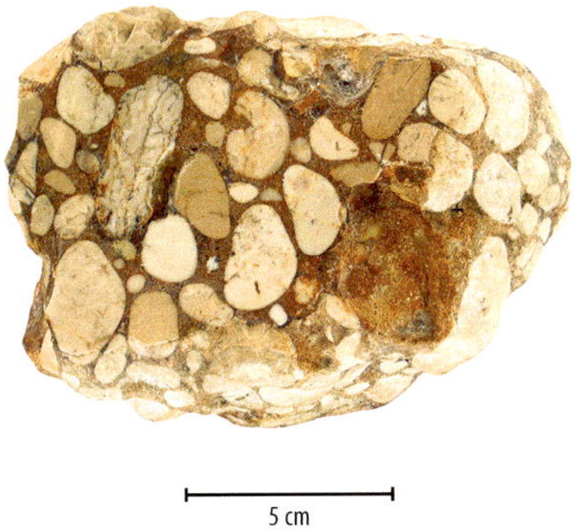

cat. no. 482

cat. no. 483

cat. no. 484

cat. no. 485

cat. no. 486, front and profile

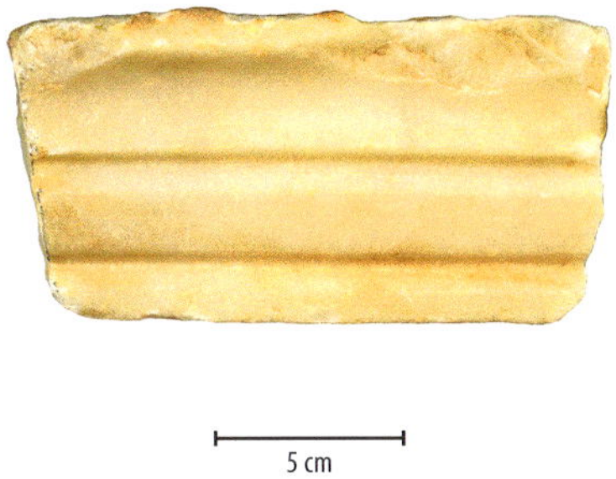

cat. no. 487

cat. no. 488

cat. no. 489

cat. no. 490

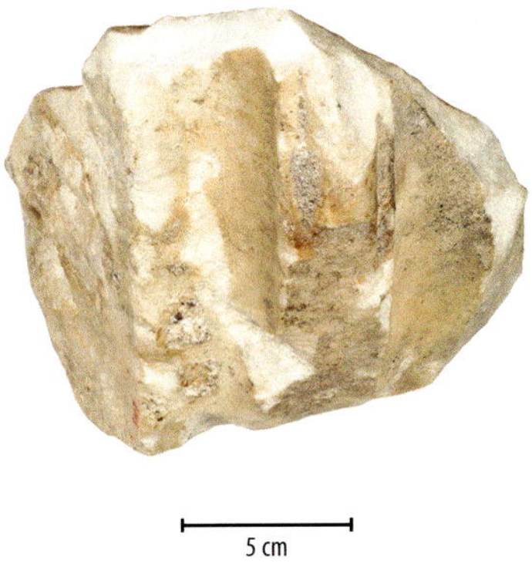

cat. no. 491, face

cat. no. 491, top

cat. no. 492

cat. no. 493

cat. no. 494

cat. no. 495

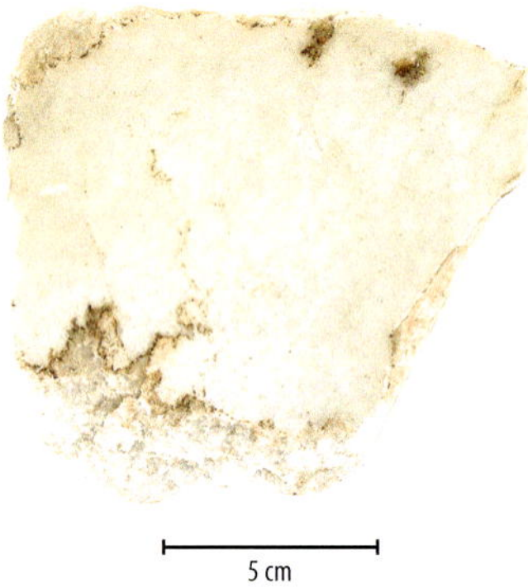

cat. no. 496

cat. no. 497

cat. no. 498

cat. no. 499

cat. no. 500

cat. no. 501

cat. no. 502

cat. no. 503, face

cat. no. 503, back

cat. no. 504, view 1

cat. no. 504, views 2 & 3

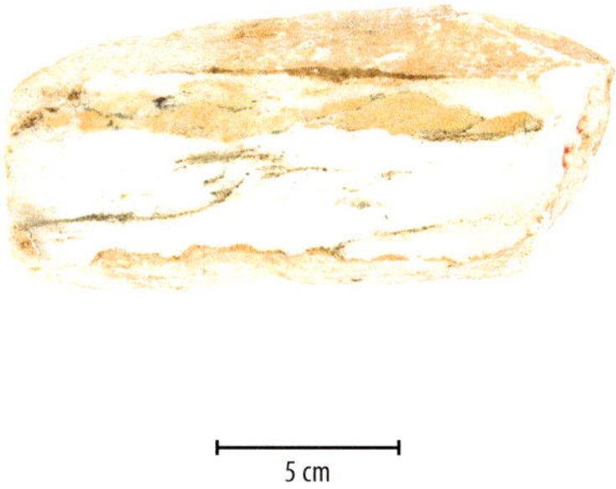

cat. no. 505

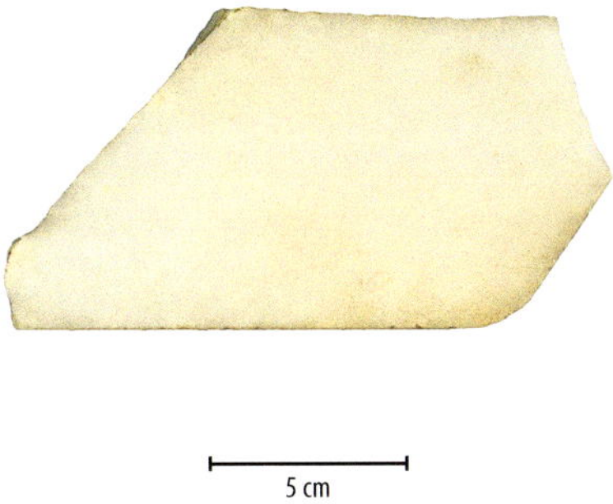

cat. no. 506

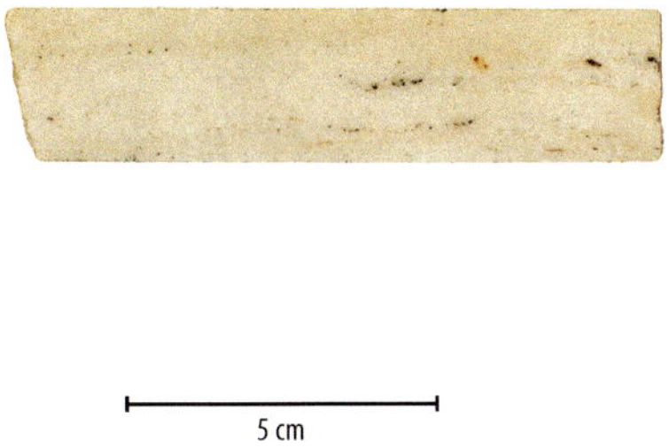

cat. no. 507

cat. no. 508

cat. no. 509

cat. no. 510

cat. no. 511

cat. no. 512

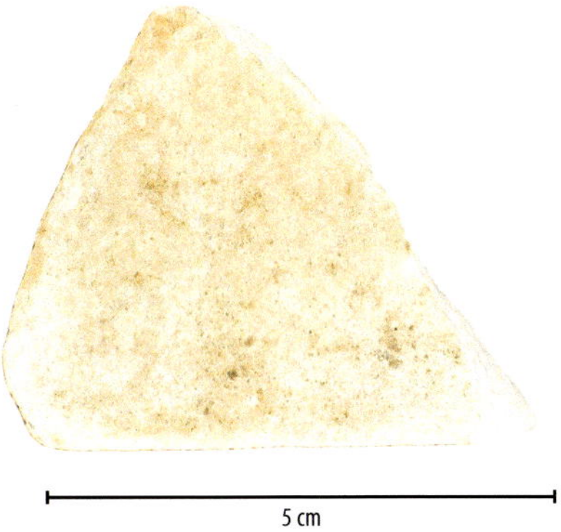

cat. no. 513

cat. no. 514

cat. no. 515, face

cat. no. 515, back

cat. no. 516

cat. no. 517

cat. no. 518

cat. no. 519

cat. no. 520

cat. no. 521, front and back

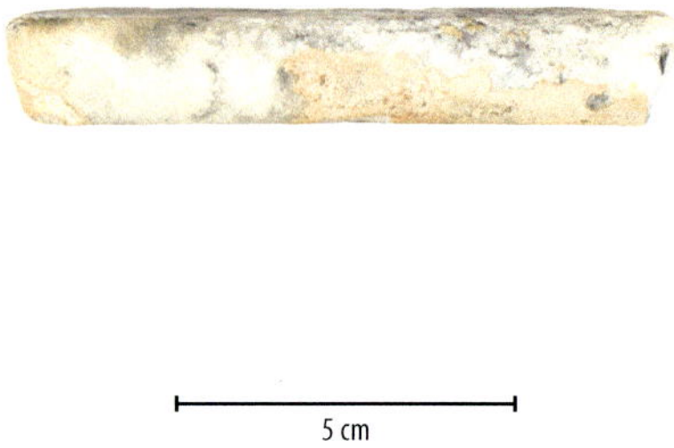

cat. no. 521, side

cat. no. 522

cat. no. 523

cat. no. 524

5 cm

cat. no. 525

5 cm

cat. no. 526

5 cm

cat. no. 527

5 cm

cat. no. 528

5 cm

cat. no. 529

5 cm

cat. no. 530

5 cm

cat. no. 531

5 cm

cat. no. 532

5 cm

cat. no. 533

5 cm

cat. no. 534

5 cm

cat. no. 535

5 cm

cat. no. 536

cat. no. 537

cat. no. 538

cat. no. 539

cat. no. 540

cat. no. 541

cat. no. 542

cat. no. 543

cat. no. 544

cat. no. 545

cat. no. 546

cat. no. 547

cat. no. 548

cat. no. 549

cat. no. 550, face

cat. no. 550, profile

cat. no. 551

cat. no. 552

cat. no. 553

cat. no. 554

cat. no. 555

cat. no. 556

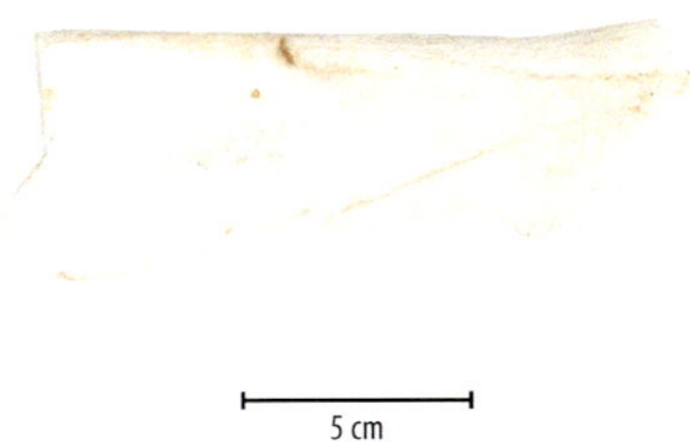

cat. no. 557

cat. no. 558

cat. no. 559

cat. no. 560

cat. no. 561

cat. no. 562

cat. no. 563

cat. no. 564

cat. no. 565

cat. no. 566

cat. no. 567

cat. no. 568

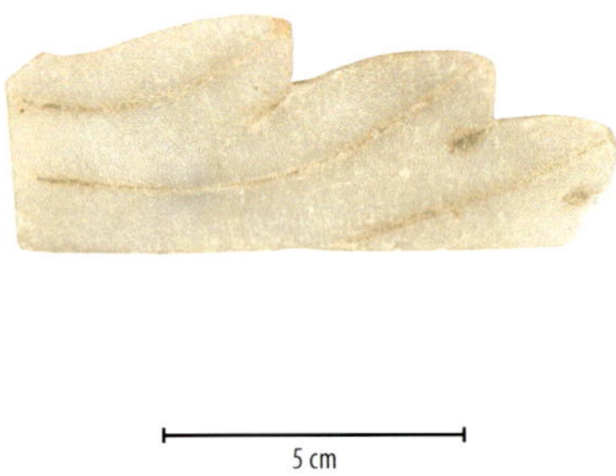

cat. no. 569

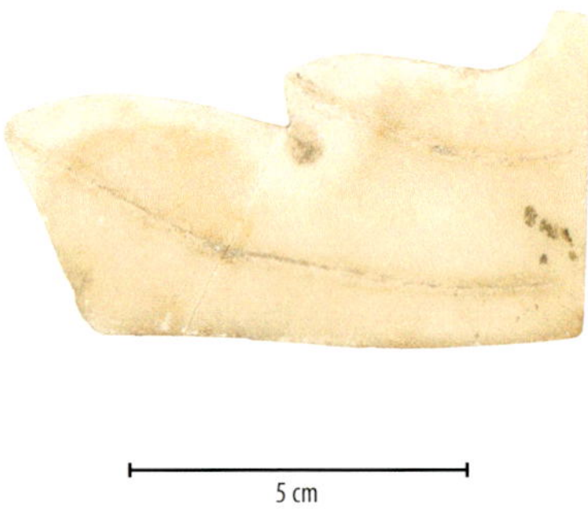

cat. no. 570

cat. no. 571

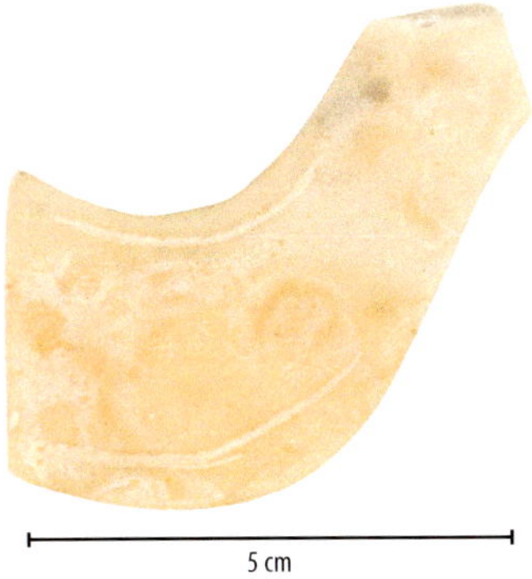

cat. no. 572

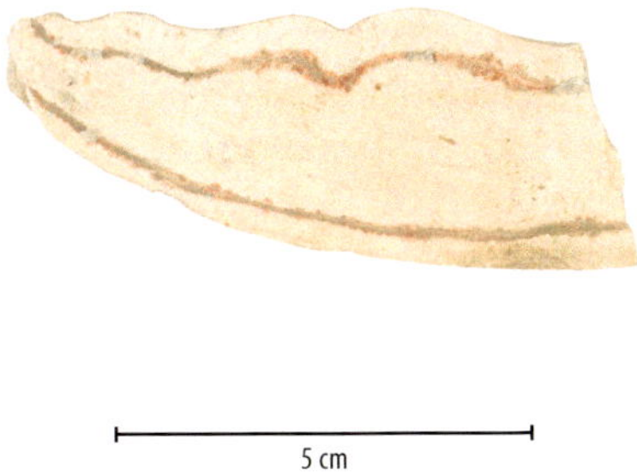

cat. no. 573

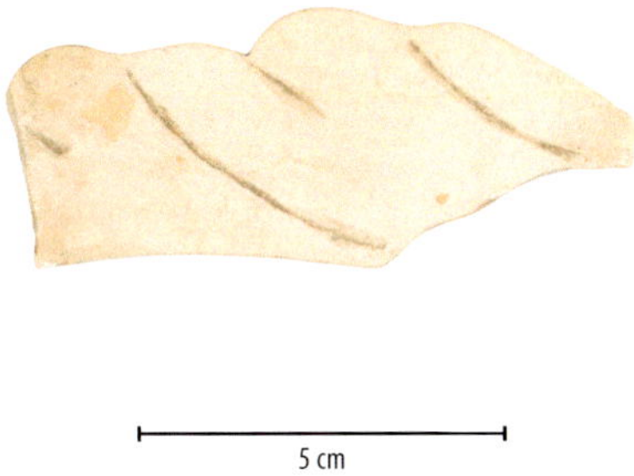

cat. no. 574

cat. no. 575

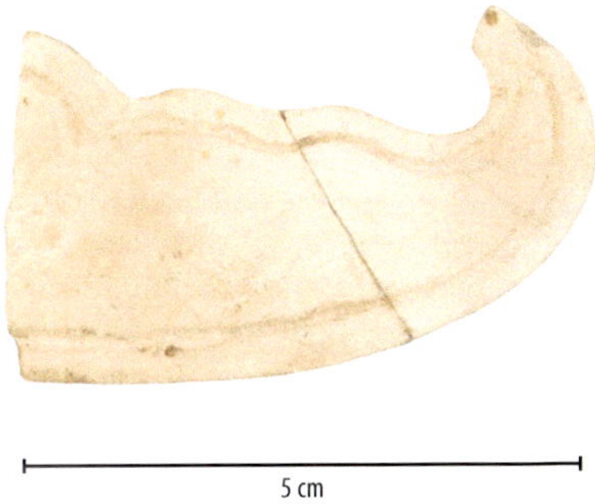

cat. no. 576

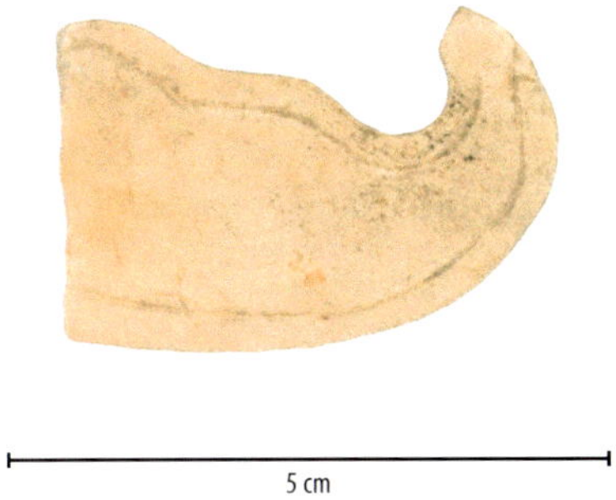

cat. no. 577

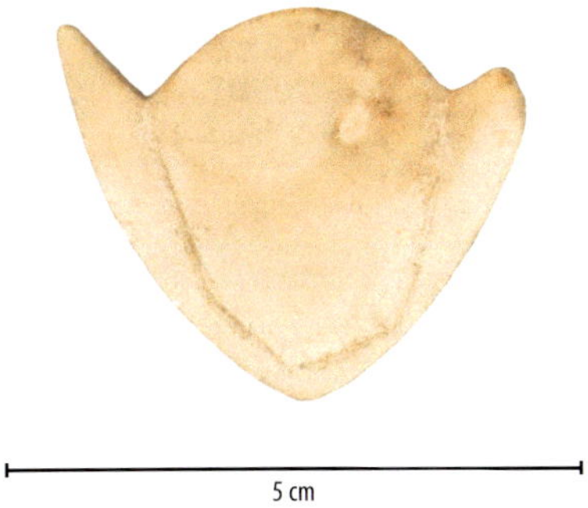

cat. no. 578

cat. no. 579

cat. no. 580

cat. no. 581

cat. no. 582

cat. no. 583

cat. no. 584

cat. no. 585

cat. no. 586, face and profile

cat. no. 587

cat. no. 588

cat. no. 589

cat. no. 590

cat. no. 591

cat. no. 592

cat. no. 593, view 1

cat. no. 593, view 2

cat. no. 594

cat. no. 595

cat. no. 596, face

cat. no. 596, tool marks

cat. no. 597

cat. no. 598

cat. no. 599

cat. no. 600

cat. no. 601

cat. no. 602

cat. no. 603

cat. no. 603, label

cat. no. 604

Plate 58

cat. no. 605

cat. no. 606

cat. no. 607

cat. no. 608

cat. no. 609

cat. no. 610, face

cat. no. 610, edge

cat. no. 611, view 1

cat. no. 611, view 2

cat. no. 612

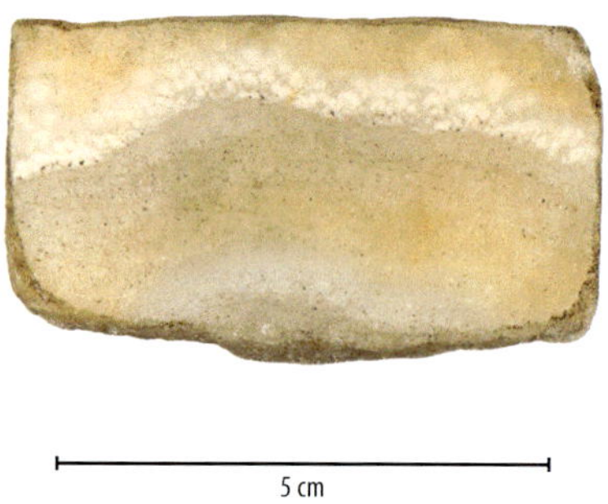

cat. no. 613

cat. no. 614

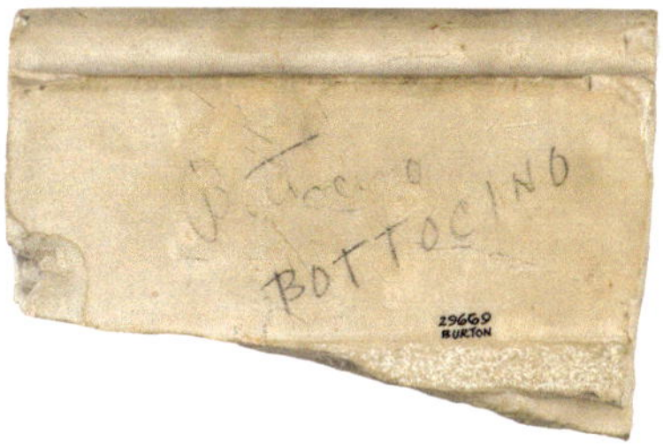

5 cm

cat. no. 615

5 cm

cat. no. 616

5 cm

cat. no. 617

5 cm

cat. no. 618

5 cm

cat. no. 619

5 cm

cat. no. 620

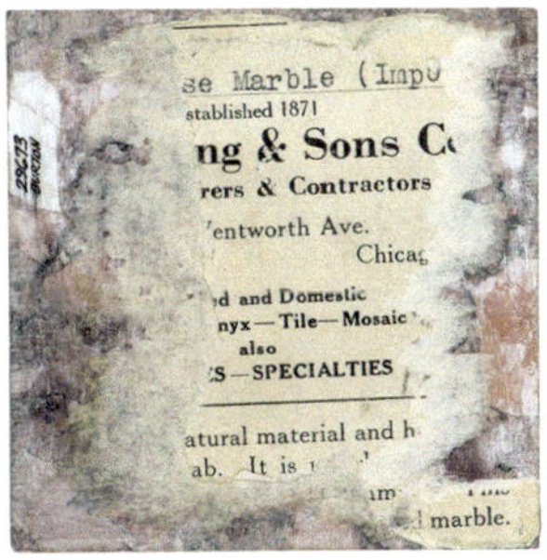

5 cm

cat. no. 620, label

5 cm

cat. no. 621

5 cm

cat. no. 622

5 cm

cat. no. 623

5 cm

cat. no. 624

5 cm

cat. no. 625

Plate 60

5 cm

cat. no. 626

cat. no. 626, label

5 cm

cat. no. 627

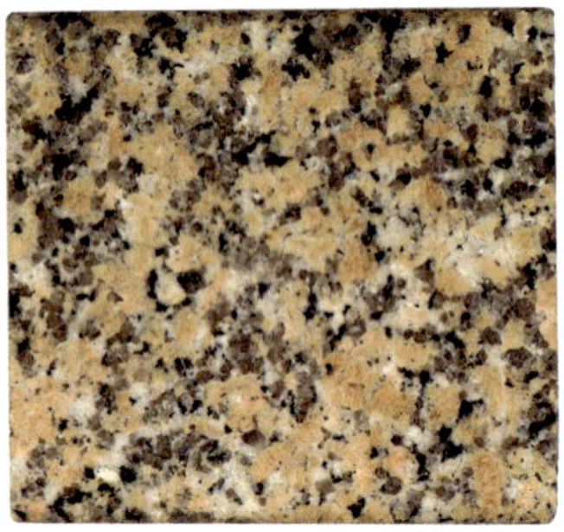

5 cm

cat. no. 628

5 cm

cat. no. 629

5 cm

cat. no. 630

5 cm

cat. no. 631

5 cm

cat. no. 632

5 cm

cat. no. 633

5 cm

cat. no. 634

5 cm

cat. no. 635

5 cm

cat. no. 636

cat. no. 637

cat. no. 638

cat. no. 639

cat. no. 640

cat. no. 641

cat. no. 642

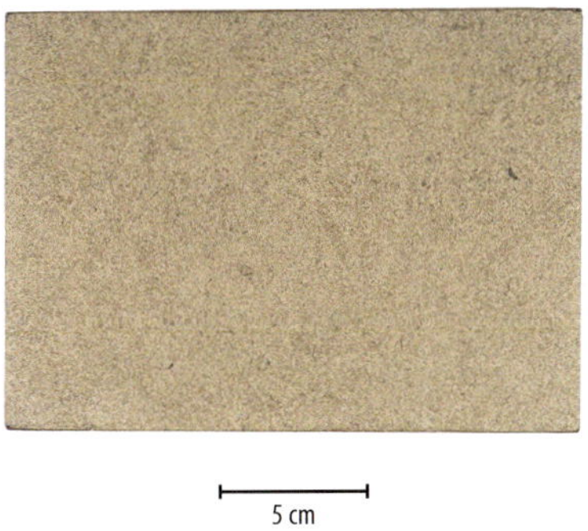

cat. no. 643

cat. no. 644

cat. no. 645

cat. no. 646

cat. no. 647

cat. no. 648

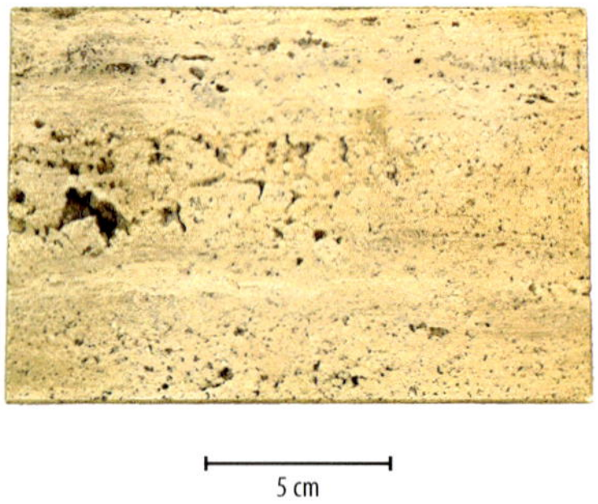

cat. no. 649

Roman Travertine
FROM
Gray Knox Marble Company
KNOXVILLE, TENNESSEE

Marble is not uniform throughout—being nature's product and varies in almost each and every slab. This piece is a general indication of above named marble. Submitted in connection our quotation November 25th

FORM NO. 322-5M-10-35

cat. no. 649, label

cat. no. 650

cat. no. 651

cat. no. 652

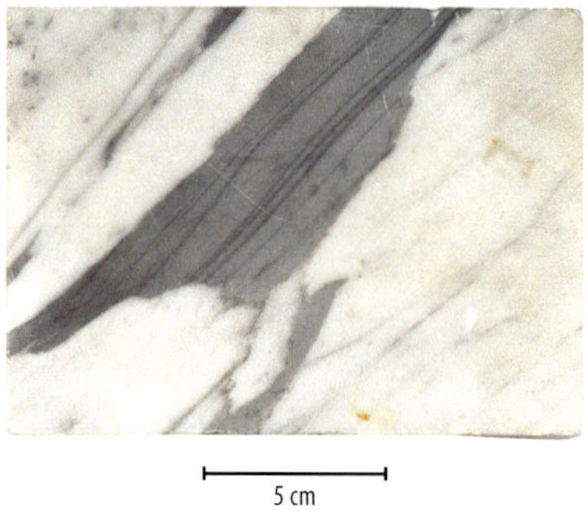

cat. no. 653

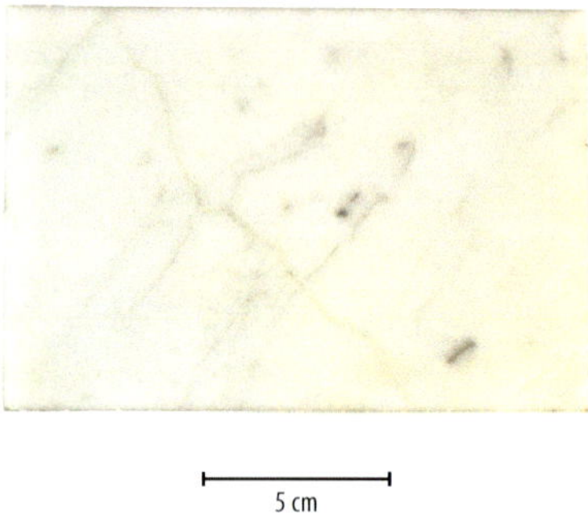

cat. no. 654

cat. no. 655

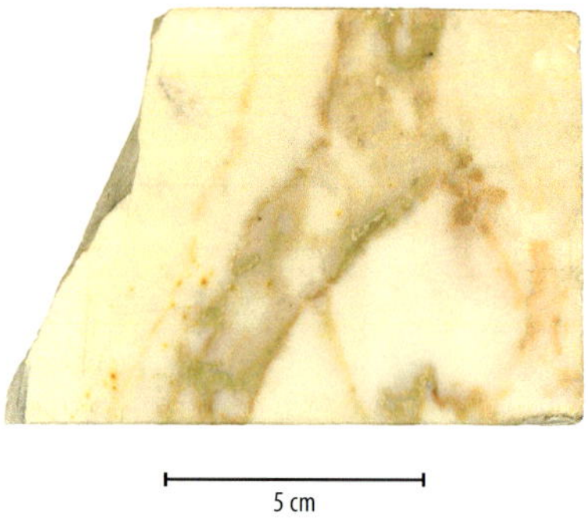

cat. no. 656

cat. no. 657

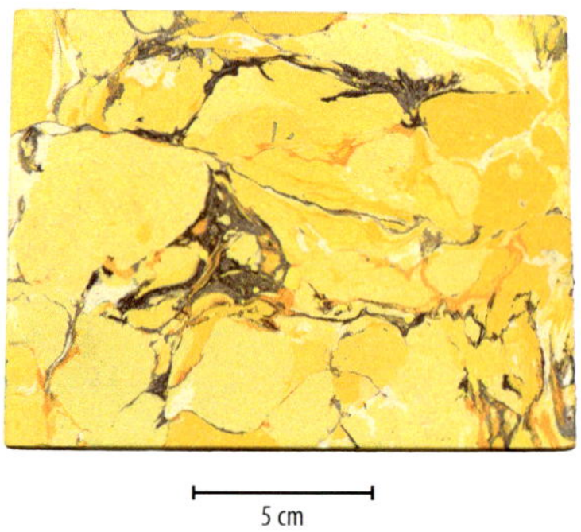

cat. no. 658

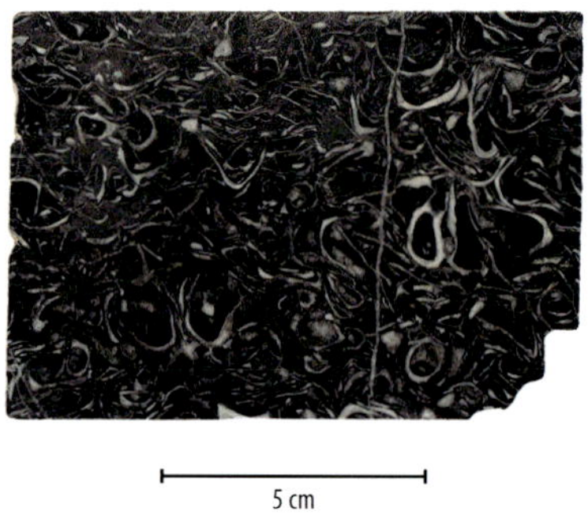

cat. no. 659

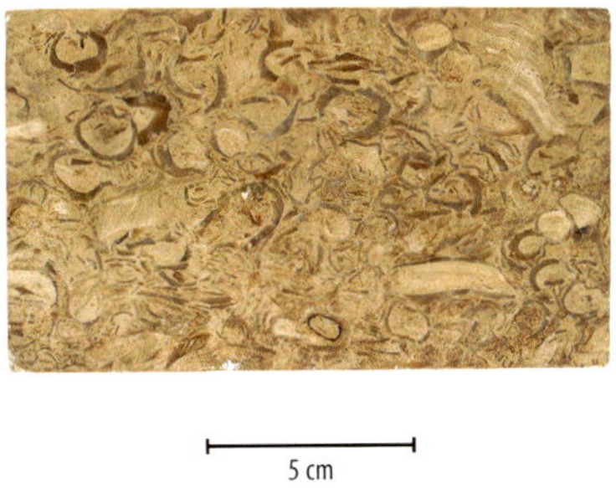

cat. no. 660

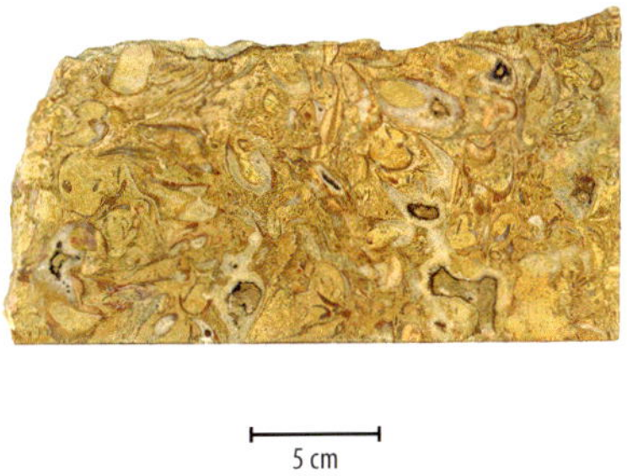

cat. no. 661

cat. no. 662

cat. no. 663

cat. no. 664

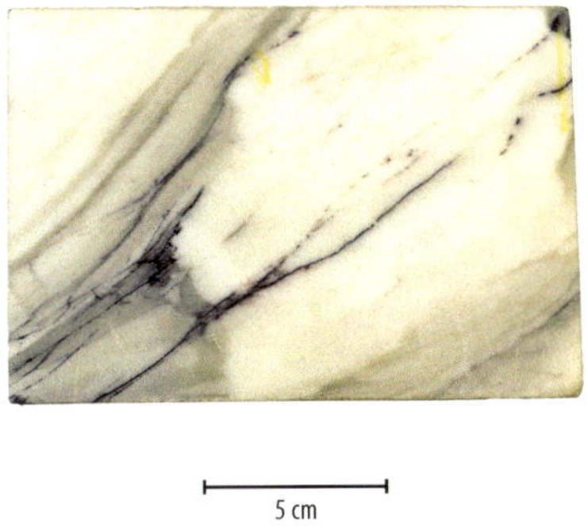

cat. no. 665

cat. no. 666

cat. no. 667

cat. no. 668

cat. no. 669

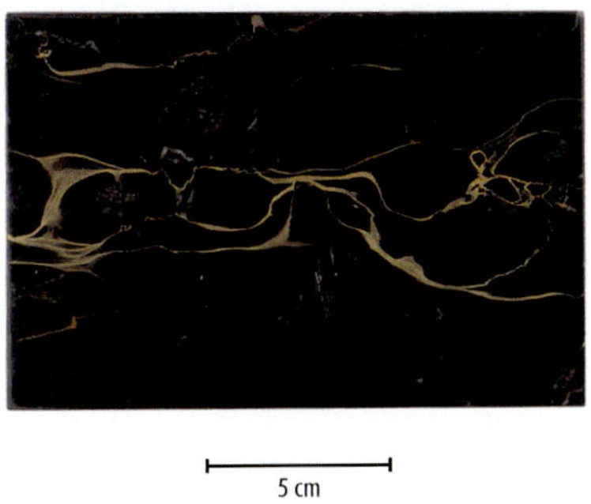

cat. no. 670

BLACK & GOLD

FROM

Gray Knox Marble Company

KNOXVILLE, TENNESSEE

Marble is not uniform throughout—being nature's product and varies in almost each and every slab. This piece is a general indication of above named marble.

FORM NO. 322-5M-10-35

cat. no. 670, label

5 cm

cat. no. 671

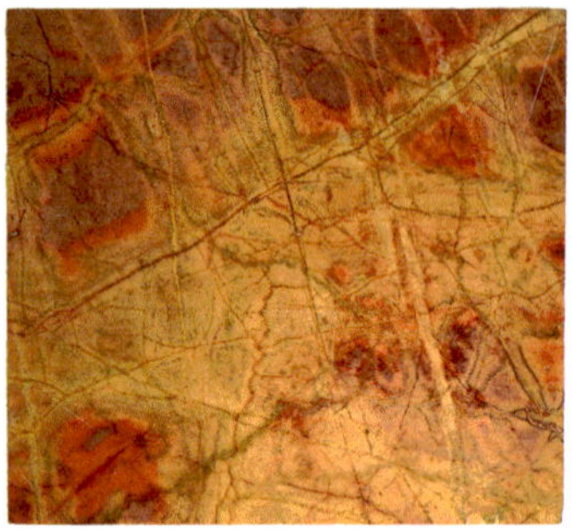
5 cm

cat. no. 672

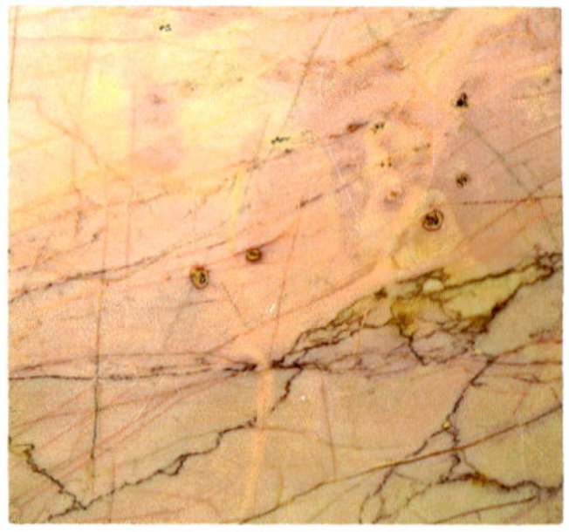
5 cm

cat. no. 673

5 cm

cat. no. 674

5 cm

cat. no. 675

5 cm

cat. no. 676

5 cm

cat. no. 677

5 cm

cat. no. 678

5 cm

cat. no. 679

5 cm

cat. no. 679, label

5 cm

cat. no. 680

5 cm

cat. no. 681

cat. no. 682, front

cat. no. 682, side

cat. no. 683

cat. no. 684, top and bottom

cat. no. 685

cat. no. 686, face and edge

cat. no. 687

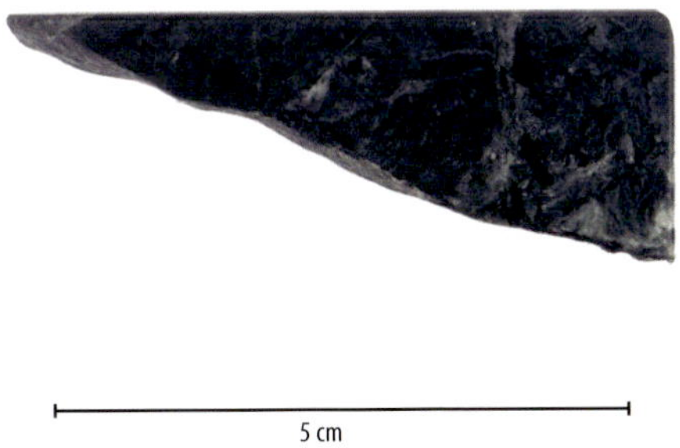

cat. no. 688

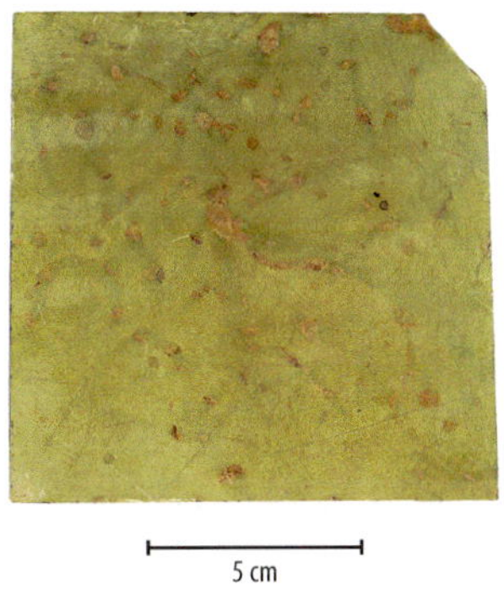

cat. no. 689

cat. no. 690

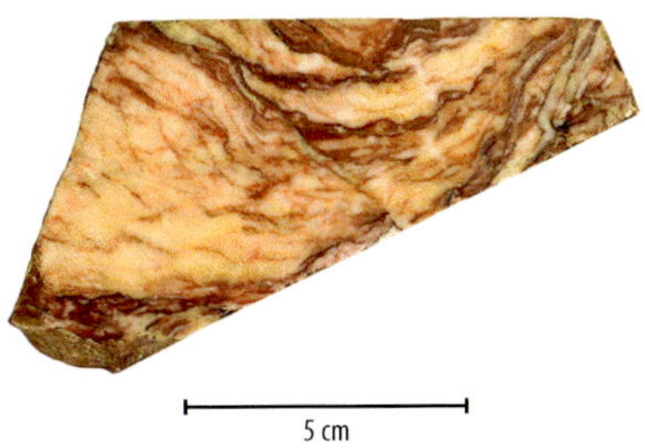

cat. no. 691, face

cat. no. 691, edge

cat. no. 692

cat. no. 693

cat. no. 694

cat. no. 695

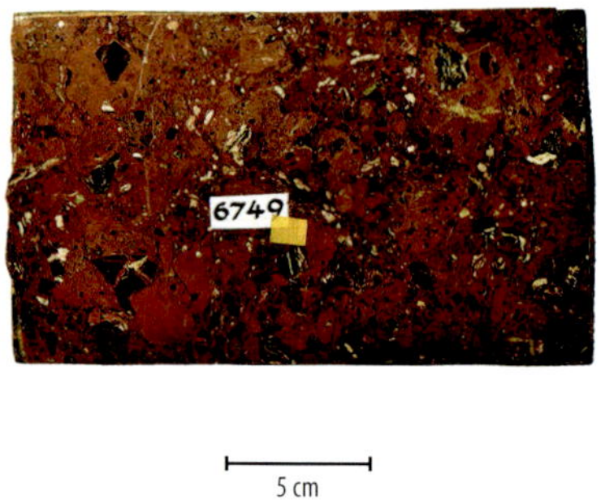

cat. no. 696

cat. no. 697

cat. no. 698

cat. no. 699

cat. no. 700

cat. no. 701

cat. no. 702

cat. no. 703

cat. no. 704